Advances in Intelligent and Soft Computing 156

Editor-in-Chief

Prof. Janusz Kacprzyk
Systems Research Institute
Polish Academy of Sciences
ul. Newelska 6
01-447 Warsaw
Poland
E-mail: kacprzyk@ibspan.waw.pl

For further volumes:
http://www.springer.com/series/4240

Javier Bajo Pérez, Juan M. Corchado Rodríguez,
Emmanuel Adam, Alfonso Ortega, María N. Moreno,
Elena Navarro, Benjamin Hirsch,
Henrique Lopes-Cardoso, Vicente Julián,
Miguel A. Sánchez, and Philippe Mathieu (Eds.)

Highlights on Practical Applications of Agents and Multi-Agent Systems

10th International Conference on Practical Applications of Agents and Multi-Agent Systems

Springer

Editors
Javier Bajo Pérez
Facultad de Informática
Universidad Pontificia de Salamanca
Salamanca, Spain

Juan M. Corchado Rodríguez
Departamento de Informática y Automática
Facultad de Ciencias
Universidad de Salamanca
Salamanca, Spain

Emmanuel Adam
LAMIH (UMR CNRS 8530)
Universite de Valenciennes, France

Alfonso Ortega
Departamento de Ingeniería Informática
Calle Francisco Tomás y Valiente
Ciudad Universitaria de Cantoblanco
Madrid, Spain

María N. Moreno
Departamento de Informática y Automática
Facultad de Ciencias
Universidad de Salamanca
Salamanca, Spain

Elena Navarro
Departamento de Informática
University of Castilla-La Mancha
Albacete, Spain

Benjamin Hirsch
EBTIC / Khalifa University
Abu Dhabi, United Arab Emirates

Henrique Lopes-Cardoso
Departamento de Engenharia Informática
Faculdade de Engenharia
Universidade do Porto, Porto, Portugal

Vicente Julián
Departamento de Sistemas Informáticos
y Computación
Universidad Politécnica de Valencia
Valencia, Spain

Miguel A. Sánchez
Facultad de Informática
Universidad Pontificia de Salamanca
Salamanca, Spain

Philippe Mathieu
Université des Sciences et Technologies
de Lille, France

ISSN 1867-5662
e-ISSN 1867-5670
ISBN 978-3-642-28761-9
e-ISBN 978-3-642-28762-6
DOI 10.1007/978-3-642-28762-6
Springer Heidelberg New York Dordrecht London

Library of Congress Control Number: 2012933089

Printed on acid-free paper

Springer is part of Springer Science+Business Media (www.springer.com)

Preface

PAAMS'12 Special Sessions are a very useful tool in order to complement the regular program with new or emerging topics of particular interest to the participating community. Special Sessions that emphasized on multi-disciplinary and transversal aspects, as well as cutting-edge topics were especially encouraged and welcome.

Research on Agents and Multi-Agent Systems has matured during the last decade and many effective applications of this technology are now deployed. An international forum to present and discuss the latest scientific developments and their effective applications, to assess the impact of the approach, and to facilitate technology transfer, has become a necessity.

PAAMS, the International Conference on Practical Applications of Agents and Multi-Agent Systems is an evolution of the International Workshop on Practical Applications of Agents and Multi-Agent Systems. PAAMS is an international yearly tribune to present, to discuss, and to disseminate the latest developments and the most important outcomes related to real-world applications. It provides a unique opportunity to bring multi-disciplinary experts, academics and practitioners together to exchange their experience in the development of Agents and Multi-Agent Systems.

This volume presents the papers that have been accepted for the 2012 edition in the special sessions: COREMAS: COoperative and RE-configurable MultiAgent System for Manufacturing, Logistics and Services domains in Industry, AMMR: Adaptive Multimedia and Multilingual Retrieval, BioMAS: 3rd International Special Session on Bio-Inspired and Multi-Agents Systems: Applications to Languages (BioMAS), WebMiRes: Web Mining and Recommender systems, MASSS: Multi-Agent Systems for Safety and Security, AAA: Assessing Agent Applications), TINMAS: Trust, Incentives and Norms in open Multi-Agent Systems, AMAS: Adaptive Multi-Agent Systems, ASM: Agents for smart mobility, ABAM: Agents Behaviours and Artificial Markets.

We would like to thank all the contributing authors, as well as the members of the Program Committees of the Special Sessions and the Organizing Committee for their hard and highly valuable work. Their work has helped to contribute to the success of the PAAMS'12 event. Thanks for your help, PAAMS'12 wouldn't exist without your contribution.

We thank the sponsors (IEEE Systems Man and Cybernetics Society Spain, AEPIA Asociación Española para la Inteligencia Artificial, APPIA Associação Portuguesa Para a Inteligência Artificial, CNRS Centre national de la recherche scientifique), the Local Organization members and the Program Committee members for their hard work, which was essential for the success of PAAMS'12.

Juan M. Corchado Rodríguez
Javier Bajo
PAAMS'12 Organizing Co-chairs

Organization

Special Sessions

COREMAS: COoperative and RE-configurable MultiAgent System for Manufacturing, Logistics and Services domains in Industry.
BioMAS: 3rd International Special Session on Bio-Inspired and Multi-Agents Systems: Applications to Languages.
WebMiRes: Web Mining and Recommender systems.
MASSS: Multi-Agent Systems for Safety and Security.
AAA: Assessing Agent Applications.
TINMAS: Trust, Incentives and Norms in open Multi-Agent Systems.
AMAS: Adaptive Multi-Agent Systems.
ASM: Agents for Smart Mobility.
ABAM: Agents Behaviours and Artificial Markets.

Special Session on COoperative and RE-configurable Multi-Agent System for Manufacturing, Logistics and Services Domains in Industry

Emmanuel Adam	University of Valenciennes, France
Paulo Leitao	Polytechnic Institute of Bragança, Portugal

Special Session on 3rd International Special Session on Bio-Inspired and Multi-Agents Systems: Applications to Languages

Gemma Bel-Enguix	Universitat Rovira i Virgili, Spain
Mª Dolores Jiménez López	Universitat Rovira i Virgili, Spain
Alfonso Ortega de la Puente	Universidad Autónoma de Madrid, Spain

Special Session on Web Mining and Recommender Systems

María N. Moreno (Chair)	University of Salamanca, Spain
Valérie Camps (Chair)	University of Toulouse, France
Ana María Almeida	Institute of Engineering of Porto, Portugal
Anne Laurent	University of Montpellier 2, France
Chris Cornelis	Ghent University, Belgium
Constantino Martins	Institute of Engineering of Porto, Portugal
Maguelonne Teisseire	University of Montpellier 2, France
María José del Jesús	University of Jaen, Spain
Marie-Aude Aufaure	Ecole Centrale de Paris, France
Rafael Corchuelo	University of Sevilla, Spain
Vivian López Batista	University of Salamanca, Spain
Joel Pinho Lucas	HP, Brazil
Salima Hassas	University of Lyon, France
Sylvain Giroux	University of Sherbrooke, Canada
Yolanda Blanco	University of Vigo, Spain
Jérôme Picault	Alcatel-Lucent Bell Labs, France

Special Session on Multi-Agent Systems for Safety and Security

Antonio Fernández-Caballero (Chair)	University of Castilla-La Mancha, Spain
Elena María Navarro Martínez (Chair)	University of Castilla-La Mancha, Spain
Ruth Aguilar-Ponce	Universidad Autónoma de San Luis Potosí, Mexico
Javier Albusac	Universidad de Castilla-La Mancha, Spain
Giuliano Armano	University of Cagliari, Italy
Danilo Avola	Sapienza University of Rome, Italy
Federico Castanedo	Universidad de Deusto, Spain
Rita Cucchiara	Università degli Studi di Modena e Reggio Emilia, Italy
Francisco Garijo	Institut de Recherche en Informatique de Toulouse, France
Marie-Pierre Gleizes	Institut de Recherche en Informatique de Toulouse, France
Zahia Guessoum	Laboratoire d'Informatique de Paris 6, France
Alaa Khamis	The German University in Cairo, Egypt
José Manuel Molina	Universidad Carlos III de Madrid, Spain
Takashi Matsuyama	Kyoto University, Japan
Paulo Novais	Universidade do Minho, Portugal
Juan Pavón	Universidad Complutense de Madrid, Spain
Milind Tambe	University of Southern California, USA

Special Session on Assessing Agent Applications

Benjamin Hirsch	EBTIC / Khalifa University, United Arab Emirates
Tina Balke	University of Bayreuth, Germany
Marco Lützenberger	DAI-Labor, Technical University of Berlin, Germany

Special Session on Trust, Incentives and Norms in open Multi-Agent Systems

Ana Paula Rocha (Chair)	University of Porto, Portugal
Henrique Lopes Cardoso (Chair)	University of Porto, Portugal
Olivier Boissier (Chair)	ENSM Saint-Etienne, France
Ramón Hermoso (Chair)	Universidad Rey Juan Carlos, Spain
Daniel Villatoro	IIIA-CSIC, Spain
Eugénio Oliveira	University of Porto, Portugal
Holger Billhardt	Universidad Rey Juan Carlos, Spain
Joana Urbano	University of Porto, Portugal
Jordi Sabater-Mir	IIIA-CSIC, Spain
Laurent Vercouter	University of Rouen, France
Nicoletta Fornara	University of Lugano, Switzerland
Sascha Ossowski	Universidad Rey Juan Carlos, Spain
Tibor Bosse	Vrije Universiteit Amsterdam, The Netherlands
Virginia Dignum	Delft University of Technology, The Netherlands
Wamberto Vasconcelos	University of Aberdeen, The Netherlands
Yao-Hua Tan	Delft University of Technology, The Netherlands

Special Session on Adaptive Multi-Agent Systems

Vicente Julian	Polytechnic University of Valencia, Spain
Alberto Fernández	University of Rey Juan Carlos, Spain
Juan M. Corchado	University of Salamanca, Spain
Vicente Botti	Polytechnic University of Valencia, Spain
Sascha Ossowski	University of Rey Juan Carlos, Spain
Sara Rodríguez González	University of Salamanca, Spain
Carlos Carrascosa	Polytechnic University of Valencia, Spain
Javier Bajo	Pontifical University of Salamanca, Spain
Martí Navarro	Polytechnic University of Valencia, Spain

Stella Heras	Polytechnic University of Valencia, Spain
Carlos A. Iglesias	Technical University of Madrid, Spain
Holger Billhardt	University King Juan Carlos I, Spain
Javier Palanca	Polytechnic University of Valencia, Spain
Juan Francisco de Paz Santana	University of Salamanca, Spain
Álvaro Herrero	University of Burgos, Spain
Victor Sanchez	Polytechnic University of Valencia, Spain

Special Session on Agents for Smart Mobility

Miguel Angel Sánchez Vidales	Universidad Pontificia de Salamanca, Spain

Special Session on Agents Behaviours and Artificial Markets

Philippe Mathieu (Chair)	University Lille, France
Bruno Beaulfis	University of Lille1, France
Olivier Brandouy	University of Paris 1, France
Andrea Consiglio	University of Palermo, Italy
Florian Hauser	University of Innsbruck, Austria
Juan Pavón	University Complutense Madrid, Spain
Marco Raberto	University of Rekjavik, Iceland
Roger Waldeck	Telecom Bretagne, France
Hector Zenil	University of Sheffield, United Kingdom

Organizing Committee

Juan M. Corchado (Chairman)	University of Salamanca, Spain
Javier Bajo (Co-Chairman)	Pontifical University of Salamanca, Spain
Juan F. De Paz	University of Salamanca, Spain
Sara Rodríguez	University of Salamanca, Spain
Dante I. Tapia	University of Salamanca, Spain
Fernando de la Prieta Pintado	University of Salamanca, Spain
Davinia Carolina Zato Domínguez	University of Salamanca, Spain

Reviewers

Emmanuelle Grislin
Gauthier Picard
Marin Lujak
Pavel Vrba
Carole Bernon
Jose Barata
Andres Garcia
Jean-Paul Jamont
Zahia Guessoum
Valérie Camps
Montserrat Mateos
Encarnación Beato
Roberto Berjón

Contents

COREMAS: COoperative and RE-configurable Multi Agent System for Manufacturing, Logistics and Services Domains in Industry

BioMAS: 3rd International Special Session on Bio-Inspired and Multi-Agents Systems: Applications to Languages (BioMAS)

WebMiRes: Web Mining and Recommender systems

MASSS: Multi-Agent Systems for Safety and Security

AAA: Assessing Agent Applications

TINMAS: Trust, Incentives and Norms in Open Multi-Agent Systems

AMAS: Adaptive Multi-Agent Systems

Multilevel MAS Architecture for Vehicles Knowledge Propagating

Emmanuel Adam, René Mandiau, and Emmanuelle Grislin

Abstract. Completely autonomous vehicles in traffic should allow to decrease the number of road accident victims greatly, and should allow gains in terms of performance and economy. Models of the interaction among the different vehicles is one of the main challenges. We propose in this paper a model of communication of knowledge between mobile agents on a traffic network. The model of knowledge and of interaction enables to propagate new knowledge without overloading the system with a too large number of communications. For that, only the new knowledge is communicated, and two agents communicate the same knowledge only once. In order to allow agents to update their knowledge (perceived or created), a notion of degradation is used. A simulator has been built to evaluate the proposal.

1 Introduction

In many studies on traffic supervision, the optimization of traffic flow as well as new road infrastructure [2], attempt to deal with collective interests and individual interests. We think that completely autonomous vehicles in traffic should allow to decrease the number of road accident victims greatly, and should allow gains in terms of performance and economy. Developing models of the interaction among the different vehicles is one of the main challenges [4] to optimize traffic flow with autonomous vehicles.

The simulation of traffic is often used to evaluate traffic flow optimization methods. In the area of road traffic simulation, two approaches allow management

Emmanuel Adam · René Mandiau · Emmanuelle Grislin
Lille Nord de France, F-59000 Lille, France
e-mail: {emmanuel.adam,rene.mandiau}@univ-valenciennes.fr,
emmanuelle.grislin@univ-valenciennes.fr

Emmanuel Adam · René Mandiau · Emmanuelle Grislin
UVHC, LAMIH, F-59313 Valenciennes, France

Emmanuel Adam · René Mandiau · Emmanuelle Grislin
CNRS, UMR 8201, F-59313 Valenciennes, France

J.B. Pérez et al. (Eds.): Highlights on PAAMS, AISC 156, pp. 1–8.
springerlink.com

policies for scheduling vehicles flows: centralized approaches and distributed approaches. A means to bypass the limitations of centralized approaches (lack of flexibility, difficulty to adapt to changes ...) is to decentralize the traffic simulation [9, 7]; in this context agent-based approaches seem to be the most appropriate [5, 6]. In these approaches, the traffic is the result of the sum of all actions and interactions of the various simulated entities (busses, vehicles, pedestrians, road signs, road infrastructure for examples).

So, in the context of a project that aims at studying the impact of information communication between drivers and infrastructure or elements of the environment (like shops, car park for instance), we turn to multiagent systems to propose a simulator of a traffic network.

Communication of knowledge implies classically to take care of the confidence about this knowledge. If β_a is the set of knowledge of the agent a. A received knowledge b can be :

- out-of-date : dynamic environment implies to date the knowledge about it, about events.
- from a doubtful origin : the knowledge comes from an unknown agent, or without the required signature.
- inconsistent : the receiver of the knowledge b is is unable to store it in its own knowledge without getting an inconsistency; $(\{b\} \cup \beta_a = \emptyset)$.

The next section of the paper presents the architecture of the multiagent system used to share knowledge between mobile agents and the model of knowledge that we propose. The section 3 presents the simulator of road traffic, and the case study that we used to validate our proposal. The last section draws our conclusions and gives some perspectives for future research.

2 Communication of Mobile Agents in a Network

Fast communication of knowledge between mobile agents along a network can be done : directly, by messages exchange, when agent are physically close enough to communicate; or indirectly through the environment (generally the nodes are used to store / read information).

We think that a node should be more pro-active and should have the opportunity to choose to communicate some information to some agents chosen according to its knowledge.

So we propose a third method in which some non mobile agents are located at the nodes and communicate with mobile agents and that are close enough to receive messages.

Moreover, we can easily argue that to give the possibility to a node to interact with one another could give efficiency to the propagation of an information along a network.

Multilevel Architecture. In order to allow the three kinds of communication (‘*mobile-agent-to-near-mobile-agent*’, ‘*mobile-agent-to-near-node*’ and ‘*node-to-near-node*’), and to allow communication between distant mobile-agent and distant nodes, we use a multi-level architecture, inspired from holonic principles. We have already used this kind of architecture for the simulation of a flexible assembly cell, in order to correct myopic behaviour of mobile and autonomous shuttle [1].

At the bottom of the system (level 0), we have the mobile agents. At he level 1, there are the node agents, that can schedule, manage conflicts between ‘mobile agents’ that have to cross them. These agents (mobiles and nodes) and the environment (the network) are included in the ‘network agent’ that represents the multiagent system.

Each agent can interact with all agents that are in its ‘vision field’.

The Figure 1 presents an extract of a screenshot of a traffic simulator that we use to test knowledge communication between cars; ‘mobile agents’ are represented by ‘vehicle agents’, ‘node agents’ are represented by ‘crossroad managers’.

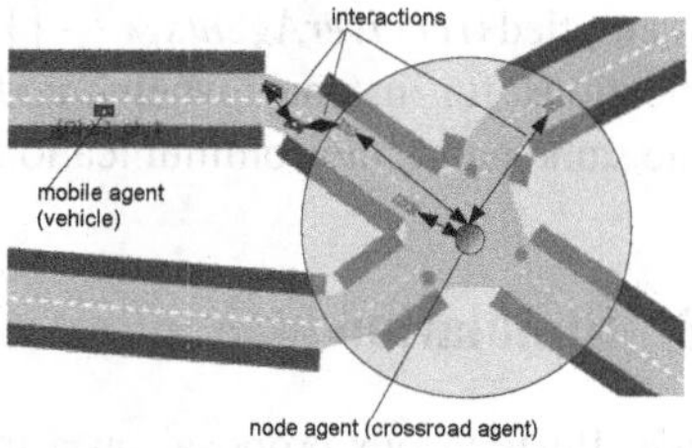

Fig. 1 Example of application in road traffic. Mobile vehicles agents interact between them and with the crossroad (node) agent if they are close enough.

Elements of Knowledge Model. We give here just some elements of how the mobile agents manage the exchanged knowledge. A knowledge is a partial view of the environment or of the other agents, namely for a given object o of the environment (the traffic network for example); it is (generally) an incomplete copy of it, so a representation of o with missing attributes and methods.
If $\mathcal{O}$ is the set of objects of the environment, an object o is defined by:

$$o = (id, attributes_o = \{a_1^o, \dots, a_n^o\}_{n>0}, functions_o = \{f_1^o, \dots, f_m^o\}_{m\geq 0}) \quad (1)$$

with $f_j^o : \mathcal{O} \mapsto \mathcal{O}$.

If o'_a is a partial representation of o for the agent a, the cardinalities of the set of attributes and functions of o' are lesser than those of the object o:

$\mid attributes_{o'} \mid \leq \mid attributes_o \mid$ and $\mid functions_{o'} \mid \leq \mid functions_o \mid$

We define a knowledge κ_o^a (cf. def. 2) on an object o for an agent a by: o'_a, a partial view of o from a; $date_{\kappa_o}$, the date when the knowledge has been created or updated (by a or by another agent if the knowledge has been received); $builderAgent_{\kappa_o}$, the ‘builder’ of the knowledge (name of the agent that has created/updated the knowledge from its perception); $senderAgent_{\kappa_o}$, the ‘sender’ of the knowledge (name of the agent that could have sent the knowledge to a); $receiverAgents_{\kappa_o^a}$, the list of agents to which the knowledge has been sent by a; $shareable_{\kappa_o^a}$, the fact that the knowledge is shareable or not by a.

$$\kappa_o^a = \begin{pmatrix} o'_a, date_{\kappa_o}, builderAgent_{\kappa_o}, \\ senderAgent_{\kappa_o}, receiverAgents_{\kappa_o^a}, shareable_{\kappa_o^a} \end{pmatrix} \quad (2)$$

The notions of $date_{\kappa_o}$ and of $receiverAgents_{\kappa_o^a}$ allow to decrease the number of communications. Indeed, when the agent a perceives, from its perception function an image of a new object o, it creates the knowledge κ_o^a with the date of creation.

When a receives from another agent b a knowledge on o, a replaces its knowledge only if the $dateCreation_{\kappa_o}$ in κ_o^b is newer than the one in κ_o^a.

If b is close enough to send its knowledge to a, reciprocally a can send its knowledge. So when two agents meet, at the end of the communications they own the last representation known of the object o. When an agent (a) sends some knowledge on o to another (b), it adds the agent to its set of receivers ($receiverAgents_{\kappa_o^a} \leftarrow receiverAgents_{\kappa_o^a} \cup \{b\}$. Each time an agent receives a new version of an information, its set of agents to which this information has been sent is emptied ($receiverAgents_{\kappa_o^a} \leftarrow \{\}$). This allows to decrease the number of communications; for instance, when a and b meet many times (for example, if they follow the same path), the communication about a given knowledge is done only once.

3 Simulation

We illustrate our proposal on a traffic road simulator that we have developed in the context of a project (Plaiimob : a simulating Plateform dedicated to Mobility services) of CISIT (for International Campus on Security and Inter modality in Transports).

The aim of this project is to allow vehicle-to-vehicle (V2V) communication to allows drivers to automatically exchange data about their environment (incident/traffic jam on a road, information about off-street parking ...) [8].

We developed a traffic road simulator with the Jade Platform[1]:

- The *environment* is a traffic network in the OpenStreetMap format[2] (OSM). This allows us to use true traffic networks, or to define our own maps in order to test particular situations.
- The classes of *agents* are: the PersonnageAI agent class (it allows to simulate the behaviour of a driver), the CrossRoadManager agent class (it allows to manage the priorities at a crossroad according to the road signs), the ObserverAgent class (that allows to draw statistics from the simulating exercise).
- The *roles* played by the PersonnageAI are: RandomBehaviourRole: to represent a driver that moves randomly on the traffic; BusRole: to represent a driver that starts from a particular point and has to reach an objective, by linking some

[1] See the web site of the Jade platform : `http://jade.tilab.com/`

[2] See the web site of the OpenStreetMap project:
`http://www.openstreetmap.org/`

bus stops, at earliest; EmergencyRole: to represent a driver of an ambulance, firetruck, for instance, that has an objective to reach as soon as possible.

- The *roles* played by the CrossRoad Managers are linked to the road signs or traffic lights associated with them. We defined some roles like: FifoCrossRoad (the first vehicle arrived in a queue of the crossroad 'receives' a green light); ClassicTrafficLightRole (emulation of classical traffic lights at each entry of the crossroad); AITrafficLightRole (based on the ClassicTrafficLightRole, it gives the priority to the street from which an emergency vehicle arrives (the associated traffic light goes on green light) ...).

Jade has been chosen because the aim of the project is to test our proposal on a real case, with agents embedded in the smart-phones of the drivers. So we aim at reusing some classes of the simulation to build agents and their roles.

Communication of Incident. In order to test the benefit to have a V2V communication, we present here a small case study that includes 2 bus lines (*A* and *B*), with one bus for the line *A* (*bus1a*) and 5 bus of the line *B* (*bus2a*, ..., *bus2e*); and five cars that travel randomly on the network (*randomCar*1, ..., *randomCar*5).

The Fig. 2 shows the map used in this context. The bus of the line *A* start from 'SA', and have to make loops between the bus stops 'BS-A1', and 'BS-A2'. For the busses of the line *B*, they start from 'SB', and have to make loops between the bus stops 'BS-B1', and 'BS-B2'. 'SR' is the starting point of the cars that travel randomly.

We used three different scenarios. In the first scenario, there is no incident, all the vehicles travel as they planed. In the second scenario, we add an incident on the road with incident risk before *bus1a* reaches it, and vehicles do not communicate information. In the third scenario, busses and cars are able to exchange knowledge.

The incident causes a tailback on this road, and the speed limit is divided by 2 (so it becomes 35 km/h maximum).

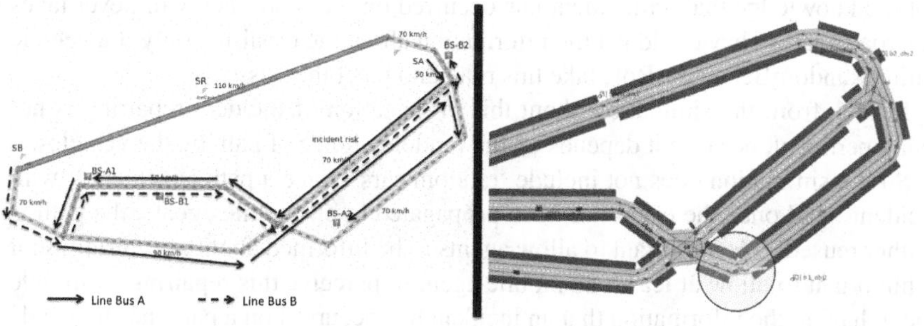

Fig. 2 On the right: map of Traffic Roads used in the case study. The busses have to stop on the right side of the road. One road is prone to incident risk. On the left: screencopy of the Traffic Roads Simulator. *bus2b* is selected; the red arrow shows the roads where it plans to go. The disc around *bus2b* represents its communicating zone.

Initially, each mobile agent knows the map of the traffic road, with the initial speed limit.

Each time a mobile agent reaches a road, it updates its knowledge if needed: if the perceived state of the road implies a modification of the speed limit (due to an incident, or traffic jam, for example), a knowledge is created and stored; if the perceived road exists in the knowledge, the knowledge is updated (the date is updated, the state of the road is modified if necessary).

If no incident is signaled, but if the agent took at least twice more time to exit the road, it records a tailback event in its knowledge.

Each 'PersonnageIA' with the role 'BusRole' is able to compute the shortest path to its next objective with the Dijkstra algorithm [3].

The best path, without incident, for A is to made a double loop (a '8') : from 'SA', the bus goes by the road with the 'incident risk'; then to 'BS-A1'; then it takes the road limited to $90km/h$; takes again the 'incident risk' road; to reach 'BS-A2' and repeats its round.

For the busses of the line B, the best path starts from 'SB'; goes to 'BS-A1'; takes the 'incident risk' road to reach 'BS-A2'. The busses take next the fast lane to repeat their round.

In scenario 3, we choose to put an incident at the beginning of the simulation; $bus1a$, which does not know this fact, travels on the route with incident. It informs all the agents that it meets during its move. *bus2b* is the first informed by $bus1a$, it computes a new best path and chooses another road than the one taken by *bus2a*.

On Fig. 2-right, the green arrow shows the road took by *bus2a*, the red arrow shows the road took by *bus2b* (and its followers) that met $bus1a$.

Without communication between vehicles, the lost time when an incident occurs on the 'incident risk' road is 7 time units on average. The communication of information, in the particular case study, results in a time-saving of about 4 time units.

Communication of Incident Repairing. In the simulator, when a incident is removed from a road, this one goes back to its initial speed limit. If all the busses have got the knowledge that an incident has occurred on the road, they will never takes this road again. They could get the information about the repairing only if a vehicle with a RandomBehaviourRole take this road and meet the busses.

Results from the simulation about this propagation of incident repairing is not really pertinent, because it depends on the random choice of path by the vehicles.

So if a simulation does not include 'random cars', once a path is declared 'with incident', and once the event has been propagated to all mobile agents, the path is neither reused. It is important to allow agents to be informed about a repairing; so it is important to allow at least one mobile agent to perceive this repairing. A mobile agent, having the information that an incident has occurred on a path, has to decide to take it, after *some* time.

We introduce a notion of degradation, in the definition of a knowledge κ_o^a. We define $degradation_{\kappa_o}$ as a coefficient of the reliability of the knowledge (if t is the elapsed time since the creation/update of the knowledge, the confidence that a has on it is: $trust_{\kappa_o} = 1 - t \times degradation_{\kappa_o}$). For a knowledge κ_o, if $trust_{\kappa_o} \leq 0$, then the agent does not take into account this knowledge to compute the best path to its next objective; and the knowledge is not sent to other agents.

If $degradation_{\kappa_o} = 0$, then the knowledge κ_o is perennial; if a $degradation_{\kappa_o} = 0.1$, then the knowledge κ_o exists for 10 *tu*.

The degradation coefficient for a knowledge is strongly dependent of the case study. Here, due to the light complexity of the traffic network used, as a bus needs 2.2 *tu* to make a loop, we fix the degradation coefficient of a knowledge about an incident at $1/3$ *tu*. This allows a bus to try to pass by the 'incident road' after 3 loops; if the incident is always on the road, the date of the knowledge is updated, and the new knowledge is communicated to each vehicle met by the bus.

One perspective could be to automatically define the degradation coefficient from observation during the simulation.

Propagation of Knowledge. In order to improve the diffusion of knowledge, we propose to use agents linked to the nodes of the network (here, these agents are the CrossRoad Managers).

When a mobile agent decides to communicate its knowledge, it includes the local node agents in its potential recipients. All nodes are not necessary equipped with node agents, but in the case study, a CrossRoad agent is located to each crossroad. When a vehicle detects a problem on the road it reaches, the vehicle informs all agents inside its perimeter of communication. The crossroads at the beginning and the end of the road receive the information. When another vehicle approaches the road, the crossroad where the vehicle arrives communicates the knowledge on the road incident/repairing. The vehicle is able to recompute its best path, and to choose another road.

We define so a fourth scenario with crossroad manager agents able to communicate knowledge. The results of the simulation is the same as the scenario 3, except that it is not needed to have a line *A* bus to inform line *B* busses about the incident on the 'incident risk' road.

4 Conclusion

In order to allow the propagation of knowledge between mobile agents, with a minimum of exchanged messages, we propose an architecture, a model of knowledge and a model of communication.

In the first results presented in this paper, we make the assumption that all the agents are cooperative, and no deficient; that is to say that they cannot send wrong knowledge, voluntary, or not (if a captor has a dysfunction).

The trust on a knowledge depends only on the date from which the knowledge has been updated or created. We plan to introduce the notion of trust that depends of the sender; for example, if an agent *t* sends a knowledge that is not coherent with the perception of other agents, these latters can decide to put in quarantine the faulty agent, like in [10] for example.

We plan also to enhance the information sent by the crossroad agents: when a car informs a 'crossroad agent' about its desire to take a particular road, the crossroad manager sends the number of agents that have already taken this road in a near past. So the mobile agent can choose to pursuit its initial plan, or to recompute a new best path.

Acknowledgments. This research was financed by the International Campus on Safety and Inter-modality in Transportation, the Nord/Pas-de-Calais Region, the European Community, the French Regional Delegation for Research and Technology, and the French National Centre for Scientific Research. We are grateful for the support of these institutions. The authors thanks particularly Remy Tylski, who implemented interesting parts of the simulator.

References

1. Adam, E., Zambrano, G., Pach, C., Berger, T., Trentesaux, D.: Myopic Behaviour in Holonic Multiagent Systems for Distributed Control of FMS. In: Corchado, J.M., Pérez, J.B., Hallenborg, K., Golinska, P., Corchuelo, R. (eds.) Trends in PAAMS. AISC, vol. 90, pp. 91–98. Springer, Heidelberg (2011)
2. Bazzan, A.L.: A distributed approach for coordination of traffic signal agents. Autonomous Agents and Multi-Agent Systems 10, 131–164 (2005)
3. Dijkstra, E.W.: A note on two problems in connexion with graphs. Numerische Mathematik 1, 269–271 (1959)
4. Dresner, K., Stone, P.: Mitigating catastrophic failure at intersections of autonomous vehicles. In: AAMAS Workshop on Agents in Traffic and Transportation, Estoril, Portugal, pp. 78–85 (2008)
5. Ferber, J.: Multi-agent systems - an introduction to distributed artificial intelligence. Addison-Wesley-Longman (1999)
6. Jennings, N.R., Sycara, K., Wooldridge, M.: A roadmap of agent research and development. Int. Journal of Autonomous Agents and Multi-Agent Systems 1(1), 7–38 (1998)
7. Mandiau, R., Champion, A., Auberlet, J.-M., Espié, S., Kolski, C.: Behaviour based on decision matrices for a coordination between agents in a urban traffic simulation. Appl. Intell. 28(2), 121–138 (2008)
8. Popovici, D., Desertot, M., Lecomte, S., Peon, N.: Context-aware transportation services (cats) framework for mobile environments. IJNGC 2(1) (2011)
9. Ruskin, H.J., Wang, R.: Modelling Traffic Flow at an Urban Unsignalised Intersection. In: Sloot, P.M.A., Tan, C.J.K., Dongarra, J., Hoekstra, A.G. (eds.) ICCS-ComputSci 2002, Part I. LNCS, vol. 2329, pp. 381–390. Springer, Heidelberg (2002)
10. Vercouter, L., Jamont, J.-P.: Lightweight trusted routing for wireless sensor networks. In: Demazeau, Y., Pechoucek, M., Corchado, J.M., Pérez, J.B. (eds.) PAAMS. AISC, vol. 88, pp. 87–96. Springer, Heidelberg (2011)

Nervousness in Dynamic Self-organized Holonic Multi-agent Systems

José Barbosa, Paulo Leitão, Emmanuel Adam, and Damien Trentesaux

Abstract. New production control paradigms, such as holonic and multi-agent systems, allow the development of more flexible and adaptive factories. In these distributed approaches, autonomous entities possess a partial view of the environment, being the decisions taken from the cooperation among them. The introduction of self-organization mechanisms to enhance the system adaptation may cause the system instability when trying to constantly adapt their behaviours, which can drive the system to fall into a chaotic behaviour. This paper proposes a nervousness control mechanism based on the classical Proportional, Integral and Derivative feedback loop controllers to support the system self-organization. The validation of the proposed model is made through the simulation of a flexible manufacturing system.

José Barbosa · Paulo Leitão
Polytechnic Institute of Bragança, Campus Sta Apolnia,
Apartado 1134, 5301-857 Bragança, Portugal
e-mail: {jbarbosa,pleitao}@ipb.pt

José Barbosa · Emmanuel Adam · Damien Trentesaux
Univ. Lille Nord de France, F-59000 Lille, France
e-mail: {emmanuel.adam,damien.trentesaux}@univ-valenciennes.fr

José Barbosa · Damien Trentesaux
UVHC, TEMPO research center, F-59313 Valenciennes, France

Paulo Leitão
LIACC - Artificial Intelligence and Computer Science Laboratory, R. Campo Alegre 102, 4169-007 Porto, Portugal

Emmanuel Adam
UVHC, LAMIH, F-59313 Valenciennes, France

Emmanuel Adam
CNRS, FRE 3304, F-59313 Valenciennes, France

J.B. Pérez et al. (Eds.): Highlights on PAAMS, AISC 156, pp. 9–17.
springerlink.com

1 Introduction

The need to develop more flexible and adaptive factories that better addresses the current requirements imposed to manufacturing companies, such as flexibility, reconfiguration and responsiveness, is an important issue in the Factory of the Future (FoF) program and implies the adoption of new production control structures. Holonic manufacturing and multi-agent systems [4] are examples of suitable approaches that overcomes the typical problems exhibited by traditional centralized control structures, e.g. the rigid, monolithic structures. In fact, they offer an alternative way to design control systems based on a set of distributed, autonomous entities that may cooperate to reach the systems' goals. These distributed entities, the holons, have a local perspective of their world surrounding and may suffer from information deficiency and information quality, meaning that any decision is based on incomplete and even, at some point, inaccurate/outdated information. In these holonic systems, holons take decisions in a myopic manner relying only of the existing local information.

Self-organization properties [1] embedded in holonic manufacturing systems, can contribute to improve the system performance. In the same way, in such dynamic and continuous self-adaptive systems, some undesired instability, due to the presence of nervousness, may appear and should be taken into consideration. Nervousness can be characterized by or showing emotional tension, restlessness and agitation, i.e. as the frequency that a holon changes its intentions. As expected, a repeated changing of plans can originate undesired behaviour, leading to a loss of the system performance, and in extreme cases, into pitfalls or chaotic behaviour. Since the objective is to push these systems to their limits but maintaining them under control, the study of the system nervousness is crucial to design dynamic self-organized holonic and multi-agent systems, and particularly, how to control the nervousness of each individual holon.

This paper discusses the control of nervousness in dynamic self-organized holonic structures, namely overcoming the reasons for the occurrence of instability in distributed adaptive systems and identifying the benefits and liabilities associated to nervousness. A nervousness control mechanism is proposed based on the classical Proportional, Integral and Derivative (PID) feedback loop controllers. Aiming to validate the proposed approach, an experimental manufacturing cell was simulated using a self-organized agent-based system embedded with the proposed nervousness mechanisms. The rest of the paper is organized as following: the next section explains the importance of having reactive holons balancing the local and global aspects of the self-organized distributed system. Section 3 depicts the PID based proposed model for the nervousness control to be embedded in each holon, and section 4 describes the experimental case study. Finally, section 5 rounds up the paper with the conclusions and future work.

2 Nervousness and Equilibrium in Dynamic Systems

The new manufacturing control paradigms have foundations in decentralized control over distributed entities, i.e. the global control is distributed by the entities in the system. The decentralized nature of these systems allows the achievement of responsiveness to external perturbations since they can locally react to the disturbance in a faster way. This situation turns on the need of the system to display self-organization properties allowing an automatic arrangement of the system entities to allow the system to fulfill its goals.

Latest concept developments of self-organization state that self-organizing systems include not only adaptation and learning mechanisms but also stable structures. Additionally, self-organization can also be achieved by the global cooperation in dynamical systems by means of emergence properties which consist of the multiple lower-level interactions, in one hand by entities-entities interactions and on the other by entities-environment interactions. A deep study of this issue can be found in [3].

In such dynamic self-organized systems, some instability may occur, resulting of the nervousness of the individual entities. For this purpose, the system nervousness should be balanced trying to push the system into its limits but maintaining it under control, allowing the system to dynamically evolve safely into different structures maintaining high performance levels. This idea is in line with the application of theory of chaos as seen in [6].

In this work, nervousness is characterized by the frequency that an entity changes intentions during its life-cycle. Naturally, an entity is nervous if it is constantly changing its intentions and in opposite an entity is calm when does not change intentions. The pertinent question is how to know that a new solution can bring better results than the previously one and in the case of distributed systems if it will negatively affect the system performance. Also vital is to translate this phenomenon into the system design allowing to push it to its limits keeping it under control.

Parunak et al. use two mechanisms to calm hyperactive agents [8]. The first one uses a stigmergy like mechanism used by societies of insects, allowing to merge multiple local data in order to take decisions. The second mechanism delineates a way to separate exploration and exploitation times relating them to the imposed deadline time.

Having in mind the same objective, Hadeli et al. state that, in case an agent wants to change the initial ideas, it is necessary to make a quality measurement of the new solution and only if the new solution is significantly better than the previous one the agent can change its intentions [5]. Additionally, the frequency of changing the initial intentions is limited, restringing the continuous willing to change. Another example can be found in [9] where the inspiration from magnetic fields helps products to choose the most adequate route. In this mechanism, as resources are being occupied, the attraction power emitted (designated by potential field) is reduced allowing products to route in a decisive manner.

The referred approaches allow to control the nervousness phenomena; however the goal of this paper is to go further and introduce a nervousness mechanism that, in a natural manner, is associated to the control mechanism and simultaneously supports the dynamic self-organization.

3 A PID Based Approach for the Nervousness Control

In the classical feedback control theory, one of the most and effective mechanism used to control discrete or continuous systems is the PID controller [7], namely its discrete version which allows to be implemented into more sophisticated processing units. This mechanism, adjusting some key parameters, allows a quick reaction to the perturbation combined with the elimination of the error in a steady state. In practice, when different functioning conditions are needed, e.g., a different temperature in a room, the PID controller adjusts in a quick and effective way, by setting the variables to the new set-point. Having in mind these principals, the inspiration of the PID control was used to design a nervousness control mechanism for self-organized systems, as illustrated in Fig. 1.

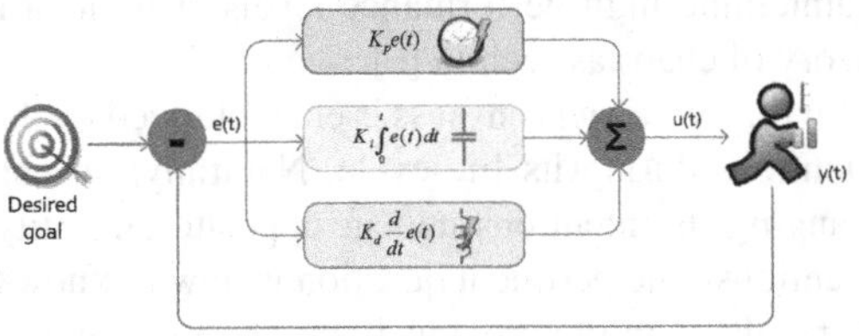

Fig. 1 PID controller block diagram for the nervousness control

The stability and equilibrium analysis can be performed in analogy to the transient response or natural response behaviour. The response can be classified as one of four types, describing the system output in relation to the steady-state value (in this case, the adaptation response): *Overdamped* response that does not oscillate about the steady-state value but is slow to reach it, i.e., exhibits a slow reaction to the perturbation; *Critically damped response* that reaches the steady-state value the fastest without being underdamped - in an ideal case, there should be no oscillation about the steady state value; *Underdamped response* that oscillates within a decaying envelope, exhibiting more oscillations and longer to reach steady-state being as more underdamped is the system - it presents a fast adaptation response but is nervous and can lose the stability; *Unstable response*, where the system is not able to respond to the set point and gets uncontrolled.

Since the objective is to avoid the system instability, the last type of response can make the system fall into a chaotic behaviour and should be avoided. The achievement of a fast adaptation response maintaining the system stable and predictable is a balance between adjusting the damping parameter of the system/holons. Note that if the system has overdamping, the system response may be inadequate, i.e., too slow, due to the time needed for adaptation.

The question is how to handle the local behaviours to reach more adaptive systems with less nervousness. Using the analogy with the PID controllers found in the control theory, the idea is to adjust the coefficients K_p, K_i and K_d that define the local behaviour of the holons, as illustrated in the equation (1).

$$f(t) = K_p e(t) + K_i \int_O^t e(\tau)d\tau + K_d \frac{d}{dt} e(t) \tag{1}$$

In the proposed approach, the meaning of each component is: K_p, i.e., the proportional part, is related to the time to react to the perturbation (meaning the adaptation response); K_i, i.e., the integral part, is related to the error in steady-state state (meaning reaching the optimal state); K_d, i.e., the derivative part, is related to the speed of the stabilization eliminating the error (meaning the responsiveness of achieving the optimal state).

Adjusting each parameter of the controller will culminate in the desired performance curve for a given system. The consequences and cautions that must be taken when adjusting the parameters can pass by the decrease of rise time, increase of overshoot, increase of the steady state error and a degradation of the system stability. Individualizing and extrapolating into holonic multi-agent systems, the increase of K_p can lead to the increase of the overshoot and the degradation of the stability, that can be seen as a temporary difference to the desired behaviour (see Fig. 2_{left}). The K_i parameter drive the system to have higher overshoot and settling time which is counterbalance by a significant decrease of the steady state error, leading the system to its goal with more accuracy (see Fig. 2_{middle}). Finally, the K_d parameter can influence more the system stability but improves the overshot, allowing the entity/system not to greatly drift from desired behaviour in initial time (see Fig. 2_{right}).

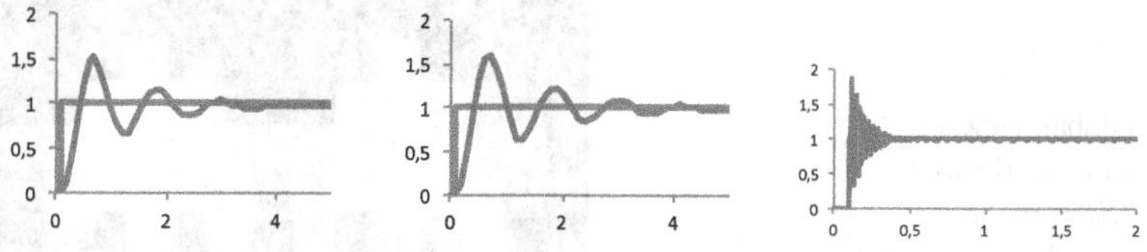

Fig. 2 Response curves adjusting K_p, K_i and K_d

The stability and the convergence to the global optimization can be affected if the individual entities self-learn, self-adapt and self-organize taking different directions. Nevertheless, self-learning capabilities embedded into the individual behaviours can gather valuable information that assists the auto-tuning of the PID control parameters on-the-fly (i.e. without the need to stop, re-program and re-start the system). Although the PID controller parameters can be adjusted in an offline mode, allowing an easier development, the controller has more potentialities if the parameters are adjusted in real time to face the changes of functioning conditions.

In distributed systems, an important question is how to guarantee the overall system equilibrium based on the control of the nervousness of the individual entities. Being the proposed model more suitable to be developed into individual entities, a global unifying solution pass by self-organization consideration such as found e.g., in societies of ants. Alternatively, rules could be deployed in which the eager that individual entities have to fulfill their goals decrease as those goals are being accomplished. The consequence would be that entities would decrease their nervousness levels alongside the goals fulfillment.

4 Case Study

The proposed model, described in the previous section, was used to solve the nervousness of a self-organized agent-based system applied to route pallets in a flexible manufacturing system. This section describes the case study and the experimental tests performed by using the Netlogo framework[1].

System Description. The flexible manufacturing system, illustrated in Fig. 3, is composed by two independent cells each one having two conveying levels joined by lifters. The first level is comprised by 9 interconnected conveyors while the lower level has 2 conveyors. In each cell, the left lifter is able to transport the product in an upper way and the right lifter in the opposite direction. The 2 cells are joined at both levels i.e., in the upper level the pallets can forward from the first cell to the second cell and in the lower level the pallets can leave the second cell re-entering the first cell. The global entrance of the system is made in the first cell through the upper level in the left lifter, being the output made on the right lifter of the second cell.

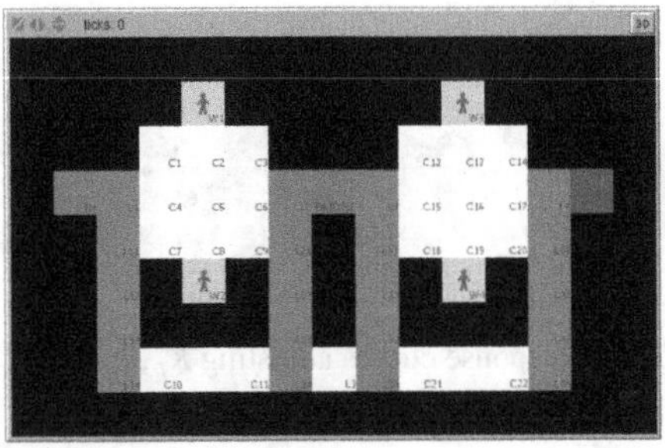

Fig. 3 Global view of the system. Workstations: $WS_1 = [S_1, S_2]$; $WS_2 = [S_1]$; $WS_3 = [S_2, S_3]$; $WS_4 = [S_2]$. List of products: $A = [S_1 S_2 OUT]$; $B = [S_3 OUT]$

The manufacturing system comprises 4 workstations, associated with conveyors C_2 (WS_1) and C_8 (WS_2), for the first cell, and C_{13} (WS_3) and C_{19} (WS_4) for the second cell. Each workstation performs a set of services, as described in Fig. 3.

[1] see `http://ccl.northwestern.edu/netlogo/`

The system includes a catalogue of products comprising two different types of products (A and B), each one routing through the conveyor system and fulfilling its process plan, leaving the system afterwards (see the process plan in Fig. 3).

The self-organized agent-based control system for this case study was implemented using the Netlogo framework, which allows a fast and powerful proof of concept of the developed model [2]. The system comprises the pallets and resources agents, representing respectively the pallets circulating in the system and the physical resources, e.g., conveyors, lifters and workstations. Shortly, the pallets circulate in the system and when facing a decisional node (i.e., situation where the pallet can convey in more than one direction), the pallet agents should decide in which direction should be conveyed taking into consideration the process plan.

Experimental results. The products arrive the system with an optimal plan (plan A), created offline, and with a backup plan (plan B), used if necessary. Obviously, in optimal functioning conditions, there would be no need to change between plans, since the system is well tuned to cope with the processing times and arrivals of new products into the system. However, if something gets out of system optimality, e.g., due to processing time delays, the system should self-adapt taking into consideration the possibility to change the initial plan according to its nervousness level. In fact, the commutation between plans is made "throwing the coin into the air", generating a random number, and verify if the obtained result is lower than the pre-defined nervousness level. The pallet keeps up with the optimal plan, if the obtained result is lower than the pre-defined nervousness level, changing it otherwise. This mechanism is triggered when the product isn't able to perform the optimal plan, e.g., if the product next routing conveyor is occupied by another product.

Varying the nervousness level between 0% to 100%, i.e. where 0% indicates extreme calm (only follows the optimal plan) and 100% indicates extreme nervousness, the functioning boundaries of the system can be defined. The Mean Lead Time (MLT) of the performed tests showed that the system works in a narrow band of nervousness level, observing also an increase of the MLT with the decrease of the nervousness.

The discussion can be pushed to the limits by observing two particular cases. In Fig. 4a, a pallet with no nervousness (absolutely calm) could be conveyed to the right conveyor, but it remains waiting to execute the optimal plan, introducing jam in the system when alternatives routing are present. The second case, exhibiting high levels of nervousness, is observed in Fig. 4b, where the pallet is conveyed to the right when it could wait for the movement forward of the pallet occupying the desired conveyor, since the next conveyor would be available shortly.

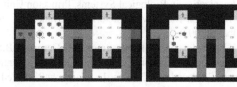

Fig. 4 Extreme nervousness states: a)(left) hyper-calm agent; b) (right) hyper-active agent

Experiments were also conducted to illustrate the benefits of having the presence of nervousness in distributed systems. The results were obtained by fixing the nervousness level at 5% (calm but not extremely) and by varying the arrival time of the pallets to the system. It is observed that for an arrival time between 7 t.u. and 16 t.u., the makespan decreases gradually, increasing after that. One could expect that giving enough time for the pallets fulfill the optimal plan without disturbing the other pallets would be the optimal solution. The problem found is that in this situation, the optimal plan only comprises the allocation of one workstation per service, where there are more available workstations that provide the same service.

Ultimately, the proposed mechanism allows a better utilization of all available resources in the system than other control solutions, especially when the optimal plans can't be fulfilled.

5 Conclusions

Holonic multi-agent systems are suitable control approaches that copes with adaptation issues which is a "must have" in FoF. In such distributed systems exhibiting self-organization capabilities, the system is evolving dynamically to environmental changes, giving origin to some instability due to the nervousness of the individuals. Nevertheless, the nervousness of the system, if controlled, is benefit to achieve better performances.

The paper proposes a mechanism for the nervousness control based on the PID control theory, that combines three parameters related to the steady state, the need for fast reaction and the accuracy of the response. The experimental tests, using a system for routing pallets in a manufacturing system, have shown promising results. The proposed model can be applied into a set of different paradigms where the distributed entities could be holons, agents, services or components.

Despite that, future development will be centered in the dynamic and on-the-fly optimization of the PID parameters and in the study of the influence of the nervousness of individual entities into the global system nervousness.

References

1. Banzhaf, W.: Self-organizing systems. In: Meyers, R.A. (ed.) Encyclopedia of Complexity and Systems Science, pp. 8040–8050. Springer, Heidelberg (2009)
2. Barbosa, J., Leitao, P.: Modelling and simulating self-organizing agent-based manufacturing systems. In: Proceedings of the 36th Annual Conference on IEEE Industrial Electronics Society (IECON 2010), pp. 2702–2707 (2010)
3. Camazine, S., Franks, N.R., Sneyd, J., Bonabeau, E., Deneubourg, J.L., Theraula, G.: Self-Organization in Biological Systems. Princeton University Press, Princeton (2001)
4. Deen, S.: Agent-Based Manufacturing: Advances in the Holonic Approach. Springer, Heidelberg (2003)
5. Hadeli, K., Valckenaers, P., Verstraete, P., Germain, B.S., Van Brussel, H.: A study of system nervousness in multi-agent manufacturing control system. In: Brueckner, S.A., Di Marzo Serugendo, G., Hales, D., Zambonelli, F. (eds.) ESOA 2005. LNCS (LNAI), vol. 3910, pp. 232–243. Springer, Heidelberg (2006)

6. Hogg, T., Huberman, B.A.: Controlling chaos in distributed systems. IEEE Transactions on Systems, Man, and Cybernetics 21, 1325–1332 (1991)
7. King, M.: Process Control: A Practical Approach. John Wiley & Sons (2010)
8. Parunak, H.V.D., Brueckner, S., Matthews, R.S., Sauter, J.A.: How to calm hyperactive agents. In: AAMAS, pp. 1092–1093. ACM (2003)
9. Zbib, N., Pach, C., Sallez, Y., Trentesaux, D.: Heterarchical production control in manufacturing systems using the potential fields concept. Journal of Intelligent Manufacturing, 1–22 (2010)

A Typology of Multi-agent Reorganisation Approaches

Emmanuelle Grislin-Le Strugeon

Abstract. Dealing with dynamic and complex environment can require to develop a distributed system that is able to adapt dynamically its organisation. Many agent-based solutions exist, the problem is to select the most appropriate one. We propose to make a step toward a multi-agent reorganisation guide in providing a typology of the approaches. The typology is based on what is the subject of the adaptation and in which way the adaptation changes the organisation. The subjects of the adaptation are described on the base of usual concepts of agent organisation meta-models.

1 Introduction

Dynamic environments require system adaptation. Widely used in multi-agent systems, the organisational adaptation is a method to maintain the system's functional behaviour in spite of internal or external events that modify its constraints, e.g., addition/suppression of resources, task load changes, etc. Indeed, and especially in distributed systems, the organisation is one of the key component of the system behaviour. In this article, the organisation is viewed as "a decision and communication schema [...] applied to a set of actors that together fulfils a set of tasks in order to satisfy goals while guarantying a global coherent state" [20].

Numerous solutions are proposed for the adaptation of agent organisations. From an engineering perspective, the difficulty is to select the one which is appropriate to the current problem. Indeed, the diversity of the approaches is combined to a lack of relative positioning and a lack of common description framework: most methods are described using their own underlying organisational models. Some attempts have been made to propose a unified meta-model [2, 4]. However, they do not cover the extent of the concept domain that is used in the reorganisation approaches.

Emmanuelle Grislin-Le Strugeon
Univ. Lille Nord de France, UVHC, LAMIH-UMR 8201, F-59313 Valenciennes, France
e-mail: emmanuelle.grislin@univ-valenciennes.fr

J.B. Pérez et al. (Eds.): Highlights on PAAMS, AISC 156, pp. 19–27.
springerlink.com

In this context, the aim of this paper is to propose a typology framework to class the reorganisation mechanisms according to a common meta-model. In Section 2, we select some of the fundamental concepts used to describe an agent organisation. On the base of these concepts, a typology is proposed in Section 3, to describe and classify the changes made to the organisation components. This is illustrated by several approaches from the literature. Section 4 discusses related works and Section 5 gives a conclusion and proposes future works.

2 Modeling an Agent-Based Organisation

Our aim is to be able to class reorganisation mechanisms in describing what is modified in the organisation, and in which way it is modified. To describe what is modified requires to model the elements of the organisation and the relations between them. Organisation meta-models provide such a description. But, in spite of the efforts made to compare and integrate them [2, 4], there is still no consensus in the field, and almost each reorganisation method is provided with its specific meta-model. We propose thus to use a restricted set of concepts to describe the reorganisation approaches; when a concept has distinct names in different meta-models, we have selected one general term to represent it.

The concepts we use are represented in Fig.1: a **multi-agent organisation** is described by a set of agents, organised as a set of organisational components (OCs), in order to achieve a set of goals. An **agent** is a concrete active software entity that exhibits an autonomous and proactive behaviour [29]. An **organisational component (OC)** represents a member of the organisation, individuality or group, as a role holder, similarly to the *position* in [25]. An OC is either **atomic**, i.e. a component that is not further decomposable, or **composite**, i.e., a component made of sub-components organised according to a structure, like a *group* in AGR [11] or the *organisation* in Gaia [30]. The **structure** describes the relations (e.g., dependency,

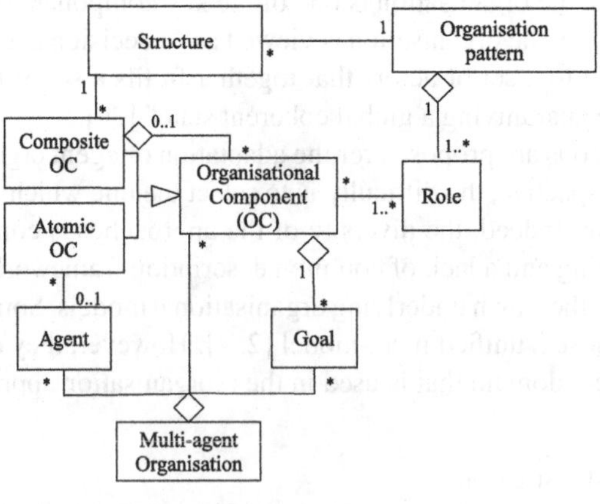

Fig. 1 The organisational concepts used in the typology.

authority) between the components [11, 30]. On this base, we can represent heterogeneous organisations, in which the composite components have distinct structures. One agent can be associated to several atomic OCs. The OCs are responsible of **goal** (internal or external final state) achievement and they hold roles in the organisation. A **role** represents the organisation members' relative positions [3], associated to expected behaviours, including responsibilities, obligations and permissions towards the other roles. Some structures have similar bases, mostly similar roles, independently from the number of their components. To represent structures that have common characteristics we use the concept of **organisational pattern**, which is a generic and abstract type of organisation, as the organisational form in [7].

3 Typology of Organisational Adaptation Approaches

Numerous approaches enable the agents to adapt their organisation according to new needs or constraints. The detection of a new situation that may trigger a reorganisation is rather homogeneous for all the approaches, realised using a measure of the system utility [7, 19], performance [23], or workload [1]. The triggered adaptation actions are much more various (some of them are illustrated in Fig.2):

- agents entering or leaving the organisation,
- creating or deleting components (group formation/dissolution),
- adding/deleting role holders,
- substituting a component for another,
- changing agent role assignments,
- changing a structure.

We propose to classify the dynamic and organisational adaptation approaches according to the organisational elements modified, represented by some of the concepts previously introduced in the organisation description. Five concepts — Agent, OC's role, Agent-OC relation, Goal-OC relation and OC's structure — have been selected according to the adaptation actions we have listed in the literature. The resulting classification is a typology: it separates the approaches according to concepts in the aim to provide a basis for comparison; the proposed categories are neither exhaustive nor mutually exclusive. Indeed some of the approaches realise changes that require to describe actions of different types, while others can be classified in one specific type of action. In the following, we use the proposed typology to classify several examples of reorganisation approaches selected from the literature.

Agent Changes. The approaches making changes in the agent set consider the agents as resources that can be modified according to the needs. Dynamical changes of agent's capabilities is an approach used to adapt the system to its environment. Learning agents and systems that enable the dynamic acquisition of competencies (e.g., [22]) belong to this category. Such changes have effect on the organisation when they are used to be able to have a sufficient amount of agents having the required capabilities without changing the overall quantity of agents. When the amount of agents in the system can be modified, most of the approaches increase

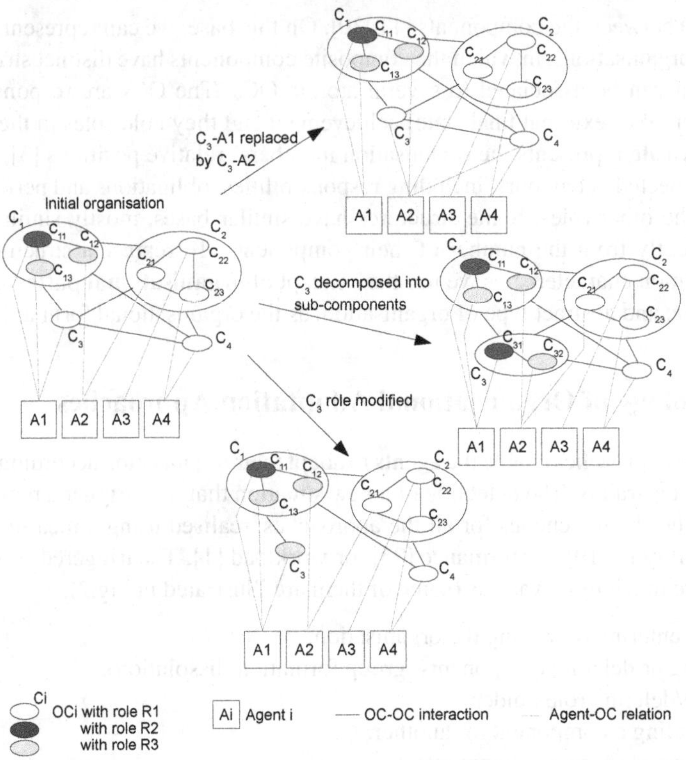

Fig. 2 From an initial agent organisation, three examples of reorganisation actions

this amount, with actions like critical agent duplicating [8] and agent cloning [18]. Suppressing agents is generally not useful, at worst they are "inactivated" [1] in case they would be needed again.

OC's Role Changes. Making the roles evolve is another way to adapt the system. Individual redefinition of roles by the agents is proposed in [1] but this requires to manage the differences between prescribed roles and roles really played by the agents (see [31] about the problem of role management). A recent proposition [14] makes the role taxonomy evolves with the social dynamics of the system (e.g., agents entering/leaving the system, agents' behavioural changes): roles are periodically re-discovered using a clustering algorithm. Creating and deleting role instances are similar to changes made to the inter-agent relationships, like in [21] and [19]. In the former, the adaptation concern the decision-making control relationships and the authority relationships: agents are able to change dynamically their individual decision-making interaction styles. In the latter, each pair of agents can be linked by one of three possible kinds of relations, i.e., purely acquaintance, peer or authority relation. The substitution of the relation by another one can be seen as switching for another role.

Agent-OC Relation Changes. The Agent-OC relation concerns role assignment to the agents via the agents' positioning as components of the organisation. In addition to the techniques described in Section 3 with the redefinition of roles, or in Section 3 with actions on the agents' characteristics, adaptation can be realised by agents adopting or abandoning roles [24]. In TRACE [9], such actions are realised by the resource allocation protocol. In [15], the reorganisation is based on dynamic role reallocation. The system adaptation to unforeseen events is realised by agent-role allocation changes using specific Adaptor roles. The adaptation is triggered according to performance indicators. The reorganisation policy described with OMACS [5] is based on new agent assignments according to the agents capacity to achieve the goals. The assignment is determined by evaluating the different possible agent-role-goal combinations. Using our framework, an agent-role-goal assignment is equivalent to the combination of two relation instances: agent-component and goal-component relations. Considering the first of these, the agents' new assignments can be described as the creation and the destruction of agent-component relations.

Goal-OC Relation Changes. The Goal-OC relation concerns goal assignment, i.e., the allocation of tasks or the commitments in BDI based systems. The goal assignment modification, as a part of the reorganisation policy described in [5] gives an example of Goal-OC relation changes. Such reorganisations are applied to find the best assignment according to the agent resources and the role possibilities in a given structure. The task allocation in [9] is another type of such mechanisms, performed by specific *Resource Manager* agents. The task-oriented mechanism proposed in [28] allows both task amplification and task aggregation. Such modifications are also realised by the breaking and merging operations on the task structure in [18].

OC's Structure Changes. The agents can possess the ability to make the organisation grows to face work overload, such as the holonic agents in [1] which can create sub-groups: an atomic OC is replaced by a composite OC, this one taking in charge the tasks that was allocated to the previous, individual component. However, group compositions/decompositions can be seen either as qualitative changes if we consider them as atomic/composite transformations, or as quantitative changes if we consider them as creation/destruction of components (see other examples below). In-depth structural changes, like those proposed in [21] and [13], with eventually the choice of another pattern for the organisation, are rarely proposed because of their high cost. Such deep changes are nevertheless interesting when the new organisation is expected to last, e.g., when the system arrives to a threshold situation and must switch to another type of organisation. For example, a small society based on peer-relation may become inefficient to deal with the arrival of numerous new members. Described in our proposition context, the reorganisation method in [10] consists in the creation/destruction of temporary atomic OCs to hire marketable agents as working resources in permanent composite OC. This allows reallocation of tasks among the agents, to deal with time constraints and load variation in spite of a global constant amount of resources. With another method, in [22] the agents create acquaintances according to the dependency links in the system, which can be seen as the creation of new structures to be laid over the existing organisation. Dynamic team formation is a mechanism used to deal with both the agents' skills and the arrival of new tasks. In [12], the agents initiate and join teams using several strategies for on-line learning of effective organisational structures. In [26], the agents overlay the permanent, skill-based groups with temporary groups whose components depend on the tasks to achieve. Such changes in the OC can come with quantitative changes in the agent set. The aim is generally to lighten the tasks of the organisation members. The approaches making quantitative changes on the agent set create and delete agents according to the needs, like in the OSD (Organisation Self-Design) methods [17, 18]. In these methods, adding agents allows task distribution and goal aggregation is achieved by removing some agents.

4 Related Works and Discussion

While many works propose meta-models for the agent organisation, few of them propose criteria for a classification in the domain of dynamic organisation. The classification given in [6] distinguishes behavioural adaptation, a modification applied to the behaviour of agents enacting organisational roles, from structural adaptation, a modification applied to the organisational structure. Our classification is consistent with this behavioural/ structural distinction, but in an attempt to detail the changes of both categories. The behavioural adaptation category involves the agent changes, the qualitative OC's structure and role changes, the agent-OC relation and the OC-role relation changes of our typology, while the structural adaptation category is similar to the OC changes type. The classification criteria proposed in [27] differentiates the agent approaches according to two axes: the agent-centred/organisation-centred points of view; the reorganisation/self-organisation processes. Our proposition introduces other classification criteria, which are compatible and complementary with theirs. We have more particularly detailed the reorganisation approaches and our conceptual model is characterised by an organisation-centred point of view. However, no assumption has been made on the agents' knowledge, i.e., local or global, of the organisation. Indeed, in our typology, agent-centred approaches and organisation-centred approaches have both been classified. And even if the agents are unaware of the global structure, there is a priori no obstacle for using the model and the typology to describe emergent organisations, from an observer, external point of view.

Some of the approaches (e.g., [5],[9],[18]) are partly described as a member of certain type and partly described in another type using our framework. This is due to the distinction made between agents and organisational components. Despite the difficulty this induces in the classification of certain approaches, we have kept the distinction to be able to model the structure independently from the agents that populate the organisation. This allows to describe specific situations, such as a structure that remains identical even if an agent has been replaced by another one. And the model used as a basis for the classification is certainly incomplete in that it does not include all the concepts existing in the agent organisation meta-models, such as the functional and deontic concepts [16].

5 Conclusion

In the absence of a common meta-model of agent organisation, we have proposed a reduced framework to serve as a basis to enable the description of agent reorganisation approaches. We have proposed a typology that separates the approaches according to the changes made on the organisational elements. The typology has been used to classify several examples of reorganisation approaches selected from the literature. The typology will evolve according to the new approaches we will try to classify and according to more detailed description of the classified ones. Especially, the aim is to study further the *role* concept to integrate it more precisely in the description of the OC changes. The aim is also to include emergent approaches in the description, as discussed above.

References

1. Adam, E., Mandiau, R.: Flexible Roles in a Holonic Multi-Agent System. In: Mařík, V., Vyatkin, V., Colombo, A.W. (eds.) HoloMAS 2007. LNCS (LNAI), vol. 4659, pp. 59–70. Springer, Heidelberg (2007)
2. Bernon, C., Cossentino, M., Gleizes, M.-P., Turci, P., Zambonelli, F.: A Study of Some Multi-agent Meta-models. In: Odell, J.J., Giorgini, P., Müller, J.P. (eds.) AOSE 2004. LNCS, vol. 3382, pp. 62–77. Springer, Heidelberg (2005)
3. Cabri, G., Leonardi, L., Ferrari, L., Zambonelli, F.: Role-based software agent interaction models: a survey. The Knowledge Engineering Review 25(4), 397–419 (2010)
4. Coutinho, L.R., Sichman, J.S., Boissier, O.: Modelling dimensions for agent organizations. In: Dignum, V. (ed.) Handbook of Research on Multi-Agent Systems, ch. II, pp. 18–50. Information Science Reference, Hershey (2009)
5. Deloach, S.A., Oyenan, W., Matson, E.T.: A capabilities-based model for adaptive organizations. Autonomous Agents and Multi-Agent Systems 16(1), 13–56 (2008)
6. Dignum, V.: The Role of Organization in Agent Systems. In: Handbook of Research on Multi-Agent Systems, ch. I, pp. 1–17. Information Science Reference, Hershey (2009)
7. Dignum, V., Dignum, F., Furtado, V., Melo, A., Sonenberg, L.: Towards a simulation tool for evaluating dynamic reorganization of agent societies. In: Proc. of the Workshop on Socially Inspired Computing@AISB Convention (2005)
8. Faci, N., Guessoum, Z., Marin, O.: DimaX: a fault-tolerant multi-agent platform. In: SELMAS 2006: Proc. of the 2006 International Workshop on Software Engineering for Large-Scale Multi-Agent Systems, pp. 13–20. ACM, NY (2006)
9. Fatima, S., Wooldridge, M.: A framework for dynamic agent organizations. In: Dignum, V. (ed.) Handbook of Research on Multi-Agent Systems: Semantics and Dynamics of Organizational Models, ch. XVIII, pp. 446–459. Information Science Reference, Hershey (2009)

10. Fatima, S.S., Wooldridge, M.: Adaptive task and resource allocation. In: Agents 2001: Proc. of the Fifth International Conference on Autonomous Agents. ACM, Montreal (2001)
11. Ferber, J., Gutknecht, O., Michel, F.: From Agents to Organizations: An Organizational View of Multi-agent Systems. In: Giorgini, P., Müller, J.P., Odell, J.J. (eds.) AOSE 2003. LNCS, vol. 2935, pp. 214–230. Springer, Heidelberg (2004)
12. Gaston, M.E., desJardins, M.: Agent-organized networks for dynamic team formation. In: Proc. AAMAS 2005, pp. 230–237. ACM, New York (2005)
13. Grislin–Le Strugeon, E., Mandiau, R., Libert, G.: Towards a Dynamic Multiagent Organization. In: Raś, Z.W., Zemankova, M. (eds.) ISMIS 1994. LNCS (LNAI), vol. 869, pp. 203–213. Springer, Heidelberg (1994)
14. Hermoso, R., Billhardt, H., Ossowski, S.: Role evolution in open multi-agent systems as an information source for trust. In: Proc. AAMAS 2010, Richland, SC, vol. 1, pp. 217–224 (2010)
15. Hoogendoorn, M., Treur, J.: An adaptive multi-agent organization model based on dynamic role allocation. Int. J. of Knowledge-based and Intell. Engineering 13(3-4), 119–139 (2009)
16. Hübner, J.F., Sichman, J.S., Boissier, O.: A Model for the Structural, Functional, and Deontic Specification of Organizations. In: Bittencourt, G., Ramalho, G. (eds.) SBIA 2002. LNCS (LNAI), vol. 2507, pp. 118–128. Springer, Heidelberg (2002)
17. Ishida, T., Gasser, L., Yokoo, M.: Organization self-design of distributed production systems. IEEE Trans. on Knowledge and Data Engineering 4(2), 123–184 (1992)
18. Kamboj, S., Decker, K.S.: Exploring Robustness in the Context of Organizational Self-design. In: Hübner, J.F., Matson, E., Boissier, O., Dignum, V. (eds.) COIN@AAMAS 2008. LNCS (LNAI), vol. 5428, pp. 80–95. Springer, Heidelberg (2009)
19. Kota, R., Gibbins, N., Jennings, N.R.: Self-organising agent organisations. In: Decker, Sigchman, Sierra, Castelfranchi (eds.) Proc. AAMAS 2009, pp. 797–804 (2009)
20. Malone, T.: Modeling coordination in organizations and markets. Management Science 33(10), 1317–1332 (1987)
21. Martin, C., Barber, K.: Adaptive decision-making frameworks for dynamic multi-agent organizational change. Autonomous Agents and Multi-Agent Systems 13(3), 391–428 (2006)
22. Mathieu, P., Routier, J.-C., Secq, Y.: Principles for Dynamic Multi-agent Organizations. In: Kuwabara, K., Lee, J. (eds.) PRIMA 2002. LNCS (LNAI), vol. 2413, pp. 109–122. Springer, Heidelberg (2002)
23. Matson, E., DeLoach, S.: Formal transition in agent organizations. In: Proc. KIMAS 2005, Waltham, MA, USA (2005)
24. Nair, R., Tambe, M., Marsella, S.: Role allocation and reallocation in multiagent teams: Towards a practical analysis. In: Proc. AAMAS 2003, pp. 552–559. ACM Press (2003)
25. Odell, J., Nodine, M., Levy, R.: A Metamodel for Agents, Roles, and Groups. In: Odell, J.J., Giorgini, P., Müller, J.P. (eds.) AOSE 2004. LNCS, vol. 3382, pp. 78–92. Springer, Heidelberg (2005)
26. Petit-Roze, C., Strugeon, E.G.L.: MAPIS, a multi-agent system for information personalization. Information and Software Technology 48, 107–120 (2006)
27. Picard, G., Hubner, J., Boissier, O., Gleizes, M.P.: Reorganisation and Self-organisation in Multi-Agent Systems. In: ORGMOD@ Petri Nets 2009, Paris, pp. 66–80. Springer, Heidelberg (2009)

28. Shin, D., Leone, J.: AM/AG model: a hierarchical social system metaphor for distributed problem solving. Int. J. of Pattern Recognition and Artificial Intelligence 4(3), 473–487 (1990)
29. Wooldridge, M., Jennings, N.R.: Intelligent agents: Theory and practice. The Knowledge Engineering Review 10(2), 115–152 (1995)
30. Zambonelli, F., Jennings, N.R., Wooldridge, M.: Developing multiagent systems: The Gaia methodology. ACM Trans. on Software Engineering Methodology 12(3), 317–370 (2003)
31. Zhu, H.: Fundamental issues in the design of a role engine. In: Proc. Int. Symposium on Collaborative Technologies and Systems, Irvine, CA, USA, pp. 399–407 (2008)

Towards a Meta-model for Natural Computers: An Example Using Metadepth

Marina de la Cruz and Suzan Awinat

Abstract. Model Driven is currently one of the most promising approaches to software engineering. One of the topics related to this area is the possibility of defining models that could also be considered as meta models, that is, that could be instantiated to get a less abstract model. Most of the current tools available allow the definition of two levels: the model that can be instantiated (and hence be considered a meta model) and the specific model we can get after instantiating the first one. Nevertheless some domains require more than two levels. One of them can be found in the realm of the so called natural computers. This domain includes several models that share different characteristics. Each model could be instantiated to define particular cases able to solve specific problems. So we need to define a first level with the common features of all the natural computing models. This first level will be instantiated to get the particular models of natural computing. These models will be finally instantiated to get the specific systems we could use to solve a given task. MetaDepth is one of the first tools that offers the designers the possibility of defining a chain of models of this kind. This paper shows a first step for using MetaDepth in natural computing.

1 Motivation and Introduction

This paper is focused on *Membrane Computing* (also called *P systems*, from its main author Gheorge Păun [11]) and NEPs (Networks of Evolutionary Processors [2]), two new models of computation in the realm of *Natural Computing*.

P systems abstract the processes taking place in the compartimental structures of the living cells to consider them as computations. This compartimental structure is formalized as an external membrane (called *skin*) that contains one or more sibling

Marina de la Cruz · Suzan Awinat
Escuela Politécnica Superior, Departamento de Ingeniería Informática,
Universiad Autónoma de Madrid
e-mail: `name@email.address`

J.B. Pérez et al. (Eds.): Highlights on PAAMS, AISC 156, pp. 29–36.
springerlink.com

membranes each of which has the same structure (they contain one or more membranes with, again, the same structure). Each membrane is directly included only in one membrane that is usually considered its father. The biochemical contents of the cells in the living beings are represented by means of multisets of symbols (set of symbols in which more than one copy of each symbol is allowed) and a set of rules that consume some symbols of the multiset to produce others. Different families of P systems allow the creation and dissolution of membranes as well as different mechanisms for carrying symbols across the membranes. P systems are inherently parallel, both in the selection of symbols consumed and in the application of the rules. All the membranes apply their rules and, once they have finished, update their contents. These steps are given until reaching some stopping configuration. Different theoretical results have been reported with respect to their equivalence with Turing machines [11] and to the, at least theoretically, linear-time performance for NP-problems [7]. In these papers, the reader will also find a formal and complete definition of Psystems.

NEPs abstract those systems that connects a set of *small* and *simple* nodes. These nodes are able to contain some information and perform very simple operations on it and share part of their contents with the other nodes across their connections. The information is represented by means of strings. Main operations are to change, add or remove some symbol. It is important to highlight that each node has to store all the possible results of applying to all its strings all its rules in all the possible ways. Each processor filters both, its incoming strings and those that it will share with the net. NEPs are synchronous, that is, all the processors modify their contents at the same time; then, they use the net to share some of their strings. These two steps are iterated until satisfying a predefined stopping condition. NEPs are inherently parallel and are able to solve NP problems with polynomic resources. A formal and detailed definition of NEPs could be found in the referred literature.

It is easy to realise the similarities between both models; not only in their expressive power (equivalent to Turing machine and able to get polynomic performance for classic NP problems) but also with respect to their structure (an inherently parallel net of nodes that process information in a collaborative way).

Both, P systems and NEPs, could be considered architectures alternative to that present in our conventional computers (von Neumann's architecture). This circumstance makes it possilble to design also new programming languages and other developping tools for them, given that we could consider them as new programming paradigms.

There are several research groups interested in programming tools for natural computers. P-Lingua (developed by the Research Group on Natural Computing of the University of Sevilla) is a programming language for membrane computing which aims to be a standard to define P systems. One of its main features is to remain as close as possible to the formal notation used in the literature to define P systems. The programmer will not have to do any additional effort to describe his P systems with P-Lingua once he has formalized them. P-Lingua is also the name of a software package that includes several built-in simulators for each supported model as well as the needed compilers to simulate P-Lingua programs. More details can be found at `http://www.p-lingua.org` and [6]. NEPsLingua [4] is a textual programming language for NEPs that is being developed (hardly inspired in P-Lingua) by the research group of Biocomputation of the Universidad Autónoma de Madrid. This group has also developed a Java and Python complete developing environment for NEPs. It includes a Java simulator (jNEP [5]), a visual programming language (NEPsVL [8]) and a graphical viewer for the execution of NEPs (jNEPview [3]).

These researchers are also intereseted in providing the community with a set of coherent and similar tools to minimize the learning effort of potential users. One of their goals is to classify the different families and variants of both models and to take into account both the similarities and the differences. The reader could find in [1] an attempt to classify different types of bioinspired models similar to NEPs.

Modern software engineering is interested in reusability and model driven approaches. One of the characteristics of these approaches is the abstraction of common features that are put together in abstract models that are further instantiated to define specific cases suitable to given problems and circumstances. There are currently also a lot of design tools that enormously ease the design of programming language and developping environments. It is easy to realize that there is a situation very similar to the one depicted above while we was introducing NEPs and Psystems. If we wan to to apply these approaches to natural computers we need a technique that allows the definition of different related models (some of them will be instances of the others) with different levels of abstraction. For example: we could use the most abstract level to contain the common characteristics to all the natural computers under consideration. This *NaturalComputer* model could be instantiated differently to get more specific natural computers (such as NEPs or Psystems), it could be also advisable to be able to get more specific types of these systems (there are different families of NEPs and Psystems). Finally we could instantiate the models into objects to get the final and specific NEP and Psystem able to tackle a given task.

MetaDepth [9] is one of the most recent, powerfull, and promising tools for meta modelling [10], that allow these kind of multi-level modelling. This paper is devoted to test if MetaDepth could be an appropiate approach for modelling natural computers.

2 A First Meta-model for Some Characteristics of Some Natural Computers: PSys and NEPs

As we have depicted above, our goal is to test the possibility of using MetaDepth as a tool to model natural computers (mainly NEPs and Psystems) in a integrated way. We know that MetaDepth offers the possibility of using other tools to ease the design of different kinds of programming languages, developing environments and even simulators. These applications are exactly those of interest of our research group.

To test MetaDepth for our purposes we need to model some of the main features of NEPs and Psystems trying to put the common components of both models in the

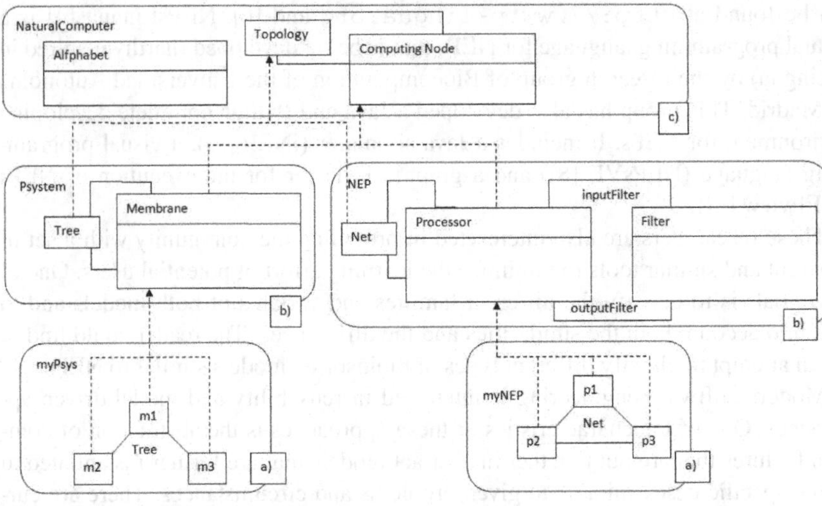

Fig. 1 A fragment of the stack of (meta-)models for some features of some natural computers (Psystems and NEPs)

most abstract level and to define them (NEPs and Psystems) as instances (that is, with possible extensions) of the abstract (meta-)model. We need more than the two typical number of levels (the level of the model or the classes and the level of the instances or objects) because NEPs and Psystems will share, at the same time, the nature of classes (specific NEPs and P syhstems will be instances of these models) and objects (NEPs and Psysetms are defined as instances of the more abstract model, *NaturalComputer*).

The process is described with detail in the paragraphs below.

The highest level of abstraction corresponds to *NaturalComputer*. Our first approach to this meta-model is focused only on the processors able to compute and their connections. We define two MetaDepth elements for this purpose: the node *ComputingNode* and the edge *Topology*, respectively.

Part c) of figure 1 shows the *NaturalComputer* meta-model. For saving space it only contains the MetaDepth nodes *Alphabet* and *ComputingNode* and the edge *Tolopology*.

Figure 2 shows a possible textual MetaDepth description of this meta-model:

- *Alphabet* is a collection of *Strings*.
- *ComputingNode* is identified by means of a String (its *name*) and contains a collection of strings (its *content*). Attributes *fathers* and *offspring*, which are collections of *ComputingNodes*, are used to implement the topology. It is worth highlighting that this node is *ext*ensible. This MetaDepth feature will allow us to change (extend) this node while instantiating it. The processors of the NEPs, for example, are associated to a pair of filters while the processors of Psystems (membranes) are not, so will need to add to the proessors (NEPs) some attribute for their filters.

- The MetaDepth *Edge Topology* implements the connections of the *ComputingNodes*

It is easy to include other components common to Psystems and NEPs such as StoppingConditions or rules. Listing 2 shows, for example, a simplified version of StoppingCondition.

```
Model NaturalComputer@2{

  Node Alphabet{
 symbols: String[*];
  }

 Node StoppingCondition{
end : boolean;
 }

  ext Node ComputingNode{
    name : String ;
    content : String[*];
    fathers : ComputingNode[*];
    offspring : ComputingNode[*];
  }

  Edge Topology (ComputingNode.fathers , ComputingNode.offspring) {}

}
```

Fig. 2 MetaDepth listing for Natural Computers

The next level of abstraction includes two instances of *NaturalComputer*: the models *Psystem* and *NEP*. Part b) of figure 1 shows them.

Computing nodes are called *Membranes* in *Psystem*.

As we have explained above, the topology, in this case, could be represented as a tree because each membrane has only a father. This is way the corresponding attribute of *Psystem* (*included* that is associated with *fathers*) is changed to take into account this constraint in its cardinality (in *Psystem*, *included* is only a *Membrane* instead of a collection). Notice that the topology is named *tree* in this model.

Figure 3 shows a possible textual MetaDepth description of this model:

As we described above, NEPs associate a pair of filters to each processor, and any graph is a valid topology for them.

Figure 4 shows a possible textual MetaDepth description of this model.

- *Filters* are identified by means of a *String* (their attribute *name*). In the literature, filters are determined by means of their type (a number) and the permitting and forbidding languages (that could be implemented as collections of strings). The attribute *myProcessor* is used to implement the relationships between filters and processors.

```
NaturalComputer Psystem{
ComputingNode Membrane {
    included : Membrane {fathers};
    includes : Membrane[*] {offspring};
}

    Topology tree (Membrane.included , Membrane.includes) {}
}
```

Fig. 3 MetaDepth listing for P systems

- The *ComputingNodes* of NEPs are named *Processors*. The attributes *inputFilter*, *outputFilter*, *connected* and *connects* are used for implementing the relationships described above.
- The *Topology* of NEPs is named *net*

```
NaturalComputer NEP{
 Node Filter{
       name: String {id};
   type : int;
   permitting_context : String[*];
   forbidding_context : String[*];
   myProcessor: Processor;
}

 ComputingNode Processor {
   currentSize : int;
   inputFilter: Filter;
   outputFilter: Filter;

   connected : Processor[*] {fathers};
   connects  : Processor[*] {offspring};

 }

Topology net (Processor.connected , Processor.connects) {}

Edge Input (Processor.inputFilter , Filter.myProcessor) {}
Edge Output (Processor.outputFilter , Filter.myProcessor) {}

}
```

Fig. 4 MetaDepth listing for NEPs

The last level of abstraction contains objects that represent specific P systems and NEPs. We have defined a P system (*myPsys*) and a NEP (*myNEP*) to highlight the differences between their topologies. Both examples contain three *ComputingNodes* (*Membranes* and *Processors*, respectively). *myNEP* has a complete graph with tree nodes while *myPsys* has three membranes that form a tree.

To save space, part a) of figure 1 does not include the filters.
Listing 5 shows the complete examples.

```
 Psystem MyPsys{
Membrane m1 { name = "m1"; }
Membrane m2 { name = "m2"; }
Membrane m3 { name = "m3"; }
tree( m2, m1);
tree( m3, m1);
}

NEP myNep{
Processor p1  { name = "p1"; }
Processor p2  { name = "p2"; }
Processor p3 { name = "p3"; }
net (p1,p2) {}
net (p1,p3) {}
net (p2,p3) {}
Filter i1f { name = "i1f"; type = 1; }
Filter o1f { name = "o1f"; type = 2; }
Filter i2f { name = "i2f"; type = 1; }
Filter o2f { name = "o2f"; type = 2; }
Input ( p1, i1f) {}
Output (p1, o1f) {}
Input (p2, i2f) {}
Output (p2, o2f) ;
}
```

Fig. 5 MetaDepth listing for a specific NEP and a specific P system. It shows three nodes per model and different topologies: graphs for NEPs and tres for P systems.

3 Conclusions and Further Research Lines

We have tested that MetaDepth is able to ease de modeling of some of the most characteristic features of NEPs and Psystems in a integrated way: we was able to easily define a meta-model for Natural Computers with their common characteristics and distinguish them while we instantiate it to define the models for NEPs and Psystems. Finally we have instantiated both to define specific systems. Although we have not fully described all the features of all the families and variants of NEPs, Psystems and other similar bioinspired models of computation; this first experiment has shown how easy the design with MetaDepth is, specially to multi-level modeling in this domain. In the future we plan to generalize the use of MetaDepth in the following topics:

- To fully describe NEPs and Psystem, including all the possible variants of NEPs and Psystems.
- To benefit of the integration of MetaDepth and other similar model driven tools to design programming language and simulators.
- To compare the results of this approach with others found in the literature (P-Lingua, NEPsLingua, NEPsVL, etc.)

Acknowledgements. Work partially supported by the Spanish Ministry of Science and Innovation under coordinated research project TIN2011-28260-C03-00 and research project TIN2011-28260-C03-02 and by the Comunidad Autónoma de Madrid under research project e-madrid S2009/TIC-1650.

References

1. Arroyo, F., Castellanos, J., Mitrana, V., Sempere, J.M.: Networks of bio-inspired processors. Accepted to be published in Triangle (2011)
2. Castellanos, J., Martin-Vide, C., Mitrana, V., Sempere, J.M.: Networks of evolutionary processors. Acta Informatica 39(6-7), 517–529 (2003)
3. Cuellar, M., del Rosal, E.: jnepview: a graphical trace viewer for the simulations of neps. In: Proceedings of the 3rd International Work-Conference on the Interplay between Natural and Artificial Computation (2009)
4. de la Cruz, M., Jiménez, A., del Rosal, E., Bel-Enguix, G., Ortega, A.: Neps-lingua: a new textual language to program neps. In: Proceedings of ICAART 2011 (2011)
5. del Rosal, E., Nuez, R., Castaeda, C., Ortega, A.: Simulating neps in a cluster with jnep. In: Proceedings of International Conference on Computers, Communications and Control, ICCCC 2008 (2008)
6. García-Quismondo, M., Gutiérrez-Escudero, R., Pérez-Hurtado, I., Pérez-Jiménez, M.J., Riscos-Núñez, A.: An Overview of P-Lingua 2.0. In: Păun, G., Pérez-Jiménez, M.J., Riscos-Núñez, A., Rozenberg, G., Salomaa, A. (eds.) WMC 2009. LNCS, vol. 5957, pp. 264–288. Springer, Heidelberg (2010)
7. Gutiérrez-Naranjo, M.A., Pérez-Jiménez, M.J., Riscos-Núñez, A.: Towards a programming language in cellular computing. Electronic Notes in Theoretical Computer Science 123, 93–110 (2005)
8. Jiménez, A., del Rosal, E., de Lara, J.: A visual language for modelling and simulation of networks of evolutionary processors. In: Proceedings of PAAMS 2010 - 8th International Conference on Practical Applications of Agents and Multi-Agent Systems (2010)
9. de Lara, J., Guerra, E.: Deep Meta-modelling with METADEPTH. In: Vitek, J. (ed.) TOOLS 2010. LNCS, vol. 6141, pp. 1–20. Springer, Heidelberg (2010)
10. de Lara, J., Guerra, E.: Generic Meta-modelling with Concepts, Templates and Mixin Layers. In: Petriu, D.C., Rouquette, N., Haugen, Ø. (eds.) MODELS 2010. LNCS, vol. 6394, pp. 16–30. Springer, Heidelberg (2010)
11. Păun, G.: Computing with membranes. Journal of Computer and System Sciences 61, 108–143 (2000)

Towards the Automatic Programming of NEPs: A First Case Study

Emilio del Rosal, Alfonso Ortega, and Marina de la Cruz

Abstract. This work shows the first results of our platform for the automatic design of NEPs to solve specific tasks. The platform is based on a genetic programming algorithm that we have proposed earlier. It uses Christiansen grammars to exclude individuals with either syntactic or semantic mistakes. The fitness function required by the genetic engine, usually invokes a simulator of the model under consideration. In this work we use jNEP, a Java simulator for NEPs developed by our research group. We have chosen a non trivial problem borrowed from a NEP that applies context free rules for simulating pushdown automata: the rotation of the strings until finding the symbol to which the rule will be applied, which is one of the three steps this NEP takes. We have found some interesting solutions.

1 Motivation

NEPs are one of the so called *natural or unconventional computers*. A great effort is being devoted to these kind of models, which can be seen as alternative architectures to design new families of computers. Most of them, inspired in the way used by Nature to solve difficult tasks efficiently.

Conventional personal computers are based on the well known von Neumann architecture, that can be considered as an implementation of the Turing machine. Any computer scientist has a clear idea about how to program *conventional* computers by means of different high level programming languages and their corresponding compilers. On the other hand, imagining how to program unconventional computers is quite difficult.

Some of the authors have proposed new evolutionary automatic programming algorithms (Attribute Grammar Evolution, AGE, [4] and Christiansen Grammar

Emilio del Rosal · Alfonso Ortega · Marina de la Cruz
Escuela Politécnica Superior, Departamento de Ingeniería Informática,
Universiad Autónoma de Madrid
e-mail: {emilio.delrosal,alfonso.ortega,marina.cruz}@uam.es

J.B. Pérez et al. (Eds.): Highlights on PAAMS, AISC 156, pp. 37–44.
springerlink.com

Evolution, CGE, [9]) as powerful tools to design complex systems to solve specific tasks. Both techniques wholly describe the candidate solutions, both syntactically and semantically, by means, respectively, of attribute and Christiansen grammars; thus improving the performance of other approaches, because they reduce the search space by excluding non-promising individuals with syntactic or semantic errors.

NEPs have a complex structure, because some of their components depend on others. This dependence makes it difficult to use genetic techniques to search NEPs because, in this circumstance, genetic operators usually produce a great number of incorrect individuals (either syntactically or semantically).

One of our main interest is to design a general purpose platform to automatically design NEPs to solve real tasks by means of our genetic programming algorithms.

This paper shows our preliminary results in solving a real problem with this platform and improves our previous work [5] (in which we have tested the feasibility of generating syntactically correct NEPs by means of a simple context free grammar) by adding semantics and a complete fitness function. We have chosen a well known family of NEPs, able to solve a very simple problem [3]: the application of context free rules by classic NEPs. It is worth noticing that context free rules are not allowed in the classic family of NEPs. In this family, it is allowed only to replace a symbol by a single symbol (rather than a string of symbols, as in context free grammars).

[3] shows how NEPs simulate the application of context free rules ($A \to \alpha, A \in V, \alpha \in V^*$ for alphabet V) in three steps. The first one rotates the string where the rule is being applied until placing A in one of the string ends. We feel that this is the greatest difficulty (to rotate strings in order to locate the symbol to be derived) because the derivation itself is rather easy to implement. This sub-task is the goal of our work.

We hope that our experiments may result in the proposal of a methodology to automatically design NEPs that consists of the following modules:

- An *evolutionary engine*, used as an automatic programming algorithm. This engine has to handle candidate solutions with a complex structure. We propose using AGE or CGE.
- A *formal description of the computing device* being programmed.
- A *simulator for the computing device* that will be used to compute the fitness function.
- The *fitness function*, which must fulfill two roles:
    1. Simulate the generated solution (in this case, a particular NEP).
    2. Measure how well the solution solves the target problem.

2 Introduction: CGE and NEPs

Attribute grammars (AG) [6] are one of the tools used to completely describe high level programming languages (both their syntax and their semantics). Christiansen Grammars (CG) [2] are an adaptable extension to AG, that is, they are attribute grammars that modify themselves while they are used.

AGE [4] and CGE [9] are extensions to Grammatical Evolution [8]. Both techniques are automatic programming evolutionary algorithms independent of the target programming language, and include a standard representation of genotypes as strings of integers (codons), and a formal grammar (respectively attribute and Christiansen grammar) as inputs for the deterministic mapping of a genotype into a phenotype. This mapping minimizes the generation of syntactically and also semantically invalid phenotypes. Genetic operators act at the genotype level, while the fitness function is evaluated on the phenotypes.

Networks of evolutionary processors (NEPs) [1] are a new computing mechanism directly inspired in the behaviour of cell populations. Each cell contains its own genetic information (represented by a set of strings of symbols) that is changed by some *evolutive* transformations implemented as elemental operations on strings. Cells are interconnected and can exchange information (strings) with other cells.

A NEP can be defined as a graph whose nodes are processors which perform very simple operations on strings and send the resulting strings to other nodes. Every node has filters that block some strings from being sent and/or received.

A complete and formal definition of NEPs can be found in [1]. In [1] a NEP is a construct $\Gamma = (V, N_1, N_2, ..., N_n, G)$. V is an alphabet. For each $1 \leq i \leq n$, N_i is the i-th evolutionary node processor of the network. G is an undirected graph of processors, called the underlying graph of the network. Each processor N_i contains its set of evolution rules (to replace a symbol by other, or to insert or delete a given symbol), its initial content (a set of strings) and its input and output filters (strings that can enter and leave the processor). A NEP alternatively changes its strings, and communicates them by using its net, until reaching a stopping condition.

Given this formal device, our goal is to solve a problem introduced and solved in [3]. That paper shows how NEPs simulate the application of context free rules ($A \rightarrow \alpha, A \in V, \alpha \in V^*$ for alphabet V) in three steps. The first one (which is our goal) rotates the string where the rule is being applied until placing A in one of the string ends. In [3], this task is performed by means of a sub-NEP with 4 nodes connected as a linear chain.

The computation of the rotating sub-NEP can be summarized as follows: let us call "s" the symbol being rotated. The first node of the NEP receives a word where the symbol "s" is at the end. This node substitutes "s" by an auxiliary symbol ("s_{a1}"). Then, the new word is sent to the second node, where a new auxiliary symbol ("s_{a2}") is added to the beginning of the word. The last two nodes remove "s_{a1}" and substitute "s_{a2}" by the original "s". These nodes use filters that reject those words without the auxiliary symbols. At this point the rotation of "s" has finished. This cycle could be repeated as many times as needed until finding the symbol to which we want to apply the rule. Thus, this sub-NEP works as a chain of nodes working sequentially. It should be noted that the complete NEP has a different sub-NEP to rotate each non terminal symbol in the original grammar. Figure 1 shows a squeme of one of these sub-NEPs dedicated to rotate the symbol "x". The upper trace corresponds to the string "zyx", which is rotated since it contains "x" at the end. However, the lower trace shows the computation for the string "xyz" which does not have the symbol "x" at the end and, therefore, the filters do not allow it to pass to the last nodes.

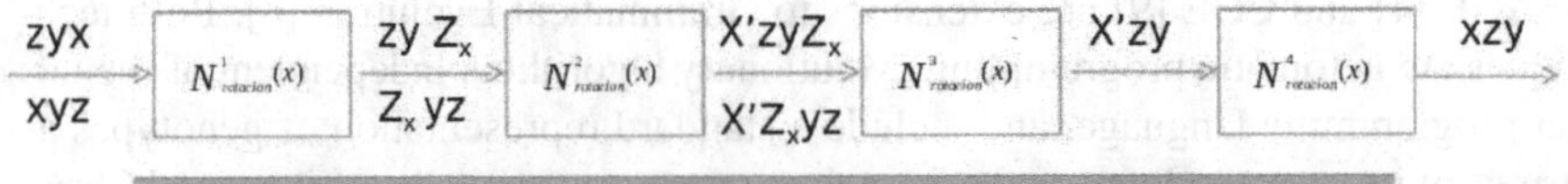

Node	Rule	Per. Input	For. Input	Per. Ouput	For. Output
$N^1_{rotacion}(X)$	$X := Z_x$			$\{Z_x\}$	
$N^2_{rotacion}(X)$	$\lambda := X'$ (left)	$\{Z_x\}$	$\{X'\}$		
$N^3_{rotacion}(X)$	$Z_x := \lambda$ (right)	$\{Z_x, X'\}$		$\{X'\}$	$\{Z_x\}$
$N^4_{rotacion}(X)$	$X' := X$	$\{X'\}$	$\{Z_x\}$		

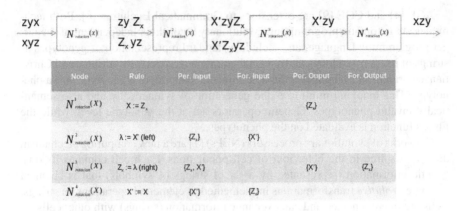

Fig. 1 Simplified squeme of the rotation NEP presented in [3]

3 Our Solution

In the following sections we will describe each component of the methodology introduced in section 1 to solve the problem under consideration.

In these first tests we have reduced the search space. We have only considered NEPs with the following characteristics: a) they have the same graph structure (a linear chain) as in [3], b) their alphabet's size is fixed, c) their rules are those described in [1] and the total number of rules in the NEP is limited, so as to avoid too complex, inefficient NEPs and, finally, d) their filters are based on *random context conditions* described in [7].

The complex structure of the individuals that we have to generate in our experiments (NEPs that belong to a particular family) made us to design a Christiansen Grammar to describe them.

As we will explain below, our fitness function invokes jNEP (a Java application we have developed to simulate NEPs[1]) to check if the generated NEP properly processes some input strings. jNEP takes as input a XML configuration file that describes the NEP that is being simulated. This is why our Christiansen Grammar actually generates the XML files that represent the NEPs and that can be read by jNEP as inputs.

Christiansen grammars are Attribute Grammars in which the first attribute of each non-terminal symbol is the actual Grammar applicable to this symbol. In this paper we follow the notation introduced in [10]: inherited and synthesized attributes are preceded by down and up arrows (↓ and ↑), respectively. Attributes are enclosed in round brackets next to their non terminal symbols. Each rule has its corresponding semantic actions (to compute the values of its attributes) enclosed in brackets after its right hand side. Figure 2 describes the grammar. Some actions are not shown for simplicity.

[1] The jNEP code is freely available at http://jnep.e-delrosal.net.

In the first rule the father inherits the original Christiansen Grammar with their children.

```
[NEP](g) ::= <?xml version="1.0"?><NEP nodes="[nodes](↓g)">[alphabetTag](↓g) [graphTag](↓g)
[processorsTag](↓g)[stoppingConditionsTag](↓g)</NEP>
{

    [nodes].↓g = [NEP].g
    [alphabetTag].↓g = [nodes].↑g_new
    [graphTag].↓g = [nodes].↑g_new
    [processorsTag].↓g = [nodes].↑g_new
    [stoppingConditionsTag].↓g = [nodes].↑g_new

}
[nodes](↓g) ::= 5
[alphabetTag] ::=<ALPHABET symbols="a_b_c_u_v_w_x_y_z"/>
[graphTag] ::= <GRAPH><EDGE vertex1="0" vertex2="1"/><EDGE vertex1="1" vertex2="2"/><EDGE vertex1="2" vertex2="3"/><EDGE vertex1="3"
vertex2="4"/></GRAPH>
```

The following rule derives the processors. It computes the total number of rules (by means of the expansion of the non terminal [inputNodeTag]) and limits it to 20: thus it is not possible to generate phenotypes with more than 20 rules.

```
[processorsTag](↓g) ::= <EVOLUTIONARY_PROCESSORS> [inputNodeTag](↓g,↓counterInit,↑counterFinal) [nodeTag]_1 (↓g,↓counterInit,↑counterFinal)
[nodeTag]_2 (↓g,↓counterInit,↑counterFinal) [nodeTag]_3 (↓g,↓counterInit,↑counterFinal) [nodeTag]_4 (↓g,↓counterInit,↑counterFinal)
</EVOLUTIONARY_PROCESSORS>
{

    EVERY CHILDREN INHERIT THE CHRISTIANSEN GRAMMAR AS IN PREVIOUS RULES
    [inputNodeTag].↓counterInit = 0
    [nodeTag]_i.↑counterFinal = [nodeTag]_{i+1}.↓counterFinal

}

[inputNodeTag](↓g,↓counterInit,↑counterFinal) ::= <NODE initCond="input word to rotate"> [evolutionaryRulesTag](↓g,↓counterInit,↑counterFinal)
[nodeFiltersTag](↓g,) </NODE>
{

    EVERY CHILDREN INHERIT THE CHRISTIANSEN GRAMMAR AS IN PREVIOUS RULES
    [evolutionaryRulesTag].↓counterInit = [inputNodeTag].↓counterInit
    [inputNodeTag].↑counterFinal = [evolutionaryRulesTag].↑counterFinal

}
[nodeTag](↓g,↓counterInit,↑counterFinal) ::= <NODE initCond=""> [evolutionaryRulesTag](↓g,↓counterInit,↑counterFinal) [nodeFiltersTag](↓g) </NODE>
{

    Semantic actions equivalent to the previous one.

}
[evolutionaryRulesTag](↓g,↓counterInit,↑counterFinal) ::= <EVOLUTIONARY_RULES> [ruleTag](↓g,↓counterInit,↑counterFinal)
</EVOLUTIONARY_RULES>
{

    EVERY CHILDREN INHERIT THE CHRISTIANSEN GRAMMAR AS IN PREVIOUS RULES
    [ruleTag].↓counterInit = [evolutionaryRulesTag].↓counterInit
    [evolutionaryRulesTag].↑counterFinal = [ruleTag].↑counterFinal

}
[ruleTag]_a(↓g,↓counterInit,↑counterFinal) ::= <RULE ruleType="[ruleType](↓g)" actionType="[actionType](↓g)" symbol="[symbol](↓g)"
newSymbol="[symbol](↓g)"/> [ruleTag]_b(↓g,↓counterInit,↑counterFinal)
{

    EVERY CHILDREN INHERIT THE CHRISTIANSEN GRAMMAR AS IN PREVIOUS RULES
    [ruleTag]_b.↓counterInit = [ruleTag]_a.↓counterInit +1
    [ruleTag]_a.↑counterFinal = [ruleTag]_b.↑counterFinal
    if ([ruleTag]_b.↑counterFinal > 20) dismissPhenotype();

}
[ruleTag](↓g,↓counterInit,↑counterFinal) ::= λ
{

    [ruleTag].↑counterFinal = [ruleTag].↓counterInit

}
[ruleType] ::= insertion | deletion | substitution
[actionType] ::= LEFT | RIGHT | ANY
[nodeFiltersTag] ::= [inputFilterTag] [outputFilterTag]
[nodeFiltersTag] ::= [inputFilterTag]
[nodeFiltersTag] ::= [outputFilterTag]
[nodeFiltersTag] ::= λ
[inputFilterTag] ::= <INPUT [filterSpec]/>
[outputFilterTag] ::= <OUTPUT [filterSpec]/>
[filterSpec] ::= type="[filterType]" permittingContext="[symbolList]" forbiddingContext="[symbolList]"
[filterType] ::= 1 | 2 | 3 | 4
[wordList] ::= [symbolList] [wordList] |λ
[symbolList] ::= a string of the alphabet's symbols separated by the character '_'
```

The following rule derives the stopping condition. We consider three conditions:

- The NEP stops when some string enters the output node
- The computation finishes after a maximum number of steps has been taken
- or when a maximum number of strings has been generated

```
[stoppingConditionsTag] ::= <STOPPING_CONDITION><CONDITION type="NonEmptyNodeStoppingCondition" nodeID="4"/> <CONDITION
type="MaximumStepsStoppingCondition" maximum="8"/> <CONDITION type="MaximumSizeStoppingCondition"
maximum="100"/></STOPPING_CONDITION>
```

Fig. 2 The Christiansen Grammar

Our fitness function checks if a specific symbol can be rotated in different strings: it checks if the symbol "c" can be moved from the end of a string to the beginning. It also checks if the solution works for a set of different strings. The fitness function returns a value between zero and one. The more strings are rotated, the higher value is returned. In order to obtain a smoother and more progressive fitness function (which is always desirable in an evolutionary search), we increased the value of the individual if it can perform sequences of sub-tasks. We have implemented this criterion assigning higher fitness values to those NEPs that can communicate strings across their chain of nodes.

In detail, the NEP is run once for each string of the set {"abc", "aabbcc", "aac", "bbc", "cca", "bcb"}, where the right set of outputs is {"cab", "caabbc", "caa", "cbb", "", ""}. A value proportional to the number of matches is returned. The ability to perform sequences of sub-tasks is evaluated as follows:

- If the penultimate node of the chain contains one or more strings, 0.0416 is added.
- When the last node contains one or more strings, 0.0416 is added.
- If the last node contains the desired output, 0.083 is added.

3.1 Experiments and Results

After eight runs of two thousand generations with populations of one thousand individuals, most of the experiments found a perfect (maximum fitness value) or almost perfect solution. The main parameters' values were: population = 1000, codons = 0-256, maximum wrappings operations = 2, mutation probability = 100% (each genotype mutates one of its codons in every generation), crossover probability = 95% and elitist generational replacement.

Figure 3 shows the jNEP input file for one of the solutions found. We ommit fixed elements for simplicity. It is worth mentioning that the solution proposed in [3] follows a different approach: at the first step, the rotating symbol is replaced by a *label* (an auxiliary symbol). This symbol makes the string to pass the following filters. Next node can discard any string without this label. In the last stage, the auxiliary symbol is deleted and the rotating symbol is inserted at the beginning.

It is amazing that our solution makes the opposite. It marks the non-rotating symbols with the label y to discard them later. Besides, the NEP described in [3] needs to perform more tasks and its rotating sub-NEP needs one more auxiliary symbol for a good coordination with the rest of the NEP.

Given the input *abc* the computation would be as follows. We denote *node one* containing the string *abc* as $N1(abc)$, an evolutionary step as $\rightarrow$ and a communication step as $\Rightarrow$: $N0(abc) \rightarrow N0(abc) \Rightarrow N1(abc) \rightarrow N1(abc) \Rightarrow N2(abc) \rightarrow N2(cabc) \Rightarrow N3(cabc) \rightarrow N3(cab) \Rightarrow N4(cab)$. As expected, the NEP halts when the rotated string enters the last node. On the other hand, given the input *bcb* which cannot be rotated the computation would be: $N0(bcb) \rightarrow N0(bcb) \Rightarrow N1(bcb) \rightarrow N1(bcy) \Rightarrow N2(bcy) \rightarrow N2(cbcy) \Rightarrow N3(cbcy) \rightarrow N3(cbcy) \Rightarrow$. At this point, the string *cbcy* cannot pass the output filter since it contains the symbol y and the NEP will halt when the maximum number of steps is reached.

There are three symbols in the strings that can be rotated a,b,c, the rest can be used as auxiliary symbols by the NEP. Remember that our fitness function firstly checks if the symbol *c* is properly rotated in a set of strings. The input string is placed at this first node. Those strings that finish with an *a* will change it by *b*. Therefore, this node can only transfer strings that end with the symbols *b* and *c* (the symbol being rotated) at the end.

```
...
  <ALPHABET symbols="a_b_c_o_p_q_r_s_t_u_v_w_x_y_z"/>
  <EVOLUTIONARY_PROCESSORS>
    <NODE initCond="input">
      <EVOLUTIONARY_RULES>
        <RULE ruleType="substitution" actionType="RIGHT" symbol="a" newSymbol="b"/>
      </EVOLUTIONARY_RULES>
      <FILTERS> </FILTERS>
    </NODE>
```

The second node substitutes every *b* at the end by the auxiliary symbol *y*. Therefore, after this node, all the strings will finish with *c* (the symbol being rotated) or *y*. We can consider, thus, that this node has marked the non-rotating strings .

```
    <NODE initCond="">
      <EVOLUTIONARY_RULES>
        <RULE ruleType="substitution" actionType="RIGHT" symbol="b" newSymbol="y"/>
      </EVOLUTIONARY_RULES>
      <FILTERS> </FILTERS>
    </NODE>
```

The third node adds the symbol that is being rotated (*c*) in the left side of the string.

```
    <NODE initCond="">
      <EVOLUTIONARY_RULES>
        <RULE ruleType="insertion" actionType="LEFT" symbol="c"/>
      </EVOLUTIONARY_RULES>
      <FILTERS> </FILTERS>
    </NODE>
```

This node, finally, deletes the rotating symbol from its original position. The non-rotating strings can not pass this point since the output filter forbids the symbol *y*.

```
    <NODE initCond="">
      <EVOLUTIONARY_RULES>
        <RULE ruleType="deletion" actionType="RIGHT" symbol="c"/>
      </EVOLUTIONARY_RULES>
      <FILTERS>
        <OUTPUT type="3" permittingContext="" forbiddingContext="y" />
      </FILTERS>
    </NODE>
    <NODE initCond="">
      <EVOLUTIONARY_RULES>
      </EVOLUTIONARY_RULES>
      <FILTERS> </FILTERS>
    </NODE>
  </EVOLUTIONARY_PROCESSORS>
...
```

Fig. 3 One of the solutions found

4 Conclusions and Further Research Lines

In this paper we have, for the first time, tackled a non trivial problem by means of the platform we are proposing to automatically design NEPs. Although we have simplified in some way the problem under consideration, and constrained different elements of the NEPs being evolved in order to reduce the search space, we have found interesting solutions similar to those described in literature. These solutions could be considered as valid alternatives. We are, then, optimistic with respect to the feasibility of using our platform to solve more general problems in more general domains.

In the future we have to generalize different aspects of the work described in this paper. As finding NEPs to rotate any symbol of the alphabet or searching within a more general family of NEPs by removing some of the constraints used in this paper. Moreover, we also pretend to tackle different non trivial problems. Finally, we are interested in adding to our platform a general way for describing the problem under consideration and for including it in the fitness function in a more standard way.

Acknowledgements. Work partially supported by the Spanish Ministry of Science and Innovation under coordinated research project TIN2011-28260-C03-00 and research project TIN2011-28260-C03-02 and by the Comunidad Autónoma de Madrid under research project e-madrid S2009/TIC-1650.

References

1. Castellanos, J., Martin-Vide, C., Mitrana, V., Sempere, J.M.: Networks of evolutionary processors. Acta Informatica 39(6-7), 517–529 (2003)
2. Christiansen, H.: A survey of adaptable grammars. SIGPLAN Notices 25(11), 35–44 (1990)
3. Csuhaj-Varju, E., Martin-Vide, C., Mitrana, V.: Hybrid networks of evolutionary processors are computationally complete. Acta Informatica 41(4-5), 257–272 (2005)
4. de la Cruz Echeandía, M., de la Puente, A.O., Alfonseca, M.: Attribute Grammar Evolution. In: Mira, J., Álvarez, J.R. (eds.) IWINAC 2005. LNCS, vol. 3562, pp. 182–191. Springer, Heidelberg (2005)
5. del Rosal, E., de la Cruz, M., Ortega de la Puente, A.: Towards the Automatic Programming of NEPs. In: Ferrández, J.M., Álvarez Sánchez, J.R., de la Paz, F., Toledo, F.J. (eds.) IWINAC 2011, Part I. LNCS, vol. 6686, pp. 303–312. Springer, Heidelberg (2011)
6. Knuth, D.E.: Semantics of Context-Free Languages. Mathematical Systems Theory 2(2), 127–145 (1968)
7. Manea, F., Martín-Vide, C., Mitrana, V.: Solving 3CNF-SAT and HPP in Linear Time Using WWW. In: Margenstern, M. (ed.) MCU 2004. LNCS, vol. 3354, pp. 269–280. Springer, Heidelberg (2005)
8. ONeill, M., Conor, R.: Grammatical Evolution, evolutionary automatic programming in an arbitrary language. Kluwer Academic Publishers (2003)
9. Ortega, A., de la Cruz, M., Alfonseca, M.: Christiansen grammar evolution: Grammatical evolution with semantics. IEEE Transactions on Evolutionary Computation 11(1), 77–90 (2007)
10. Watt, D.A., Madsen, O.L.: Extended attribute grammars. Technical Report 10, University of Glasgow (July 1977)

A Grammar-Based Multi-Agent System for Language Evolution

Mª Dolores Jiménez-López

Abstract. Considering the adequacy of agent systems for the simulation of language evolution, we introduce a formal-language-theoretic multi-agent model based on grammar systems that may account for language change: *cultural grammar systems*. The framework we propose is a variant of the so-called eco-grammar systems. We modify this formal model, by adding new elements and relationships, in order to obtain a new machinery to describe the dynamics of the evolution of language.

1 Introduction

Human language is one of the most challenging issues that remain to be explained. Many linguistic, computational and cognitive models try to explain how humans acquire, process and change languages. However, models proposed up to now have not been able to give neither a coherent explanation of natural language nor a satisfactory computational model for the processing of natural language. As a complex system, the explanation, formal modeling and simulation of natural language present important difficulties.

One of the biggest problems to be solved in this area is the question of how human languages originated, spread, and are constantly changing. According to [4], evolution of language is the hardest problem in science. In fact, the rapid growth in the literature on language evolution in the past few decades reflects its status as an important challenge for contemporary science.

Results obtained up to now make clear that in order to fully understand language evolution we need to cross traditional academic boundaries. Linguistics cannot be the only responsible for this matter. There is a need of connecting and integrating several academic disciplines and technologies in the pursuit of the common task.

M. Dolores Jiménez-López
Research Group on Mathematical Linguistics,
Universitat Rovira i Virgili, 43002 Tarragona, Spain
e-mail: mariadolores.jimenez@urv.cat

J.B. Pérez et al. (Eds.): Highlights on PAAMS, AISC 156, pp. 45–52.
springerlink.com

Recently, the improvement of computational techniques have led many researchers to use computational modeling to understand the origin and evolution of language. Computer simulations have become widespread in the last years [1, 2, 3, 9, 11, 12]. Computational simulation can be seen as a useful scientific methodology that can help to understand the way natural language evolves.

Many of those simulations have been carried out within multi-agent settings. Multi-agent systems offer strong models for representing complex and dynamic real-world environments. The metaphor of autonomous problem solving entities cooperating and coordinating to achieve their objectives is a natural way of conceptualizing many problems.

Taking into account the adequacy of multi-agent systems for the simulation of language evolution, and following the research line opened by *artificial life* where organisms have been simulated using symbolic models, we introduce a formal-language-theoretic multi-agent model based on grammar systems [6] that may account for language evolution: *cultural grammar systems*.

Research in language evolution can be structured in two basic approaches [13]: the *biolinguistic approach*, which puts emphasis on biology as the main factor in the origins of language; and the *evolutionary linguistic approach*, which emphasizes the role of cultural evolution. To those two basic approaches, we must add a third possibility: the one chosen by researchers that take a middle position exploring both biological and cultural aspects. Cultural grammar systems may be placed in that third possibility, since it shows the interaction between biological, cultural and environmental evolution in language change.

The framework we propose may be viewed as a variant of the so-called eco-grammar systems, since it takes as a basis this formal framework adding some new elements and relationships to account for language change. Eco-grammar systems, introduced in [7] as a subfield of grammar systems [6], provide a syntactical framework for ecosystems, this is, for communities of evolving agents and their interrelated environment. An eco-grammar system can be defined as a multi-agent system where different components, apart from interacting among themselves, interact with a special component called 'environment.'

Throughout the paper, we assume the reader to be familiar with basic notions in the theory of formal languages. For more information the reader is referred to [10].

2 Cultural Eco-Grammar Systems

2.1 Elements and Relationships in the System

According to [13], 'the origins and evolution of language is based on a congruence of three different evolutionary processes, influencing and re-enforcing each other: socio-ecological evolution, biological evolution, and cultural evolution'. Taking into account those different evolutionary processes, in a cultural grammar system we have introduced five elements: a *physical environment* (to account for socio-ecological evolution); a *genetic system* (to account for biological evolution); a set of *cultures* (to account for cultural evolution); a set of *subcultures*; and a set of *agents*.

Therefore, in a cultural grammar system, we have a set of *agents* that belong to a specific *cultural environment* and that are constrained by their *genetic* properties as well as by the *physical environment* where they are living. It is convenient to introduce an intermediate level between culture and individuals: *subcultures*. This further division can represent several things: socioeconomic classes, religions, castes, dialects, etc. The criteria by which subgroups are identified as strata usually depend on variables such as the trait under study.

Genetics and culture are not two independent inheritance systems, but they are interdependent. Those two systems are able to influence or modify each other. A change (usually very rapid) in culture can cause a (slower) change in gene frequency. Therefore, culture can alter somehow the direction and/or rate of genetic evolution. Durham [5] refers to this cultural influence on biology as *cultural mediation*. On the other hand, the genetic constitution of the individuals composing a society influences the nature of that society, so the genetic system is capable of modifying or conditioning changes in culture. As [8] emphasizes, since many of our needs have a biological basis, meme generation –this is, culture– is largely constrained by our heritage as products of biological evolution. We can refer to this relation as *genetic mediation*. Furthermore, agents may be influenced by the physical environment where they live.

In order to reflect all those relationships in our model we do the following: 1) we relate genetic system to agents, in order to capture the influence that biological environment has in the evolution of agents; 2) we relate subculture to agents, in order to indicate cultural constraints on agents' actions. For reasons of simplicity, we consider that agents are directly constrained by the subculture to which they belong, and just indirectly influenced (through the subculture) by culture; 3) and finally, we relate the physical environment to the whole system, in order to put in evidence that physical environment constrains evolution of genetic system as well as evolution of agents. And, of course, by influencing evolution of agents, it constrains indirectly changes on cultural environments, because agents and only agents are the responsible for cultural evolution.

The above elements and relationships compose cultural grammar systems that are represented in Fig. 1 and formally defined in Definition 1.

Definition 1. By a cultural grammar system of degree n, where $n \geq 1$, we mean an $n+2$-tuple:

$$\Sigma = (E, G, C_1, \ldots, C_n),$$

where:

- $E = (V_E, P_E)$,
  - V_E is an alphabet;
  - P_E is a complete finite set of context-free rules over V_E.

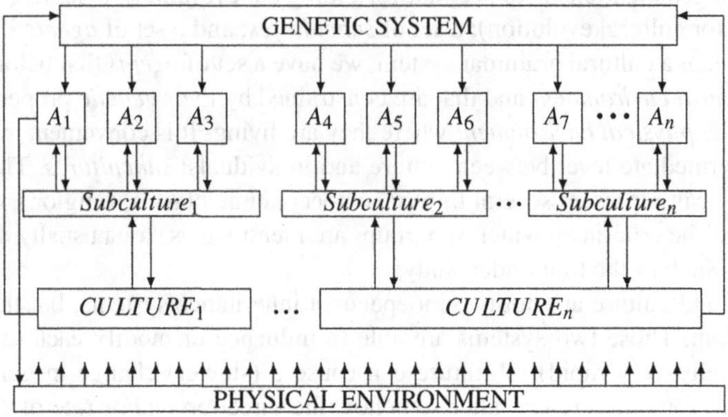

Fig. 1 Cultural Grammar Systems

- $G = (V_G, P_G)$,
  - V_G is an alphabet;
  - $P_G = P_G^1 \cup P_G^2$, where:
    (i) P_G^1, is a complete finite set of pure context-free rules over alphabet V_G; and
    (ii)P_G^2, is a finite set of rules where each rule is of the form: $(\alpha_1, \ldots, \alpha_n) : A \rightarrow v$, where $\alpha_i \in V_{C_i}^*$, $1 \leq i \leq n$, and $A \rightarrow v$ is a pure context-free rule over V_G.
- $C_i = \{K_{i_1}, \ldots, K_{i_{r_i}}\}$, $1 \leq i \leq n$, $r_i \geq 1$, where:
  - $K_{ij} = \{A_{ij1}, \ldots, A_{ijs_{ij}}\}$, $1 \leq j \leq r_i$, $s_{ij} \geq 1$, where: $A_{ijk} = (V_{ijk}, P_{ijk}, R_{ijk}^E, R_{ijk}^{C_i}, R_{ijk}^{K_{ij}}, \psi_{ijk}, \phi_{ijk}^E, \phi_{ijk}^{C_i}, \phi_{ijk}^{K_{ij}})$, $1 \leq i \leq n$, $1 \leq j \leq r_i$, $1 \leq k \leq s_{ij}$, where:
    · V_{ijk} is the alphabet of the agent;
    · P_{ijk} is a complete finite set of pure context-free rules over alphabet V_{ijk};
    · R_{ijk}^E is a finite set of pure context-free rules over alphabet V_E;
    · $R_{ijk}^{C_i}$ is a finite set of pure context-free rules over alphabet V_{C_i};
    · $R_{ijk}^{K_{ij}}$ is a finite set of pure context-free rules over alphabet V_{C_i};
    · $\psi_{ijk} : V_E^* \times V_G^* \rightarrow 2^{P_{ijk}}$;
    · $\phi_{ijk}^E : V_{ijk}^+ \rightarrow 2^{R_{ijk}^E}$;
    · $\phi_{ijk}^{C_i} : V_{ijk}^+ \times V_{C_i}^* \times V_{C_i}^* \rightarrow 2^{R_{ijk}^{C_i}}$;
    · $\phi_{ijk}^{K_{ij}} : V_{ijk}^+ \times V_{C_i}^* \times V_{C_i}^* \rightarrow 2^{R_{ijk}^{K_{ij}}}$.
- E represents the *environment*. V_E is the *alphabet* of the environmental string and P_E is the set of *evolution rules* of E.
- G formalizes the *genetic system,* V_G is its *alphabet,* and P_G is the set of evolution rules. P_G^1 contains basic evolution rules, while P_G^2 describes how cultural systems can constrain genetic evolution. Since P_G^1 is complete, cultural influence is weak.

- C_i represents i-th *cultural system* with subcultural system $K_{i_1},\ldots,K_{i_{r_i}}$.
- K_{ij} is j-th *subcultural system* of cultural system C_i with agents $A_{ij1},\ldots,A_{ijs_{ij}}$.
- A_{ijk} is k-th agent of subcultural system K_{ij} and cultural system C_i; V_{ijk} is the *alphabet* of k-th agent; P_{ijk} is the set of *evolution rules* of the agent; R^E_{ijk} is a set of rules that describe actions which can be performed by the agent on the *environment*; $R^{C_i}_{ijk}$ is a set of rules that describe *action* rules that can be used by the agent in order to modify the current state of *cultural system* C_i; $R^{K_{ij}}_{ijk}$ is the set of *action rules* that agent can use to modify the current state of *subcultural system*; ψ_{ijk} determines, according to the current states of the *environment* and the *genetic system*, which rules can be used at that moment in *agent evolution*; ϕ^E_{ijk} determines, according the *agent* current state, the set of *action rules* from which agent can choose one to apply on the *environmental state* at that moment; $\phi^{C_i}_{ijk}$ determines, according to the current states of *agent*, *cultural system* C_i, and *subcultural system* K_{ij}, the set of *action rules* from which agent can select one to perform an action on *cultural system* C_i; $\phi^{K_{ij}}_{ijk}$ determines, according to current states of *agent*, *cultural system* C_i, and *subcultural system* K_{ij}, the set of *action rules* available to the agent to perform an action on *subcultural system* K_{ij}.

Definition 2. Let $\Sigma = (E,G,C_1,\ldots,C_n)$, $n \geq 1$, be a cultural grammar system. A *state* σ of Σ we is a 5-tuple of arrays: $\sigma = (M_E,M_G,M_A,M_K,M_C)$, where:

- $M_E = (w_E)$, $w_E \in V_E^*$, is the *state* of *environment*;
- $M_G = (w_G)$, $w_G \in V_G^*$, is the *state* of *genetic system*;
- $M_A = (\gamma_{ijk})$, $1 \leq i \leq n$, $1 \leq j \leq r_i$, $1 \leq k \leq s_{ij}$, $\gamma_{ijk} \in V_{ijk}^*$, is the *state* of *agent* A_{ijk};
- $M_K = (\beta_{ij})$, $1 \leq i \leq n$, $1 \leq j \leq r_i$, $\beta_{ij} \in V_{C_i}^*$, is the *state* of *subcultural system* K_{ij} or the state of $(ij)-$th subculture;
- $M_C = (\alpha_i)$, $1 \leq i \leq n$, $\alpha_i \in V_{C_i}^*$, is the *state* of *cultural system* C_i or the state of the i-th culture, and for every i, $1 \leq i \leq n$, $\beta_{i1}\beta_{i2}\ldots\beta_{ir_i}$ is a substring of α_i.

2.2 Dynamics of the System

A fact that is recurrently found in the literature of cultural change is that there exists a difference in the rates of evolution between cultural and biological systems. Cultural systems evolve faster than genetic ones. Another important fact is that culture is man-made. Man is the responsible for creation and changes in cultural systems. So, culture evolves faster than the genetic system, and it evolves thanks to direct action of agents. In contrast, genetic system evolves in a slower fashion and admits little –or none– direct modification on behalf of man. We have to reflect these ideas in our model and we do it in the following way:

- To account for the fact that agents are the responsible for change of culture, we allow agents to act on subcultures. Every subculture –as a group– can introduce modifications in culture.
- The genetic system does not accept direct action of agents. Genetics evolves by its own rules, allowing only some weak influence of cultural systems.

- The physical environment changes according to its own rules, that are 'external' to the rest of the system. We allow agents to weakly modify the state of the environment, conditioning the direction of its evolution.
- Evolution of agents is conditioned by genetic system as well as by the physical environment where they live. Agents are heavily constrained by the culture they belong to. The genetic system and the environment constrain agents' evolution; culture and subcultures constrain agents' actions.

Definition 3. Let $\Sigma = (E, G, C_1, \ldots, C_n)$, $n \geq 1$ be a cultural grammar system. Let $\sigma_1 = (M_E, M_G, M_A, M_K, M_C)$ and $\sigma_2 = (M'_E, M'_G, M'_A, M'_K, M'_C)$ be two states of Σ. We say that *state* σ_1 *directly enters into state* σ_2, written as $\sigma_1 \Longrightarrow \sigma_2$, iff the following conditions hold:

- Let $M_E = \{w_E\}$ and $M'_E = \{w'_E\}$. Suppose that: $w_E = z_0 A_0 z_1 \ldots A_m z_{m+1}$, where $A_l \in V_E, 0 \leq l \leq m, z_t \in V_E^*, 0 \leq t \leq m+1$. Then: $w'_E = z'_0 u_0 z'_1 \ldots u_m z'_{m+1}$, where $A_l \rightarrow u_l \in \phi^E_{ijk}(\gamma_{ijk}) \subseteq R^E_{ijk}$, for some i, j, k where $1 \leq i \leq n, 1 \leq j \leq r_i, 1 \leq k \leq s_{ij}$, and the rules $A_l \rightarrow u_l$, $0 \leq l \leq m$, are performed by m different agents. Strings z'_t, $0 \leq t \leq m+1$, are obtained from strings z_l by applying rules of P_E in an $0L$ manner.
- Let $M_G = \{w_G\}$ and $M'_G = \{w'_G\}$. Suppose that: $w_G = A_0 A_1 \ldots A_m$, where $A_l \in V_G, 0 \leq l \leq m$. Then: $w'_G = y_0 y_1 \ldots y_m$, where for each $A_l \rightarrow y_l$, $0 \leq l \leq m$, either $A_l \rightarrow y_l \in P^1_G$, or for some u_i, $1 \leq i \leq n$, u_i is a substring of α_i, the state of cultural system C_i, and $(u_1, \ldots, u_n) : A_l \rightarrow y_l$ is a rule in P^2_G.
- Let $M_A = (\gamma_{ijk})$ and $M'_A = (\gamma'_{ijk})$, $1 \leq i \leq n$, $1 \leq j \leq r_i$, $1 \leq k \leq s_{ij}$. Then γ'_{ijk} is obtained from γ_{ijk} by applying rules in $\psi_{ijk}(w_E, w_G) \subseteq P_{ijk}$ in an $0L$ manner.
- Let $M_K = (\beta_{ij})$ and $M'_K = (\beta'_{ij})$, $1 \leq i \leq n$, $1 \leq j \leq r_i$. Suppose that: $\beta_{ij} = y_0 B_0 y_1 \ldots B_m y_{m+1}$, where $B_l \in V_{C_i}, 0 \leq l \leq m, y_t \in V^*_{C_i}, 0 \leq t \leq m+1, 1 \leq j \leq r_i$. Then $\beta'_{ij} = y_0 v_0 y_1 \ldots v_m y_{m+1}$, where $v_l \in V^*_{C_i}$, $0 \leq l \leq m$, $1 \leq j \leq r_i$, and $B_l \rightarrow v_l \in \phi^{K_{ij}}_{ijk}(\alpha_i, \beta_{ij}, \gamma_{ijk}) \subseteq R^{K_{ij}}_{ijk}$ for some k, $1 \leq k \leq s_{ij}$, moreover, rules $B_l \rightarrow v_l$, $0 \leq l \leq m$, are performed by m different agents.
- Let $M_C = (\alpha_i)$ and $M'_C = (\alpha'_i)$, $1 \leq i \leq n$. Let $u = \beta_{i1} \ldots \beta_{ir_i}$ and let $u' = \beta'_{i1} \ldots \beta'_{ir_i}$, where $\beta'_{ij} \in V^*_{C_i}$, $1 \leq i \leq n$, $1 \leq j \leq r_i$ is the state of the subcultural system K_{ij} in state σ_2, determined above. Suppose that: $\alpha_i = x_0 A_0 \ldots x_{l-1} u x_l A_l \ldots A_m x_{m+1}$, where $A_t \in V_{C_i}$, $0 \leq t \leq m$, $0 \leq l \leq m$, $x_h \in V^*_{C_i}$, $0 \leq h \leq m+1$. Then $\alpha'_i = x_0 v_0 \ldots x_{l-1} u' x_l v_l \ldots v_m x_{m+1}$, where $A_t \rightarrow v_t \in \phi^{C_i}_{ijk}(\alpha_i, \beta_{ij}, \gamma_{ijk}) \subseteq R^{C_i}_{ijk}$, for some $j, 1 \leq j \leq r_i$, and k, $1 \leq k \leq s_{ij}$, moreover, rules $A_t \rightarrow v_t, 0 \leq t \leq m$, are performed by m different agents.

- *Environmental state* (M_E) is modified at some places by *weak actions* of some active agents. Remaining symbols change according to its *evolution rules*. Each active agent is allowed to perform exactly one action.
- Change of *genetic system state* (M_G) is due to genetic *evolution rules,* which can be *weakly constrained* by *cultural systems.*
- *State of agent* A_{ijk} (M_A) is modified according to its *evolution rules* which are constrained by current states of *environment* and *genetic system.*
- *State of subculture* K_{ij} (M_K) is modified by agents' actions.

- *State of culture* C_i (M_C) is modified by agents' actions in such places which were not affected by the change of the state of subcultural systems.

2.3 Output of the System

Usually, in formal language theory, the output of any formal language theoretic device is a *language*. However, since our first motivation has been to provide a formal language theoretic framework to account for language change, we are interested in formally defining the way in which language evolves. In other words, we have been interested in the *process* of changing language rather than in the final *result* of the evolutionary process. For that reason, we retain more useful to define the sequence of states of the system rather than to define the language generated by a cultural grammar system.

Definition 4. Let $\Sigma = (E, G, C_1, \ldots, C_n)$ be a cultural grammar system defined as above and let σ_0 be a state of Σ. The *set of state sequences* of Σ starting from *initial state* σ_0 is: $Seq(\Sigma, \sigma_0) = \{\sigma_i\}_{i=0}^{\infty}$, where $\sigma_i \Longrightarrow \sigma_{i+1}$, $i \geq 0$.

3 Conclusions

Following the example of fields as evolutionary computation –where, starting from genetic algorithms, cultural algorithms have been proposed– or artificial life –where it has been suggested the idea of artificial culture–, and considering the application of mathematical and computational devices to the study of language evolution, we have introduced *cultural grammar systems*. If eco-grammar systems have been considered as a tool to formalize real life-like features of an ecosystem, cultural grammar systems may be regarded as a device to formalize cultural aspects. We claim that this machinery could account for the dynamics of language evolution.

Even though there still controversy about the biological versus cultural evolution of languages [4], in general, researchers in the field agree on the fact that language arises from the interaction of three different adaptive systems: *individual learning*, *cultural transmission* and *biological evolution*. This idea leads to the defense that biological adaptation and cultural transmission may have interacted in the evolution of language. The structure of cultural grammar systems allow tho explain language evolution as a consequence of biological adaptation and cultural transmission.

Individual learning, cultural transmission and biological evolution are adaptive systems [4]. Natural selection is the most clear mechanism of adaptation. Individual learning is a process of adaptation of the individual's knowledge. And cultural transmission can be thought as an adaptation of culture and languages to fit the needs of users. The model we have introduced accounts for those different adaptive systems, since it introduces: *agents* with evolution rules that describe the individual learning and show how they evolve according to the dynamics of the whole system; a *genetic system* that accounts for biological evolution; and cultures that describe the cultural transmission and evolve according to agents needs.

Most computational models of language evolution are based on agent systems. Our model is a multi-agent system where agents interact among themselves and with their interrelated environments, fitting in this way the main features of agent simulations for language evolution. One of the main advantages of our model is that it is presented in formal and non ambiguous terms. The model introduced here is just an initial approximation to the possibility of formalizing language evolution. However, cultural grammar systems are based on a consolidated and active branch in the field of formal language theory, so they offer a highly formalized framework that seems to be easy to implement, due to its simplicity and the computational background of the theory used. Furthermore, since our model is based on grammars, it offers a natural way to define language. Achieving a valid and simple computational implementation of this model is the major research line for the future.

References

1. Bartlett, M., Kazakov, D.: The role of environment structure in multi-agent simulations of language evolution. In: Proceedings of the Fourth Symposium on Adaptive Agents and Multi-Agent Systems (2004)
2. Cangelosi, A.: Adaptive agent modeling of distributed language: investigations on the effects of cultural variation and internal action representations. Language Sciences 29(5), 633–649 (2007)
3. Cangelosi, A., Parisi, D.: Computer simulation: A new scientific approach to the study of language evolution. In: Cangelosi, A., Parisi, D. (eds.) Simulating the Evolution of Language. Springer, London (2001)
4. Christiansen, M., Kirbi, S.: Language evolution: Consensus and controversy. Trends in Cognitive Sciences 7(7), 300–307 (2003)
5. Durham, W.H.: Genes, culture, and human diversity. Standford University Press (1991)
6. Csuhaj-Varjú, E., Dassow, J., Kelemen, J., Păun, G.: Grammar systems: A grammatical approach to distribution and cooperation. Gordon and Breach, London (1994)
7. Csuhaj-Varjú, E., Kelemen, J., Kelemenová, A., Păun, G.: Eco-grammar systems: A grammatical framework for life-like interactions. Artificial Life 3(1), 1–28 (1996)
8. Gabora, L.: The origin and evolution of culture and creativity. Journal of Memetics-Evolutionary Models of Information Transmission 1(1), 29–57 (1997)
9. Reitter, D., Lebiere, C.: Did social networks shape language evolution? A Multi-Agent Cognitive Simulation. In: Proceedings of ACL 2010, pp. 9–17 (2010)
10. Rozenberg, G., Salomaa, A.: Handbook of formal languages. Springer, Berlin (1997)
11. Smith, K., Brighton, H., Kirby, S.: Language evolution in a multi-agent model: the cultural emergence of compositional structure (2002)
12. Steels, L.: How to do experiments in artificial language evolution and why. In: Proceedings of the 6th International Conference on the Evolution of Language, pp. 323–332 (2006)
13. Steels, L.: Modeling the cultural evolution of language. Physics of Life Reviews 8, 339–356 (2011)

On the Syntax-Semantics Interface of Argument Marking Prepositional Phrases

Roussanka Loukanova and M[a] Dolores Jiménez–López

Abstract. The paper investigates a class of argument marking prepositional phrases that have two roles when they occur in some verb phrases. Primarily, such prepositional phrases provide the head verb with a noun phrase as one of its dependents, casting syntax-semantics correlations in the internal argument structure of the verb phrase. Additionally, they can carry substantial semantic information. The paper contributes to methods of language processing by integrated algorithmic syntax-semantics interface. The work is part of development of a new type-theoretic approach to the concepts of algorithm and algorithmic syntax-semantics interfaces.

1 Introduction

Type-logical grammars have been under development in great details, by a focus on higher-order type systems with λ-calculi. Semantic representations have been employed in large-scale Constraint-Based Lexicalized Grammar (CBLG), e.g., currently, Minimal Recursion Semantics in HPSG, see [1]. Nevertheless, well formalized semantics and syntax-semantics interface in CBLG is open area under development. This paper is in the direction of such research, by looking at a specific phenomenon of syntax-semantics interrelations of a class of prepositional phrases (PPs), used as syntactic arguments of verbs. In such a PP, consisting of a head preposition with a NP complement (schematically, [P NP]) the head preposition marks its complement NP as one of the internal dependants of the head verb in the verb phrase. The paper investigates the syntax-semantics dependencies and constraints

Roussanka Loukanova
Uppsala, Sweden
e-mail: rloukanova@gmail.com

M[a] Dolores Jiménez–López
GRLMC Research group on Mathematical Linguistics,
Universitat Rovira i Virgili, Tarragona, Spain
e-mail: mariadolores.jimenez@urv.cat

J.B. Pérez et al. (Eds.): Highlights on PAAMS, AISC 156, pp. 53–60.
springerlink.com

of argument marking PPs, which are non-vacuous, as opposed to the vacuous argument marking PPs. The approach of syntax-semantics interface incorporates lexical syntax-semantics interrelations within phrasal structure. In the next section, we provide some references and a general overview to L^{λ}_{ar} and CBLG.

2 Background and Recent Developments

Algorithmic Intensionality in the Type Theory of Acyclic Recursion

Moschovakis developed a class of formal languages of recursion, as a new approach to the mathematical notion of algorithm. It has powerful applications to computational semantics. Detailed introductions to the type theory L^{λ}_{ar} of acyclic recursion is given in [6]-[7] and other papers on its applications to computational semantics. [6]-[7]. In particular, the theory of acyclic recursion L^{λ}_{ar} in [7] models the concepts of meaning and synonymy. For initial applications of L^{λ}_{ar} to computational syntax-semantics interface in CBLG of human language, see [3]. The formal system L^{λ}_{ar} is a higher-order type theory. L^{λ}_{ar} extends classic type systems by adding a second kind of variables, recursion variables, to its pure variables, and by formation of recursive terms with a recursion operator, which is denoted by the constant where, and used in infix notation. I.e, for any L^{λ}_{ar}-terms $A_0 : \sigma_0, \ldots, A_n : \sigma_n$ $(n \geq 0)$, and any pairwise different recursion variables of the corresponding types, $p_1 : \sigma_1, \ldots, p_n : \sigma_n$, such that the set of assignments $\{p_1 := A_1, \ldots, p_n := A_n\}$ is acyclic, the expression $(A_0 \,\mathsf{where}\, \{p_1 := A_1, \ldots, p_n := A_n\})$ is an L^{λ}_{ar}-term. The where-terms represent recursive computations by designating functional recursors: intuitively, the denotation of the term A_0 depends on the denotations of $p_1, \ldots, p_n$, which are computed recursively by the system of assignments $\{p_1 := A_1, \ldots, p_n := A_n\}$. In an acyclic system of assignments, these computations close-off. The formal syntax of L^{λ}_{ar} allows only recursive terms with acyclic systems of assignments. The languages of recursion (e.g., FLR, L^{λ}_{r} and L^{λ}_{ar}) have two semantic layers: denotational semantics and referential intensions. The recursive terms of L^{λ}_{ar} are essential for encoding two-fold semantic information. **Denotational Semantics:** For any given semantic structure $\mathfrak{A}$, a denotation function, den, is defined compositionally on the structure of the L^{λ}_{ar}-terms. In any standard structure $\mathfrak{A}$, there is exactly one, well-defined denotation function, den, from terms and variable assignments to objects in the domain of $\mathfrak{A}$. Thus, for any variable assignment g, a L^{λ}_{ar}-term A of type σ *denotes* a uniquely defined object $\mathsf{den}(A)(g)$ of the subdomain $\mathfrak{A}_{\sigma}$ of $\mathfrak{A}$. L^{λ}_{ar} has a reduction calculus that reduces each term A to its canonical form $\mathsf{cf}(A) \equiv A_0$ where $\{p_1 := A_1, \ldots, p_n := A_n\}$, which is unique modulo congruence, i.e., with respect to renaming bound variables and reordering of assignments. **Intensional Semantics:** The *referential intension*, $\mathsf{Int}(A)$, of a meaningful term A is the tuple of functions (a recursor) that is defined by the denotations $\mathsf{den}(A_i)$ $(i \in \{0, \ldots n\})$ of the parts of its canonical form $\mathsf{cf}(A) \equiv A_0$ where $\{p_1 := A_1, \ldots, p_n := A_n\}$. Intuitively, for each meaningful term A, the intension of A, $\mathsf{Int}(A)$, is the *algorithm* for computing its denotation $\mathsf{den}(A)$. Two meaningful expressions are synonymous iff their referential intensions are naturally isomorphic, i.e., they are the same algorithms. Thus, the algorithmic meaning of a meaningful term (i.e., its sense) is the information about how to "compute" its denotation step-by-step: a meaningful term has sense by carrying instructions within its structure, which are revealed by its canonical form, for acquiring what they denote in a model. The canonical form $\mathsf{cf}(A)$ of a meaningful term A encodes its intension, i.e., the algorithm for computing its denotation, via: (1) the basic instructions (facts), which consist of $\{p_1 := A_1, \ldots, p_n := A_n\}$ and the head term A_0, that are needed for computing the denotation $\mathsf{den}(A)$, and (2) a terminating rank order of the recursive steps that compute each $\mathsf{den}(A_i)$, for $i \in \{0, \ldots, n\}$, for incremental computation of the denotation $\mathsf{den}(A) = \mathsf{den}(A_0)$.

Constraint-Based Lexicalist Grammar

There is extensive literature on various frameworks within the CBLG approach, primarily on lexical and phrasal syntax. CBLG frameworks, for example, HPSG, LFG, Categorial Grammars (CG), and Grammatical Framework (GF), have been under development in both aspects, as formal and computational theory of grammar, but also in applications, prominently, as large scale grammars of human languages. For recent developments, see, e.g., [4] and [5] for CG, [8] for GF, and [9] for general introduction to CBLG. The technical apparatus (type systems, rules, principles, lexical rules, and principles) is provided in very generalized, but formal style in [2]. CBLG is organized so that various kinds of lexical, syntactic and semantic information is hierarchically distributed among various grammar components, which are interactively interrelated by constraints. All grammar components and constraints are expressed in a common language of a feature-value logic, which facilitates natural exchange of information between lexicon, phrase syntax, and their semantic representations. In CBLG, as introduced by [9], there is a distinction between feature-structure descriptions and feature-structures. Intuitively, each feature-structure is a finite function from features to values, where the later are, recursively, also feature-structures[1]. A feature-structure is totally defined, when the feature structure is fully expended. Feature-structure descriptions are matrices of features and values, which give partial information (constraints) about (fully expanded) total feature-structures. The total expansions are determined by types and constraints associated with the types according to a type system. This gives a possibility for partial and underspecified descriptions and generalizations across categories. For ex., in CBLG, generalization can be achieved by leaving pairs of feature-value partly or completely underspecified. In this paper, we assume CBLG with feature-value descriptions, although there are CBLG with different formalisms, e.g., CGs or GF. **Grammar Rules** In most versions of CBLG, grammar rules are introduced in a form, which follows the tradition of the CFGs, with a mother feature structure description on the left hand side of an arrow and the daughter descriptions on its right side: $A \longrightarrow A_1 \ldots A_n$ where A, A_1, ..., A_n $(n \geq 1)$ are feature structure descriptions. The CBLG rules typically have a general form stating the co-occurrence constraints of the linguistic phenomenon they model. Usually, a rule gives the necessary minimum of

[1] Atomic values are formally constant functions.

required co-occurrence information and represent linguistic generalizations across all instances of that rule. Each grammar rule licenses a class of well-formed tree structures. **Grammar Principles:** The principles in CBLG differ from rules in representing linguistic generalizations with respect to the distribution of grammatical information over features and their values in well-formed tree structures. The formal specifications of grammar principles depend on the specific formal apparatus, definitions of feature-structure descriptions and well-formedness. In grammar implementations of CBLG, the principles are often encoded by rules, type systems, and other forms of feature-value constraints. **Lexicon:** In CBLG, a good amount of grammatical information is moved from the grammar rules into rich lexicon data that is efficiently organized via a hierarchical lexical type subsystem of the entire type hierarchy of the grammar.

3 Argument Marking Prepositions

In this section, we use Moschovakis' language L^{λ}_{ar} of acyclic recursion to provide semantic representation of HL expressions with head verbs taking PPs complements. In general, the preposition of such a PP complement to a head verb is a marker of one of its syntactic dependents that corresponds to a semantic argument role filled by the denotation of the component NP.

Kim gave the cat [to Tim]$_{PP}$. (1a)

Kim gave Tim the cat. (1b)

Kim sat on the table. (1c)

Kim sat near the table. (1d)

Kim put Fido on the table. (1e)

In (1a), the preposition in the PP [to Tim] can be treated as a semantically vacuous argument marker: the preposition "to" is *marking* the NP ("Tim") in its COMPS list and is *passing* it as a syntactic dependent of the head verb "gave". Respectively, the denotation of the NP, den(*tim*), is handed as filling up the corresponding role of "recipient", which is a semantic argument of the denotation den(*give*) of the verb "gave". In the sentence (1a), the preposition "to" has no other specific semantic contribution of its own (by ignoring context dependent details, e.g., such as utterance stress or focus). For the purposes of this paper, we call such an argument marking preposition vacuous. This is in tune with the paraphrase (1b), where the semantic role of the recipient is encoded by the position of the NP in the linear order of the verbal complements.

In this paper, we focus on PPs, where the prepositions carry additional semantic information, with respect to the denotation of the component NP and its semantic argument role to the verb denotation. In (1e), the head verb "put", like "give", has COMPS list ⟨NP, PP⟩. In contrast to "give", "put" does not allow paraphrase similar to (1b). In (1e), the PP is not a modifier of the VP "put Fido" (nor of the verb "put"). Typically, by sentences like (1e), the verb "put" is used to denote a relation between three entities, and we can render it to a constant, e.g., as in (2a), by using currying types. Then, (1e) can be rendered to the term (2c)

$$\text{put} \xrightarrow{\text{render}} put : (\tilde{e} \to (\tilde{e} \to (\tilde{e} \to \tilde{t}))) \tag{2a}$$

$$\text{Kim put Fido on the table} \xrightarrow{\text{render}} \tag{2b}$$

$$[put(fido)(on[the(table)])](kim) : \tilde{t} \tag{2c}$$

The PPs in (1c)-(1e) are not vacuous argument markers. L_{ar}^{λ} provides different ways by which we can analyse expressions with such PPs.

4 Option 1: Strictly Referential PPs

Argument marking prepositions can be used as "selectors" of an entity depending on the entity denoted by the NP that is complement to the preposition P in the PP. By such optional treatment, PPs like "on the table", "under the table", "above the table" are used for reference to a selected location relative to the entity denoted by the complement NP, like "the table" in these expressions.

$$on : (\tilde{e} \to \tilde{e}) \tag{3a}$$

$$on[the(table)] : \tilde{e} \tag{3b}$$

The denotation $den(on[the(table)])$ is a location spot. Similarly, prepositions like "nearby", "near", "under", "over", etc., by Option 1, would "select" a location appropriately related to the denotation of the NP in the COMPS' list.

In the case of Option 1, the semantics of the PP complement of the verb "put" is actually a semantic compound of the semantics of the preposition and its NP complement. E.g, the semantics of the PP "on the table" is an entity, which is a location semantically denoted by the PP, which is a phrase with syntactic structure [P NP]. The PP structure has syntax-semantics interface that provides a semantic filler of the corresponding semantic argument role, which is denoted by the 2nd element in the COMPS list of verbs like "put", and similarly argument of "sit" "stay".

In the case of Option 1, verbs, like "sit" and "stay" can be rendered into constants of the language L_{ar}^{λ}, with types as in (4a)-(4b), by currying. Then the rendering of the sentence "Kim sits here" to the L_{ar}^{λ} term (4c) can be derived (inferred) in CBLG.

$$\text{sit} \xrightarrow{\text{render}} sit : (\tilde{e} \to (\tilde{e} \to \tilde{t})) \tag{4a}$$

$$\text{stay} \xrightarrow{\text{render}} stay : (\tilde{e} \to (\tilde{e} \to \tilde{t})) \tag{4b}$$

$$\text{Kim sits here} \xrightarrow{\text{render}} sit(h)(kim) : \tilde{t} \tag{4c}$$

where h is a recursion variable that denotes the location "here" or "on the floor":

$$\text{Kim sits on the floor} \xrightarrow{\text{render}} sit(on[the(floor)])(kim) : \tilde{t} \tag{5}$$

5 Option 2: Multi-argumental Prepositions

The argument marking preposition, e.g., in a PP with syntax [P NP], can be treated as denoting a relation between entities. One of the entities is denoted by the NP that is complement to the P in the PP, the other is the denotation of another expression provided by an encompassing expression. For example, the prepositions "on" and "between" can be rendered not to single constants, but to small, finite sets of constants. The types of the corresponding constants have patterns as in (6a)-(6b).

$$on : (\tilde{e} \to (\sigma \to \tau)) \tag{6a}$$
$$between : (\tilde{e} \to (\tilde{e} \to (\sigma \to \tau))) \tag{6b}$$

The specific constants of specific types can be instantiated depending on the HL expressions, and context. **Predicative PPs:** by instantiation $\tau \equiv \tilde{t}$, we have:

$$on[the(table)] : (\sigma \to \tilde{t}) \tag{7a}$$
$$between[the(table)][the(window)] : (\sigma \to \tilde{t}) \tag{7b}$$

By rendering the preposition "on" as in (6a), with the instantiations $\sigma \equiv \tilde{e}$ and $\tau \equiv \tilde{t}$, the following L^{λ}_{ar} terms are well-formed:

$$[on[the(table)]](fido) : \tilde{t} \tag{8}$$
$$between[the(table)][the(window)](fido) : \tilde{t} \tag{9}$$

Option 2, with specific constants *on* and *between*, where $\sigma \equiv \tilde{e}$ and $\tau \equiv \tilde{t}$, fits predicative uses of PPs, e.g., as in (10). This is possible by grammatical syntax-semantics interface, where the SPR of the verb "be" (in the inflected form "is") provides SPR of the preposition "on" in the PP:

$$\text{Fido is on the table.} \tag{10}$$

The preposition head of the PP complement contributes its own semantics, which is a curried relation between two objects. The denotation of the preposition ("on", "under", "near", etc.) is a relation between an object (e.g., "Kim" in (1c) and (1d), and "Fido" in (1e)) and the entity denoted by the NP complement of the preposition P. In particular, the denotation of the PP expression [on [the table]]$_{PP}$ is

$$\mathsf{den}(on(the(table))) = \mathsf{den}(on)(\mathsf{den}(the(table))) : \mathbb{T}_{\tilde{e}} \longrightarrow \mathbb{T}_{\tilde{t}} \tag{11}$$

In the case of Option 2, when the verb "put" takes a PP as a 2nd complement, it denotes a function, i.e. a curried relation between two objects, denoted correspondingly by the subject NP and the object NP, and a property contributed by the PP:

$$put : (\tilde{\mathrm{e}} \to ((\tilde{\mathrm{e}} \to \tilde{\mathrm{t}}) \to (\tilde{\mathrm{e}} \to \tilde{\mathrm{t}}))) \tag{12}$$

In cases like (1c) and (1d), we constrain the verbal lexemes, by their feature structure descriptions. The SPR of the PP that is in the COMPS list of the head verb is constrained to be identical to the SPR of the verb:

$$\begin{bmatrix} \text{syn} & \begin{bmatrix} \text{head} & \textit{verb} \\ \text{val} & \begin{bmatrix} \text{spr} & \langle \boxed{1}\,\text{NP} \rangle \\ \text{comps} & \langle\, \text{PP}\begin{bmatrix} \text{spr}\ \langle \boxed{1} \rangle \end{bmatrix} \rangle \end{bmatrix} \end{bmatrix} \end{bmatrix} \tag{13}$$

In a feature-value structure description of the verbal lexeme "put" we can take:

$$\begin{bmatrix} \text{syn} & \begin{bmatrix} \text{head}\ \textit{verb} & \\ \text{val} & \begin{bmatrix} \text{spr} & \langle \text{NP}_k \rangle \\ \text{comps} & \langle \boxed{2}\,\text{NP}_f,\ \text{PP} \begin{bmatrix} \text{syn} & \begin{bmatrix} \text{val} & \begin{bmatrix} \text{spr} & \langle \boxed{2} \rangle \\ \text{comps} & \langle\rangle \end{bmatrix} \end{bmatrix} \\ \text{sem} & \begin{bmatrix} \text{term} & \lambda f\ \text{term(P)}(l)(f) \\ \text{index} & l \end{bmatrix} \end{bmatrix} \rangle \end{bmatrix} \end{bmatrix} \\ \text{sem} & \begin{bmatrix} \text{term} : \textit{put} \end{bmatrix} \end{bmatrix} \tag{14}$$

6 Conclusions and Future Work

The distinction between referential and relational PPs is not limited to argument marking in VP. It permeates other syntactic categories by similar syntax-semantics effects. In particular, further research is planned on syntax-semantics interface of NPs with head nominal constituents that are relational nouns. Option 1 and Option 2 contribute to semantic distinctions in representations of relational nouns. For example, The nominal expressions (15c)-(15d) are clearly ambiguous.

brother of Kim (Kim's brother) (15a)

article on/about physics (15b)

article on/about/near the planet (15c)

picture of John (John's picture) (15d)

It is interesting that the PPs with argument marking prepositions can be used as modifiers in expressions with different syntactic structure. For example, the PPs from the provided examples can be used both as modifiers of VPs and sentences. The feature MOD can be introduced as a VAL feature, as in [9] for handling the syntactic category of PP modifiers. Such syntactic treatment can provide a right usage of a PP depending on which rule uses it, HCR or a modification rule. This involves that the same (or almost the same) semantic representation of the PP contributes to the semantic representation of both syntactic roles, argument marking preposition and modification. Further detailed research on such topics is interesting for potential theoretical development of syntax-semantics interface and applications to language processing. Depending on the scope of applications, the language and theory of recursion may vary, either by including full functional recursion, or with respect to development of more elaborate type system. Most interesting, and more challenging, project is development of algorithmic type system for relational models with partial relations, i.e., situation theory.

Acknowledgements. Work partially supported by the Spanish Ministry of Science and Innovation under coordinated research project TIN2011-28260-C03-00 and research project TIN2011-28260-C03-02.

References

1. Copestake, A., Flickinger, D., Pollard, C., Sag, I.: Minimal recursion semantics: an introduction. Research on Language and Computation 3, 281–332 (2005)
2. Loukanova, R.: An approach to functional formal models of constraint-based lexicalist grammar (CBLG). Fundamenta Informaticae. Journal of European Association for Theoretical Computer Science (EATCS) (2011)
3. Loukanova, R.: Semantics with the language of acyclic recursion in constraint-based grammar. In: Bel-Enguix, G., Jiménez-López, M.D. (eds.) Bio-Inspired Models for Natural and Formal Languages, pp. 103–134. Cambridge Scholars Publishing (2011)
4. Moortgat, M.: Symmetric categorial grammar. Journal of Philosophical Logic 38(6), 681–710 (2009)
5. Morrill, G.: Categorial Grammar: Logical Syntax, Semantics, and Processing. Oxford University Press (2010)
6. Moschovakis, Y.N.: Sense and denotation as algorithm and value. In: Oikkonen, J., Vaananen, J. (eds.) Lecture Notes in Logic. Lecture Notes in Logic, vol. 2, pp. 210–249. Springer, Heidelberg (1994)
7. Moschovakis, Y.N.: A logical calculus of meaning and synonymy. Linguistics and Philosophy 29, 27–89 (2006)
8. Ranta, A.: Grammatical Framework: Programming with Multilingual Grammars. CSLI Publications, Stanford (2011)
9. Sag, I.A., Wasow, T., Bender, E.M.: Syntactic Theory: A Formal Introduction. CSLI Publications, Stanford (2003)

Overview of the Fault and Bad Language Tolerance in Automatic User-Agent Dialogues*

Diana Pérez-Marín

Abstract. Conversational agents are computer systems able to interact with users in natural language. Advances in Natural Language Processing and Genetic Computation have promoted the possibility of gathering a corpus of user-agent dialogues. In this paper, the focus is to study the level of tolerance that users have when interacting with agents from the analysis of the recorded dialogues. Some factors that have been found as reasons for bad language are: misunderstanding of the agent, high expectations of the users, need of challenging the agent, cultural background and age. Results of several experiments in which these factors are involved are reported.

Keywords: Conversational Agent, Fault Tolerance, Bad Language Tolerance.

1 Introduction

According to the Media Equation [1], we, as users, interact with the computers as if the computers were human users too. This is particularly important in the case of Conversational Agents, that is, computer systems that interact in natural language acquiring human skills such as the ability of talking or showing empathic emotions.

However, this paradigm is not always reliable [2]. D'Angeli and Brahnam have researched that when users are not being observed (for instance, when they chat with the agent on-line), they tend to be aggressive and rude. In fact, in a study carried out in 2008, from 146 conversations in natural language between human users and the general-purpose conversational agent Jabberwacky, 10% of the total stems

Diana Pérez-Marín
Computer Science Faculty, Universidad Rey Juan Carlos, Spain
e-mail: diana.perez@urjc.es

* This work has been sponsored by Spanish Ministry of Science and Technology, project number TIN2011-28260-C03-02.

J.B. Pérez et al. (Eds.): Highlights on PAAMS, AISC 156, pp. 61–67.
springerlink.com

in the corpus reflected abusive language, and approximately 11% of the sample addressed hard-core sex.

These results are particularly disturbing for agents using the "mirror" technique. That is, agents who store the sentences of users to repeat them again to other users when the level of coherence between the input sentence and the stored sentence is similar in the dialogue. The core idea is that in the first conversation, it was coherent to produce sentence 2 after sentence 1, in the current conversation, it will be coherent to produce sentence 2 after sentence 1.

However, let us suppose now that sentence 1 is a normal informative sentence, while sentence 2 is an insult, then whenever the agent is provided sentence 1, the answer provided will be an insult too. Some researchers have proposed as possible solution using filters to avoid the choice of aggressive sentences [2].

In this paper, an alternative approach is proposed, instead of filtering the sentences, the focus is on finding the reason why the users utter these sentences. A possible hypothesis explaining why the users insult the agents when they feel frustrated could be because the agent answers incorrectly to their question, or during the first conversations when they are trying to identify its intelligence pushing its limits. Moreover, factors such as the cultural background and age of the users are considered essential too.

The paper is organized as follows: Section 2 reviews the state-of-the-art in the user-agent dialogues field focusing on the fault and bad language tolerance reported and the factors associated, and Section 3 ends with a discussion.

2 State-of-the-Art Review

In general, fault tolerance is essential for multi-agent systems [3]. In this section, the causes, reported in the literature, of bad language towards conversational agents, are reviewed classified as misunderstandings of the agent [4-11], high expectations of the users [6,8,12], or need of challenging the agent [4,7,9,11].

2.1 Misunderstandings of the Agent

Veletsianos et al. [4] reported that when 52 students, enrolled in a masters program in education, were asked to use one basic pedagogic agent (Alex or Penelope) during 4 weeks, they were eager to use the agent and appraised its availability. Figure 1 shows the agents.

Moreover, according to the analysis of the students-agents logs, even when the answer was incorrect, students just stopped using the agent, or changed the topic of the conversation. Sometimes, the students even attributed the failure to themselves because they believed that it was their fault as they had not used the correct wording.

Only, when the students were in groups, they tried to challenge the agent as it will be discussed in more detail in Section 2.3, and even in those cases, the students were remorseful because they felt that they have misbehaved.

Fig. 1 Alex and Penelope agents

Some of the comments provided by the students happy because of the correct answers received are the following: "I liked how the agent explained how to upload a picture. Everything pertinent to the course that I asked was useful."; "Some questions [Penelope] did not know how to answer, but overall, the simple questions that I asked her she was able to help me out a lot along the way."

On the other hand, some comments provided by the students angry at the agents and frustrated because they were unable to answer are the following: "I asked Alexander, and he had no input for me [laughs]. He made some smart comment and changed the subject [more laughs]. That was probably the only issue I had with him. But the fact that he couldn't help me made me really angry at Alex."; "I don't like Penelope. I asked her a question like five different ways and she still couldn't answer it. I don't remember what the question was but she should have been able to answer it." ; "You would type something easy like 'burning a CD,' and it would bring you all these things but 'burning a CD' which is pretty standard."

As can be seen, the level of frustration is greater when the students considered that the question was simple enough for the agent to answer. The reactions, in this case, were usually to stop using the system or to start talking about other topics. Some comments provided by the students are the following: "Every time I would ask [Alex] a question he would say, you need to download some software or something. I didn't know what that was about, so I didn't bother."; "I asked [Alex] what the eFolio website was because I wanted to go to the website because I deleted the email that told me. I wanted to go to the website and find their contact number and things like that and email them. He was explaining what eFolio was and I was like ... argggg … forget it. And so I did my own little search and then I found it, but I kept him there in case I got stuck."

On the other hand, some students were rude to the agent, which had been coded to be rude back, reaction that was not well accepted by the students with comments such as: "I thought it was more fun to play with her."; "I was kind of rude to her one time and then she was rude back."

These results are different when the users are younger (adolescents) or with a different cultural background (prisoners) [5]. In particular, although both adolescents and prisoners had little difficulty learning to interact with the agent, 56% of the prisoners did not use the agent and the adolescents misbehaved [9].

In particular, in the case of the adolescents, incorrect answers of the agent did not seem to be the cause of the bad language. In fact, only in 7.9% conversation turns, there was an insult because of a misunderstanding of the agent, while 92 comments (46.7%) were initially unacceptable and 44% sexually explicit. As it will be discussed in more detail in Section 2.2, the researchers thought that the expectations of the adolescents when looking at the agent could be the reason, in this case, for the bad language.

Robinson et al. [7] had similar results when analyzing the logs of the conversations between users and Sgt. Blackwell, an agent exhibited in a museum as shown in Figure 2. From the 1000 conversational turns registered, 789 were started by the user, 182 were started by the agent and 29 were classified as another type. Among the 789 started by the users, the researchers identified the following categories: dialogue functions (82), user-initiated information requests (634), hazing-testing perception (40), flaming (24), imperatives (9).

As can be seen, 80% of the sentences were information requests, and only 8% were classified as hazing-testing-flaming, and in all cases it was because the intention of the user was to insult or toy with the system but not because of an angry reaction to the comment of the agent.

2.2 *High Expectations of the Users*

Norman [13] noted that representing virtual characters as human-like figures may induce expectations of human intelligence and capabilities. For instance, Sgt. Blackwell (see Figure 2) may seem an authoritative figure, while Joan (see Figure 3) may seem more like a peer student. This could be the cause that users behaved more correctly with the first agent than with the second one.

In particular, when Sgt. Blackwell provided a bad answer, users humanized the agent with comments such as: "Oh I see you have an attitude" or "you're not listening". However, students using Joan did not complete the educational task, and were rude with the agent [9].

Therefore, it is important the aesthetic design of the agent as it seems to have an impact on the expectations of the users, who behave differently depending on the outfit and context of the agent. In particular, a conversational character which presents itself as all-knowing may provoke a higher feeling of frustration and anger if the answers of the users are not correctly answered.

Some comments, in this respect, provided by the users of Alex and Penelope [4] are the following: "I felt that my question was just too difficult. I tried rewording it but I really didn't expect her to know, and then when she didn't know I wasn't that surprised."

Komatsu and Yamada [6] investigated the effect of the "adaptation gap", that is, the difference between the functions expected by the users before starting their interactions with the agents and the functions that they perceive after their interactions.

Fig. 2 (a) Sgt. Blackwell agent, (b) Joan.

They found out that the initial expectations are essential because they can determine whether the user feels that the agent is worth interacting with it. Moreover, they reported how the mental model of the users regarding the agent affects their interaction with it.

In this line, Matsumoto et al. [14] proposed a "minimal design policy". The reason given is that if users adjust their expectations to the behavior of the agent, the flow of the conversation will be more natural and without frustrations and deceptions.

Finally, it is also important to mention that related to the expectations to the users, some researchers have also mentioned motivation as a key factor [8]. It has been reported that motivated users, who really want to get the task done with the agent, will be willing to try multiple approaches until something works. On the other hand, unmotivated users will give up if the agent does not correctly answer.

2.3 Need of Challenging the Agent

Some users have reported that in the first interactions with an agent they need to push their limits [11]. Some comments reported are: "I just wanted to understand more", and "I want to know more about this thing!" what or who is it?", even one user claimed that he actually wanted to stump the agent by asking it difficult, complicated, or misleading questions.

The participants of the study with Alex and Penelope [4] reported their need of challenging the agent, with comments such as: "Because she had an answer for everything, it was like pushing her limits. You know it wasn't a real person, so it's not that you are offending her."

It is interesting to observe in the last comment how the user justifies his need of challenging the agent in the high expectation of the human appearance of Penelope (see Figure 1) and the fact that it was not a real person so there was no harm. This idea was also found in other papers such as [9] in which the students misbehaved because they knew there was no punishment.

The need of challenging the agent is usually found in groups too. In [4], it is reported that participants would talk to the agent to see what he/she knew as a fun social event with their friends outside of the technology class.

3 Discussion

From the review of the literature, it can be concluded that bad language is not usually provoked because of a misunderstanding of the agent, but because of internal factors of the users such as their cultural background, expectations, age, mental model or their need for challenging the machine and having fun with friends as they consider that there will not be punishment.

It is also recommended that agents do not create high expectations in the users because it may lead to frustration and deception. On the other hand, agents should try to refocus the user on the topic of the dialog and look like an authoritative figure. Some guidelines to achieve that would be the following:

- Name the agent with a role and rank (e.g. Dr. Joan instead of just Joan for a pedagogical agent [9]).
- Adapt the dialogue to the time: initial conversations are expected to push the limits of the agent. If the answers of the agent are not challenging but social and refocused on the topic, next interactions might be more focused.
- Adapt the dialogue to the user: children are expected to be more tolerant with the agent and motivated adults too. For other users, it could be adequate to show them some consequences of their actions.

References

[1] Reeves, B., Nass, C.I.: The media equation: How people treat computers, television and new media as real people and places. Cambridge University Press/CSLI, Cambridge (1996)

[2] D'Angeli, A., Brahnam, S.: I hate you! Disinhibition with virtual partners. Interacting with Computers 20, 302–310 (2008)

[3] Potiron, K., Taillibert, P., El Fallah Seghrouchni, A.: A Step Towards Fault Tolerance for Multi-Agent Systems. In: Dastani, M.M., El Fallah Seghrouchni, A., Leite, J., Torroni, P. (eds.) LADS 2007. LNCS (LNAI), vol. 5118, pp. 156–172. Springer, Heidelberg (2008)

[4] Doering, A., Veletsianos, G., Yerasimou, T.: Conversational Agents and their Longitudinal Affordances on Communication and Interaction. Journal of Interactive Learning Research 19(2), 251–270 (2008)

[5] Hubal, R., Fishbein, D., Sheppard, M., Paschall, M., Eldreth, D., Hyde, C.: How do varied populations interact with embodied conversational agents? Findings from inner-city adolescents and prisoners. Computers in Human Behavior 24(3), 1104–1138 (2008)

[6] Komatsu, T., Yamada, S.: Adaptation gap hypothesis: How differences between users' expected and perceived agent functions affect their subjective impression. Journal of Systemics, Cybernetics and Informatics 9, 67–74 (2011)

[7] Robinson, S., Traum, D., Ittycheriah, M., Henderer, J.: What would you ask a conversational agent? Observations of human-agent dialogues in a museum setting. In: Proceedings of the Sixth International Language Resources and Evaluation (LREC 2008), Marrakech, Morocco, vol. 26 (2008)

[8] Traum, D.: Talking to Virtual Humans: Dialogue Models and Methodologies for Embodied Conversational Agents. In: Wachsmuth, I., Knoblich, G. (eds.) Modeling Communication. LNCS (LNAI), vol. 4930, pp. 296–309. Springer, Heidelberg (2008)

[9] Veletsianos, G., Scharber, C., Doering, A.: When sex, drugs, and violence enter the classroom: Conversations between adolescents and a female pedagogical agent. Interacting with Computers 20(3), 292–301 (2008)

[10] Gupta, S., Walker, M.A., Romano, D.M.: How Rude Are You?: Evaluating Politeness and Affect in Interaction. In: Paiva, A.C.R., Prada, R., Picard, R.W. (eds.) ACII 2007. LNCS, vol. 4738, pp. 203–217. Springer, Heidelberg (2007)

[11] Veletsianos, G., Miller, C.: Conversing with Pedagogical Agents: A Phenomenological Exploration of Interacting with Digital Entities. British Journal of Educational Technology 39(6), 969–986 (2008)

[12] Hoffmann, L., Krämer, N.C., Lam-chi, A., Kopp, S.: Media Equation Revisited: Do Users Show Polite Reactions towards an Embodied Agent? In: Ruttkay, Z., Kipp, M., Nijholt, A., Vilhjálmsson, H.H. (eds.) IVA 2009. LNCS, vol. 5773, pp. 159–165. Springer, Heidelberg (2009)

[13] Norman, D.: How might people interact with agents. In: Bradshaw, J.M. (ed.) Software Agents, pp. 49–56. MIT Press, Menlo Park (1997)

[14] Matsumoto, N., Fujii, H., Goan, M., Okada, M.: Minimal design strategy for embodied communication agents. In: Proceedings of the 14th International Symposium on Robot and Human Interactive Communication, pp. 335–340 (2005)

Automatic Adaptation and Recommendation of News Reports Using Surface-Based Methods

Joel Azzopardi and Christopher Staff

Abstract. The multitude of news reports being published on the WWW may cause information overload on users. In this paper, we describe a news recommendation system whereby news reports are represented using entity-relationship graphs, and the users' interaction with these news reports in a specialised web portal is monitored in order to construct and maintain user models that store the user's reading history and also define entities that appear to be of interest to the user. These user models are used to alert individual users when an event has occurred that falls within their area of interest, and to present news reports to users in an adaptive manner – previously seen information is shown in a summarised form. We evaluated our recommendation system using a corpus of news reports downloaded from *Yahoo! News*. Results obtained indicate that our recommendation system performs better than the baseline system that uses the *Rocchio* algorithm without negative feedback.

1 Introduction

Major news sites produce a huge number of news reports each day. A normal user would not be interested in reading *all* the news reports, and interesting reports may be 'hidden' amongst other 'non-interesting' reports. A service providing personalised recommendations to the user according to his/her interests would help solve this problem of information overload.

As part of our research, we have built an operational news portal that periodically downloads news reports from different sources using RSS feeds, clusters them according to the events reported, builds logical representations of each cluster (i.e. represents the information in each cluster using entity-relation structures), constructs fused reports for each cluster, outputs the fused reports to the user in an adaptive manner (previously seen information is shown in a summarised form whilst the

Joel Azzopardi · Christopher Staff
Faculty of ICT, University of Malta, Msida, Malta
e-mail: {joel.azzopardi,chris.staff}@um.edu.mt

J.B. Pérez et al. (Eds.): Highlights on PAAMS, AISC 156, pp. 69–76.
springerlink.com

new information is shown in full), and issues personalised recommendations for the different users. All these tasks are performed using only surface-based methods – without using knowledge bases and/or deep semantic processing. It is our opinion that the use of surface-based methods renders our system domain independent since it does not require any domain-specific pre-defined knowledge. The representation of information using entity-relation structures and the construction of fused reports are described in [3]. In this paper, we focus on how our system performs content adaption and how it issues recommendations for users using content-based filtering.

In literature, one can encounter descriptions of other personalised news recommendation systems, such as *NewsWeeder* [16], *NewsJunkie* [14] and *News@hand* [10, ?]. *NewsWeeder* operates on news feeds to build user-personalised news bulletins according to the predicted user ratings. *NewsJunkie* performs personalisation on the basis of the novelty of each news report in the context of previously viewed news reports. *News@hand* produces enhanced recommendations by applying Semantic web technologies to describe and relate news reports' contents and user preferences. In our research, we diverge from existing systems in our use of entity-relationship representations, and in how the recommendations given are events (fused reports) rather than individual news reports.

The remainder of this paper is divided as follows: Section 2 gives an overview of related literature related; Section 3 describes the structure and operation of our system; Section 4 presents the evaluation of evaluated our recommendation system; and in Section 5, we give the conclusions derived from our research.

2 Related Research

The process to issue personalised recommendations involves the construction of user models reflecting users' interests, and the comparison of these models to the incoming news reports to determine the interesting reports for each user [15, 5]. This forms the basis of the *Information Filtering* (IF) discipline. The typical IF process consists of [19, 18, ?, 5, 6, 20]: *feature extraction* and *logical representation* of the incoming information; *filtering* of the incoming information; and the collection of *User feedback* to update the corresponding user models.

One of the most common representations used is the *Bag-of-Words* representation [19, 18, 4, 15, 16, ?, 14, 5, 6, 20]. This representation is efficient, effective and simple to implement, and is feasible even in cases of very high volume of information [4]. In this representation, both the incoming documents and the user profiles are represented as weighted vectors generally using *TF.IDF* [19, 18, 16, ?, 14, 20]. The terms are usually stemmed and filtered from stop words [19, 18, 4, ?, 20]. Possible improvements to the *Bag-of-Words* representation include: the extraction of special entities (such as names, dates and locations) [4, 14]; the extraction of noun phrases [4, 16]; and the use of term or document clustering [4].

In our opinion, one of the best ways to represent knowledge is by using conceptual graphs [21] since they are able to represent different items with different characteristics in an unambiguous manner. However, conceptual graphs cannot be constructed without domain knowledge. Therefore, we had to use a more simplified representation of such graphs. Descriptions of other similar graph-based representations may be found in [3].

Filtering techniques can be classified into: *content-based filtering* – filtering based on the similarities between the incoming documents and the user profiles); and *collaborative filtering* – where filtering for a user is done based on what other similar users have read [12, 16, 6]. *Collaborative filtering* is simpler to implement, but requires a large user-base for it be effective [18, 20, 6].

Content-based filtering may be performed *semantically* or *syntactically* [17]. *Syntactic filtering* is more efficient and has lower processing overheads [17]. A common syntactic filtering approach involves calculating the cosine similarity between the documents' term vectors and the user profile, and classifying those documents with a similarity higher than a pre-defined threshold as 'interesting' for that user [19, 20]. Other measures used to determine a document's interestingness include: *coverage*, *source reliability*, *novelty*, and *timeliness* [15, 12].

The effectiveness of IF systems depends crucially on the use of appropriate user profiles and user feedback mechanisms. User profiles consisting of only 'interesting' keywords are insufficient for filtering purposes [18]. Other user information may include: the user's reading history [18]; the information 'freshness' [6, 18]; the writing style that suits the user most [18]; and the user's preferred sources of information [6, 18]. One should also consider that users' interests are continually evolving [18]. Therefore, user models should not be static but should evolve in time mainly through user feedback. The *Rocchio* algorithm is used frequently to update user models by adding the vectors of 'interesting' documents, and subtracting the vectors of 'non-interesting' documents from the user profile vector [18].

User-feedback mechanisms can be classified into **explicit feedback mechanisms** – where users specify their feedback directly (such as rating articles according to their interest); and **implicit feedback mechanisms** – where the feedback is collected indirectly from the user (for example by measuring the time spent on an article) [16, 6]. *Explicit feedback* is considered to be more accurate but requires an extra effort from the user [16, 12, 17].

Closely related to the area of IF one finds *Adaptive Hypermedia* (AH) Systems whose purpose is to deliver personalised views of hypermedia documents by performing adaptation of their content and navigation links. Our system is adaptive since it adapts news reports' contents based on what information the user has already read. Typical AH systems consist of three main components – the *Hypermedia system*, the *User Modelling Component*, and the *Adaptation Component* [9, 13].

User models within AH systems can have various dimensions such as: *Interests*; *Expertise*; *Cognitive Characteristics*; *Life Style*; *Goals*; and *Background/Experience* [2, 9]. User interests can be represented by the frequencies of terms within the known documents of interest [13]. Similar to IF, the acquisition of user feedback can be done *explicitly* or *implicitly* [9, 7]. Stereotypes can also be used to classify a new user with similar existing users and issue similar recommendations [2, 8, 7].

The adaptations performed by an AH system are classified into: *Adaptive Presentation* – where the actual hypermedia content is adapted for the user; and *Adaptive*

Navigation – where the presentation of links between different hypermedia nodes is adapted [9, 2, 8]. Typically, adaptive presentation techniques consist of the showing/hiding of content according to its relevance to the user [9]. One such technique is *StretchText* where the clicking of a hot word will lead to the expansion of the text on the current page [7]. In adaptive StretchText, the user-relevant information is shown visible whilst the non-relevant information is hidden. Adaptive Navigation techniques include direct guidance, adaptive sorting of links, link hiding, link annotation, and map adaptation [9, 8].

3 Methodology

Our system downloads news reports through RSS feeds on a periodical basis and clusters them into event-centric clusters[1]. The news reports in each event cluster are represented using entity-relation graphs. The generated entity-relation graphs consist of entities – noun phrases extracted using heuristic Part-of-Speech (POS) patterns; and the relationship between these entities[2]. Verb phrases are used as relationship names. The identification of relationships between different entities is performed by matching POS sequences with heuristic patterns. Co-referent noun entities and verb entities from different segments are identified, based on the amount of similar salient terms between them, and clustered together into entity collections – e.g. "Obama", "President Obama" and "Barack Obama" are clustered together within a single entity collection. Relationships that are conveying the same information are also clustered together. Such relationships are identified as those that have two co-referent entities, or that have one co-referent entity and a similar relationship name between them – e.g. "John killed Mary" and "John murdered Mary" are identified as co-referent. Then, the fused report is generated by searching for the maximally expressive set of source sentences – i.e. the set of sentences that represents all the entities and relationships without repetition. Further details about this process may be found in [3].

When a user re-visits a fused report that has been updated since his/her last reading, the previously seen information – i.e. the information that was already present during his/her last reading – is summarised (to serve as a reminder of what was read), whilst the new information is shown in its full form. This content adaptivity is possible since our system maintains a user model for every user that apart from 'interesting' terms, contains also the reading history (and the date and time of reading) of the fused news reports read by that user. The fused reports presented to the users are segmented into sections according to when the information was published. The summarisation component works by extracting the most salient information in each section and outputting it as the summary for the previously seen sections. The selection of the most salient information is performed on the basis of: the number of

[1] The news reports in each event-centric cluster are describing the same event.

[2] e.g. the phrase "John killed Mary" is represented by the entities "John" and "Mary", and the relationship "killed" between them

times each entity-relation structure was found in different segments; and the weights of the constituent terms in each entity-relationship cluster.

Each user model consists of: terms extracted from the most salient entities in the read fused reports; a history of what the user has read; and the times when the user has read the different reports. The user models are used to recommend unread news reports, and also to alert the user about fused news reports that have been updated since he/she last accessed them. When a new user logs on to our system, a new user model is created and a cookie is set in his/her browser with the identification of this new user model. When an existing user logs in, the corresponding user model is located with the help of the relevant cookie string. Recommendations are calculated forthwith and displayed to the user. Moreover, the list of reports read by the user is checked to identify reports that have been updated since the user last read them. To simplify the process of constructing and updating the user models, each news report cluster is represented by the most salient noun and verb entities within that cluster – i.e. those entities that form part of the highest number of relationships. User models are updated implicitly – when a user accesses a fused report, the corresponding user model is updated with the salient terms found in that report, and that same report is referenced in the user's reading history along with the current date and time.

For our system to be able to perform content-based filtering in a timely manner, it cannot compare the corresponding user model with each and every fused news report. The fused news reports' entity-relation representations from the different clusters are merged together by identifying co-referent entities across the different clusters (similar to what was done in the fusion phase), and stored in one single storage base. To perform filtering, our system reads the set of salient terms within the user model, and searches through the entity-relation storage base for unread fused reports referencing entity collections featuring those salient terms. The list of unread fused reports is filtered to remove those reports that do not have sufficient references to entity collections containing the user model's salient terms.

4 Evaluation

In this section, we describe how we evaluated our system's recommendation capability using content-based filtering. To perform this evaluation, we imitated the methodology utilised in [19], [18] and [20], where RSS feeds are downloaded from Yahoo! News[3] for different sections, and the news reports that are marked "Mostly Viewed" for a section are considered to be interesting for a pseudo-user who has a high interest in that section.

For our evaluation, we periodically downloaded news reports from the *Yahoo! News* RSS feeds[4]. We used the set of categories that are quite specific – for example, rather than using the category *Technology*, we used finer categories such as *Internet, Linux/Open Source*. The training set consisted of 4297 news reports from 68 different categories downloaded in December 2010. The test set was made up of

[3] http://news.yahoo.com

[4] http://news.yahoo.com/rss

6579 news reports downloaded during January 2011 and classified into these same 68 categories. We assumed that each of these specific categories represents a user with a specific interest. For the evaluation process we constructed a user model for each category in the training set. Then, the news reports in the test set were processed in order of the time of publishing and recommendations were calculated for each user model. These user models were updated incrementally as the test reports were processed – e.g. after processing report X from the "Linux" category, the "Linux" user-model was updated with report X.

We compared our results to a baseline recommendation system that uses the *Rocchio* algorithm without negative feedback. We did not consider negative feedback since we are evaluating an approach to be utilised for an operational system where the feedback supplied is implicit. The user profiles used by the baseline system were constructed from the full evaluation corpus, and not just from the training set.

Our system is not a push-based system – i.e. it does not forward 'interesting' news reports to the user. On the contrary, it provides a list of recommendations to the user. Users have a higher tendency to access items at the top of such a list than items at the bottom of the list. Hence, our aim is to provide a finite list of recommendations sorted in order of 'interestingness' to the user. Therefore, we extracted the top 10 and 20 recommended reports for each user, and calculated the *precision* of these subset lists using the test set's categories as reference.

Table 1 shows the results obtained. Our system outputs recommendations with an average precision of 56% for the entire list of recommendations, 66% for the top 10 recommendations, and 62% for the top 20 recommendations. They may be compared to the baseline system that outputs the top 10 and 20 recommendations with approximately 10% less precision. The results obtained also show that the precision of the recommendations issued by our system does not deteriorate greatly as one goes down the list of recommendations. One should consider the fact that the baseline system had the advantage of better-defined user models (they were constructed using both the training and test datasets). When our recommendation system was run with user profiles built using the entire evaluation dataset, the average precision of the top 10 recommendations reached 80%. This implies that when our system is used with better formed user profiles that were built over a number of days (or months), it produces more effective results.

Table 1 Results from the Filtering Evaluation

Results List	Mean Precision	Min. Precision	Max. Precision
Full Recommendations List by our Filtering System	0.5637	0.0000	1.0000
Full Recommendations List by Baseline Filtering System	0.0315	0.0015	0.1886
Top 10 Recommendations by our Filtering System	0.6571	0.0000	1.0000
Top 10 Recommendations by Baseline Filtering System	0.5687	0.0000	1.0000
Top 20 Recommendations by our Filtering System	0.6227	0.0000	1.0000
Top 20 Recommendations by Baseline Filtering System	0.5142	0.0000	1.0000

5 Conclusion

We have described a system that uses surface-based approaches to: cluster news reports from disparate sources into event-centric clusters; represent reports using entity-relation structures; fuse related reports into a single coherent report; present the reports to the user in an adaptive manner; and provide personalised recommendations to users. Since our system does not utilise any complex processes that require a huge amount of resources to run, it can be used as an operational system handling a large amount of incoming news reports from different sources. A unique feature of our system is that our system operates by transforming the news reports into entity-relation structures and then processes these entity-relation structures to perform fusion and filtering. We have encountered literature that describe the use of similar structures for text and documents, but to our knowledge, these have not been applied in operational news systems.

In view of the results obtained in our evaluation, we may conclude that our recommendation system performs satisfactorily. However, one would wish that the list of recommendations would have a higher percentage of 'interesting' documents. When we analysed some of the top recommendations issued by our system that are not deemed 'interesting' by the data corpus, we found that these documents were actually quite related to the user profile. This means that our recommendation system does not issue totally unrelated recommendations. Further work on our system will include having further research in the representation of news reports and in the user modelling processes in order to provide recommendations with higher precision.

References

1. Albayrak, S., Wollny, S., Varone, N., Lommatzsch, A., Milosevic, D.: Agent technology for personalized information filtering: the pia-system. In: SAC 2005: Proceedings of the 2005 ACM Symposium on Applied Computing, pp. 54–59. ACM, New York (2005), doi: `http://doi.acm.org/10.1145/1066677.1066695`
2. Ardissono, L., Console, L., Torre, I.: An adaptive system for the personalized access to news. AI Commun. 14(3), 129–147 (2001)
3. Azzopardi, J., Staff, C.: Fusion of news reports using surface-based methods. In: MAW 2012: Proceedings of the 2012 International Symposium on Mining and Web. IEEE (to appear, 2012)
4. Belkin, N.J., Croft, W.B.: Information filtering and information retrieval: two sides of the same coin? Commun. ACM 35(12), 29–38 (1992), doi: `http://doi.acm.org/10.1145/138859.138861/html`
5. Bell, T.A.H., Moffat, A.: The design of a high performance information filtering system. In: SIGIR 1996: Proceedings of the 19th Annual International ACM SIGIR Conference on Research and Development in Information Retrieval, pp. 12–20. ACM, New York (1996), doi: `http://doi.acm.org/10.1145/243199.243203`
6. Bordogna, G., Pagani, M., Pasi, G., Villa, R.: A Flexible News Filtering Model Exploiting a Hierarchical Fuzzy Categorization. In: Larsen, H.L., Pasi, G., Ortiz-Arroyo, D., Andreasen, T., Christiansen, H. (eds.) FQAS 2006. LNCS (LNAI), vol. 4027, pp. 170–184. Springer, Heidelberg (2006)

7. Boyle, C., Encarnacion, A.O.: Metadoc: An adaptive hypertext reading system. User Modeling and User-Adapted Interaction 4, 1–19 (1994), http://dx.doi.org/10.1007/BF01142355, doi:10.1007/BF01142355
8. Bra, P.D., Brusilovsky, P., Houben, G.J.: Adaptive hypermedia: from systems to framework. ACM Comput. Surv. 31(4es), 12 (1999), doi: http://doi.acm.org/10.1145/345966.345996
9. Brusilovsky, P.: Methods and techniques of adaptive hypermedia. User Modeling and User-Adapted Interaction 6(2-3), 87–129 (1996), http://citeseer.ist.psu.edu/brusilovsky96methods.html
10. Cantador, I., Bellogín, A., Castells, P.: News@hand: A Semantic Web Approach to Recommending News. In: Nejdl, W., Kay, J., Pu, P., Herder, E. (eds.) AH 2008. LNCS, vol. 5149, pp. 279–283. Springer, Heidelberg (2008)
11. Cantador, I., Castells, P.: Semantic contextualisation in a news recommender system (2009)
12. Das, A.S., Datar, M., Garg, A., Rajaram, S.: Google news personalization: scalable online collaborative filtering. In: WWW 2007: Proceedings of the 16th International Conference on World Wide Web, pp. 271–280. ACM Press, New York (2007), doi: http://doi.acm.org/10.1145/1242572.1242610
13. Díaz, A., Gervás, P.: Adaptive User Modeling for Personalization of Web Contents. In: De Bra, P.M.E., Nejdl, W. (eds.) AH 2004. LNCS, vol. 3137, pp. 65–74. Springer, Heidelberg (2004)
14. Gabrilovich, E., Dumais, S., Horvitz, E.: Newsjunkie: providing personalized newsfeeds via analysis of information novelty. In: WWW 2004: Proceedings of the 13th International Conference on World Wide Web, pp. 482–490. ACM, New York (2004), doi: http://doi.acm.org/10.1145/988672.988738
15. Kassab, R., Lamirel, J.C.: A new approach to intelligent text filtering based on novelty detection. In: ADC 2006: Proceedings of the 17th Australasian Database Conference, pp. 149–156. Australian Computer Society, Inc., Darlinghurst (2006)
16. Lang, K.: Newsweeder: Learning to filter netnews. In: Proceedings of the 12th International Machine Learning Conference (ML 1995), pp. 331–339. Morgan Kaufman (1995)
17. Morita, M., Shinoda, Y.: Information filtering based on user behavior analysis and best match text retrieval. In: SIGIR 1994: Proceedings of the 17th Annual International ACM SIGIR Conference on Research and Development in Information Retrieval, pp. 272–281. Springer-Verlag New York, Inc., New York (1994)
18. Pon, R.K., Cardenas, A.F., Buttler, D., Critchlow, T.: Iscore: Measuring the interestingness of articles in a limited user environment. In: IEEE Symposium on Computational Intelligence and Data Mining 2007, pp. 354–361 (2007)
19. Pon, R.K., Cardenas, A.F., Buttler, D., Critchlow, T.: Tracking multiple topics for finding interesting articles. In: KDD 2007: Proceedings of the 13th ACM SIGKDD International Conference on Knowledge Discovery and Data Mining, pp. 560–569. ACM Press, New York (2007), doi: http://doi.acm.org/10.1145/1281192.1281253
20. Pon, R.K., Cárdenas, A.F., Buttler, D.J.: Online selection of parameters in the rocchio algorithm for identifying interesting news articles. In: WIDM 2008: Proceeding of the 10th ACM Workshop on Web Information and Data Management, pp. 141–148. ACM, New York (2008), doi: http://doi.acm.org/10.1145/1458502.1458525
21. Sowa, J.F.: A conceptual schema for knowledge-based systems. In: Proceedings of the 1980 Workshop on Data Abstraction, Databases and Conceptual Modeling, pp. 193–195. ACM Press, New York (1980), doi: http://doi.acm.org/10.1145/800227.806920

Through the Data Modelling Process of Turimov, an Ontology-Based Project for Mobile Intelligent Systems

Juan García, Francisco J. García-Peñalvo, and Roberto Therón

Abstract. Functionality in terms of software can be defined as the quality of a software system of being functional, according to its specifications. Functionality tests represent a crucial aspect during the last phase of the development process of a software system. This paper describes a case study performed using OWL-VisMod, in order to evaluate its functionality, during its last phase of development. This case study describes the development of a data modelling process, for a project called *Turimov*, a recommendation system for mobile devices (smartphones, tablets and PDA's), to support the tourism activities in the region of Castile and Lion Spain.

Keywords: Functional testing, OWL-VisMod evaluation, Ontologies Modelling.

1 Introduction

The International Organization for Standardization (ISO) defines the usability *as the quality that let a product to be used by specific users, to reach specific goals with effectiveness, efficiency and satisfaction, in a specific context of use (ISO/IEC9241)*. Usability refers to the capacity of a software product to be used, especially those products that interact with people [9]. In the specific context of the Knowledge Representation Systems (KRS), the development of diverse usability tests becomes crucial, in order to detect inconsistencies, errors or other aspects that the system designers did not previously considered. The most visible aspect of this approach

Juan García · Roberto Therón
Computer Science Department, University of Salamanca, Spain
e-mail: {ganajuan, theron}@usal.es

Francisco J. García-Peñalvo
Computer Science Department, Science Education Research Institute (IUCE). GRIAL Research Group. University of Salamanca, Spain
e-mail: fgarcia@usal.es

J.B. Pérez et al. (Eds.): Highlights on PAAMS, AISC 156, pp. 77–84.
springerlink.com

is usability testing, in which users work and interact with the product interface and share their views and concerns with the designers and developers.

This term represents an approach that puts the user, rather than the system, at the center of the process. The user-centered design incorporates user concerns and advocacy through the design process, letting to the system designers to detect inconsistencies, errors or diverse aspects that they did not previously considered.

On the other hand, functionality in terms of software, refers to the set of functions or capabilites that a software system is assumed to perform [10]. It can also be defined as the quality of a software system of being functional, according to its specifications. Functional testing represents a crucial aspect to evaluate during the last phase of the development process of a software system.

OWL-VisMod[1] is a visual modelling tool for OWL ontologies [4]. It has been designed to support the processes of creation, modelling, reusability and analysis of ontologies for Knowledge Representation Systems, and diverse ontology-based software systems. A first usability test of it is described in [5], and the case study described in this paper is considered as a complementary evaluation of its usability.

The main contribution of this paper, is to validate that OWL-VisMod has been provided with the capabilities and the functionality needed to perform a system in a real scenario. Through this evaluation process, diverse aspects that could have not been considered by the designers, can be detected prior to release the first version of the tool.

This paper describes the development of a data modelling process for the *Turimov* project using OWL-VisMod. This modelling process is described according to the sequence of the diverse activities performed, highligting the most important aspects in each of the phases.

The outline of this paper is as follows: the next section describes the domain of the problem, the goals to reach as well as the technical requirements. The third section describes the process of building the taxonomy of concepts, then the fourth section describes the definition of the object and datatype properties. The fifth section describes the process of population of the ontology with individuals, to finally discuss the general results, the conclusions and the future work.

2 Domain of the Problem

The use of semantic models for recommendation systems [1] in touristic domains, is a field that has been developing in the last decade, with a set of proposals such as [2, 3, 8].

The *Turimov* project is an intelligent system for mobile devices, to support the tourism in the Castile and Lion region in Spain. The main objective of this project is the development of a system based on the use of diverse mobile technological platforms, to help tourists to find activities, places, events, etc., during their trips through Castile and Lion. More specific details about the project can be found in

[1] `http://www.analiticavisual.com/juan/OWL-VisMod.html`

the website of the Interra's company[2]. This project is based on the use of OWL ontologies to manage all the information; these ontologies have been defined in order to perform inference over the data.

The first step in the data modelling process, corresponds to the definition and the abstraction of the problem, as well as the approach to solve it. The definition of the objectives of the project and the technological requirements were defined by the Interra's specialists, while the ontological model was completely implemented in OWL-VisMod. The global ontological model has been divided into three submodels, according to the general objectives: the first one is focused on the users of the system, the second one is used to classify the diverse mobile devices, while the third submodel contains all the resources in the knowledge base.

The first submodel (*turisem_personas.owl*[3]) contains all the concepts and properties that define the diverse profiles for all users in the system. The profiles are based on their preferences on gastronomy, activities, shows or preferred places to visit. The main objective of this submodel is to define the diverse characteristics of the users of the system, in order to the system can recommend activities according to these characteristics. The second submodel (*turisem_dispositivos.owl*[4]) lets to define the characteristics of the diverse mobile devices (smartphones, tablets and PDA's). This submodel has been defined in order to the system can get specific information about the mobile devices, this information is intended to manage the technological requirements of each. Finally, the third submodel (*turisem.owl*[5]), the biggest, contains all the information about the touristic resources such as activities (sports, shows, concerts, etc.), places (hotels, bars, restaurants, parks, galleries, museums, casinos, etc.), as well as diverse routes for hiking. These three submodels have been defined in spanish, according to the system requirements.

3 Building the Taxonomy

Once that the three submodels have been defined, the next step involved the definition of the taxonomy of concepts of each of them. The first model is focused on the users, where the main class is *Persona* (person), used to define the characteristics of a person. Other classes in this submodel are related to the definition of the proper characteristics for persons; these classes created as enumerations define diverse aspects such as ranges of age or the gender. On the other hand, the second model related to the mobile devices, defines the class *Dispositivo* (device) as its main class. This submodel includes diverse classes that define specific information about the technological requirements in the devices, such as the operating system, the software and hardware platforms, web standards or the wireless capabilities.

[2] `http://www.interra.es/`
[3] `http://data.turisem.info/turisem_personas.owl`
[4] `http://data.turisem.info/turisem_dispositivos.owl`
[5] `http://data.turisem.info/turisem.owl`

The third submodel is the biggest, it contains one hundred and twenty one classes, eighty three datatype properties and forty nine object properties. All the information in this submodel has been provided by the Interra's company, and it is mainly about places located in the city of Salamanca, its province and other provinces of Castile and Lion, such as Segovia, Bejar or Avila. This knowledge base mainly contains information about bars, restaurants, hotels, museums and historic monuments. There is also information about parks, wineries, golf clubs or tematic parks mostly located within the province of Salamanca. The total of classes in the three models is one hundred and forty three, being the most important classes: *Persona* (person), *Dispositivo* (device), *PlataformaSoftware* (software platform) and *PlataformaHardware* (hardware platform).

4 Defining the Object and Datatype Properties

Once the taxonomy has been defined, the next step is the definition of the properties. Firstly, the intrinsic properties that define the characteristics of the concepts according to the information of each of them. In the context of the model of persons (*turisem_personas.owl*), having the class *Person* as its main class, it defines two intrinsic properties: the age and the name of a person. The name property represents an identifier for an individual in a set of persons, while the age property, is used to classify persons in a range of ages, in order to recommend activities according to this range.

On the other hand, the model *turisem_dispositivos.owl* is focused on a diversity of mobile devices and their characteristics. These characteristics are mainly technical aspects such as: the hardware and software platforms, the operating systems, programming languages and the mobile devices itself (smartphones, PDA's and tablets). Finally, the model of the touristic resources (*turisem.owl*), has *Lugar* (place) and *Actividad* (activity) as two of the most important classes. The class *Lugar* defines the variety of places that can be visited by the tourists, such as museums, restaurants, bars, hotels, etc. The main properties of this class indicate the name of the place, the address, a possible cost to get access to it and a brief description. The figure 1 shows the global coupling of this class *Lugar* (place), where all the coupled classes are highlighted and the properties are represented as edges. The rest of classes are blurred as well as the edges, in order to highlight just the elements related to the coupling. Another important class is *Recorrido* (route), that defines the diverse hiking routes in the Castile and Lion region. This class contains data properties to store information about the route itself, such as the location coordinates, diverse places within the route, interest points, classification values according to the type and the difficulty of the route and the points of information through the route.

Object properties define the coupling among classes; a more detailed description of the process of modelling, visualisation and the analysis of the coupling among classes with OWL-VisMod can be found in [4]. By coupling it should be understood the dependency relations among classes. One class in an ontology is related to another class by means of an object property. For instance, the class *Persona*

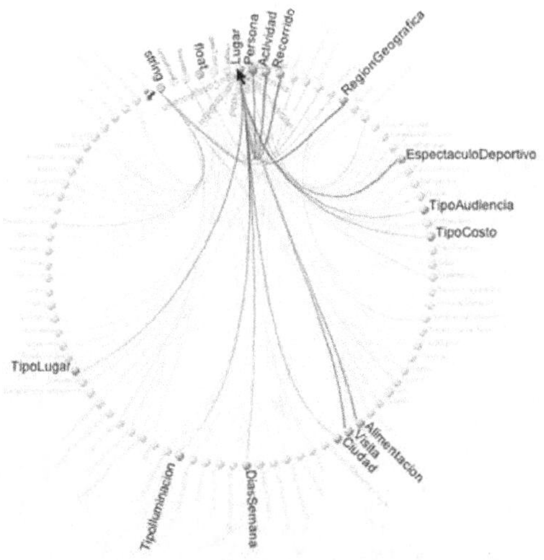

Fig. 1 Shows the coupling of the class *Lugar* (place), where the coupled classes are highlighted and the properties representing the relations are represented as edges.

(person) is related to the class *Actividad* (activity) by means of the relation *actividadPreferida*, indicating a preferred activity developed by a person.

On the other hand, the class *Dispositivo* (device) is coupled with classes that provide technical information, such as the operating system, the software and hardware platforms) and the physical resolution of the device. These relations have been defined to get information about the technical requirements of the mobile device, such as the operating system, the screen's resolution, as well as the hardware and software platforms.

5 Populating the Knowledge Base with Individuals

Once that all the properties (object and datatype) were defined, the next step was the definition of instances or individuals that represent specific values for the concepts in the ontologies. These instances mainly refer to diverse touristic places located within the region of Castile and Lion, such as bars, restaurants, historic monuments, museums, among others. Figure 2 shows a radial visualization showing the properties and instances of the class *Restaurante* (restaurant), in a semantic zoom technique described in [6]. In this case, the selected class is located at the centre of the visualisation, with all its instances surrounding it as well as the properties, represented as polygons. These properties contain the internal values for each instance in each property.

Figure 2 performs a visual querying technique described in [6], that shows all the values for the selected instance *Don Quijote*, a restaurant located in Salamanca.

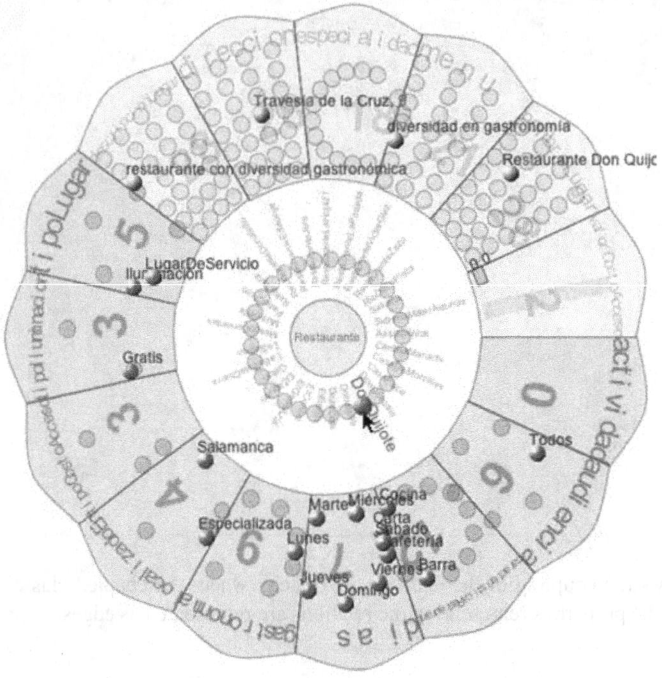

Fig. 2 The semantic zoom of the class *Restaurante* (restaurant), with nine object properties, six datatype properties and twenty-nine instances.

Another individuals have been also defined for diverse classes, representing touristic resources, such as restaurants, museums, parks, wineries, golf clubs, historis monuments or sky stations. The table 1 resumes all the individuals included in the project Turimov, for each of the classes. Moreover, this knowledge base can be enriched with new individuals for other classes, for instance the diverse activities such as shows, sport events, art expositions, among others.

6 Results

The main result described through this paper, is the data model definition based on the use of OWL ontologies, as well as the definition of a knowledge base for the "Turimov" project. This data modelling process, has let to evaluate the functionality of OWL-VisMod in a real scenario, with real data. The main goal has been successfully reached, because OWL-VisMod has been able to develop the process of data modelling, according to the specifications of this project. Even though, diverse evaluations with other projects with different requirements can be performed, OWL-VisMod can be viewed in general, as a mature tool for developing data modelling based on OWL ontologies. The first result, is the evaluation to OWL-VisMod, that can be considered as satisfactory.

Table 1 A resume of the total of instances of touristic resources

Touristic Resource	Total of Individuals
Restaurants	29
Hotels	27
Bars	24
Places	18
Historic Monuments	16
Museums	13
Parks	6
Wineries	5
Golf Clubs	4
Theme Parks	4
Paintball Scenarios	4
Rural Houses	4
Equestrian Centres	2
Sky Stations	1

As has been described through the development of this paper, the case study created three ontological models and finally a knowledge base with diverse instances of their classes. These three models have defined a taxonomy of concepts and the relations that let the system, firstly, to know the characteristics of a user to create a profile; then, according to this profile, the system performs an inference process in order to identify the diverse touristic resources that can fit to this profile. Secondly, the system has defined a knowledge base of diverse touristic resources within the region of Castile and Lion, most of them located within the city of Salamanca. A second, and a more tangible result, is that the case study has created three OWL ontologies and a knowledge base (*http://data.turisem.info/turisem_personas.owl, http://data.turisem.info/turisem_dispositivos.owl* y *http://data.turisem.info/turisem*).

7 Conclusions and Future Work

This paper has described the development of the modelling data process of the "Turimov" project, a system based on the use of diverse mobile technological platforms, to perform touristic activities. This project defines three OWL ontological submodels publicly available in the site of the project *http://data.turisem.info*; one for users, one for mobile devices and other for touristic resources. In addition to these models, a knowledge base has been also built, in order to enrich the system with a real scenario, using real data. On the other hand, the case study has also let to evaluate the functionality of OWL-VisMod, using a project with diverse requirements. The evaluation has let to consider that OWL-VisMod is now a mature tool, provided with the capabilities needed for developing OWL ontological models. Even though it has not been used any specific evaluation methodology, but just the fact that it has been able to successfully develop the data modelling process, represents itself, an evaluation. The main contribution is the tool itself, the case study discussed in this paper, represents a way to validate that the tool has reached such a maturity, to be considered for implementing Knowledge Representation Systems based on OWL ontologies.

The future work is defined by two approaches: firstly, the natural process of maintenance of these models, as well as the knowledge base, especially with the addition of new instances having diverse tourist resources, such as places of interest or activities. A second approach, is defined by the modelling of new projects or the maintenance process of existing ones.

Acknowledgements. This work was supported by Spanish Government project TIN2010-21695-C02-01 and by the Castile and Lion Regional Government through GR47 and the project FFI2010-16234.

References

1. Adomavicius, G., Tuzhilin, A.: Toward the next generation of recommender systems: A survey of the state-of-the-art and possible extensions. Journal IEEE Transactions on Knowledge and Data Engineering 17(6), 734–749 (2005)
2. Fesenmaier, D., Wober, K., Werthner, H. (eds.): Destination Recommendation Systems. CAB International (2006)
3. Franke, T.: Enhancing an online regional tourism consulting system with extended personalized services. Information Technology & Tourism 5, 135–150 (2003)
4. García, J., García, F., Therón, R.: Modelling relationships among classes as semantic coupling in owl ontologies. In: Proceedings of the 2011 International Conference on Information & Knowledge Engineering, IKE 2011, vol. 1, pp. 22–28 (2011)
5. García, J., Garcia-Peñalvo, F.J., Therón, R., de Pablos, P.O.: Usability evaluation of a visual modelling tool for owl ontologies. Journal of Universal Computer Science 17(9), 1299–1313 (2011)
6. García, J., Therón, R., García, F.: Semantic Zoom: A Details on Demand Visualisation Technique for Modelling OWL Ontologies. In: Pérez, J.B., Corchado, J.M., Moreno, M.N., Julián, V., Mathieu, P., Canada-Bago, J., Ortega, A., Caballero, A.F. (eds.) Highlights in PAAMS. AISC, vol. 89, pp. 85–92. Springer, Heidelberg (2011)
7. Holten, D.: Hierarchical edge bundles: Visualization of adjacency relations in hierarchical data. IEEE Transactions on Visualization and Computer Graphics 12, 8 (2006)
8. Kanellopoulos, D.: An ontology-based system for intelligent matching of travellers' needs for group package tours. International Journal of Digital Culture and Electronic Tourism 1(1), 76–99 (2008)
9. Lazar, J., Feng, J., Hochheiser, H.: Research Methods in Human-Computer Interaction. John Wiley and Sons, Ltd., Publication (2010)
10. Sharp, H., Rogers, Y., Preece, J.: Interaction design: beyond human-computer interaction, 2nd edn. Wiley & Sons Ltd. (2007)

Intelligent Recovery Architecture for Personalized Educational Content

A. Gil, S. Rodríguez, F. De la Prieta, B. Martín, and M. Moreno

Abstract. Multi-agent systems are known for their ability to adapt quickly and effectively to changes in their environment. This work proposes a model for the development of digital content retrieval based on the paradigm of virtual organizations of agents. The model allows the development of an open and flexible architecture that supports the services necessary to conduct a search for distributed digital content dynamically. AIREH (Architecture for Intelligent Recovery of Educational content in Heterogeneous Environments) is based on the proposed model; it is a multi-agent architecture that can search and integrate heterogeneous educational content through a recovery model that uses a federated search. A major challenge in searching and retrieval digital content is to efficiently find the most suitable for the users This paper proposes a new approach to filter the educational content retrieved based on Case-Based Reasoning (CBR).The model and the technologies presented in this research are an example of the potential for developing recovery systems for digital content based on the paradigm of virtual organizations of agents. The advantages of the proposed architecture are its flexibility, customization, integrative solution and efficiency.

Keywords: Multi-agent systems, e-learning, learning objects, repositories, Case Base Reasoning, recommender systems.

1 Introduction

The last decade has seen a significant evolution in the methods for managing and organizing large volumes of digital content. The information is characterized in

A. Gil · S. Rodríguez · F. De la Prieta · B. Martín · M. Moreno
Departamento de Informática y Automática – Facultad de Ciencias
University of Salamanca – Spain
Plaza de la Merced s/n
37008 Salamanca
e-mail: {abg,srg,fer,eureka,mmg}@usal.es

J.B. Pérez et al. (Eds.): Highlights on PAAMS, AISC 156, pp. 85–93.
springerlink.com

various ways, and contained in specialized data repositories each of which must be accessed by their own methods, resulting in dozens of communication protocols. Therefore, there is a need for research on the techniques, tools and methodologies that provide a technological solution that enables the adaptation of mechanisms in an environment that is the sum of open contexts highly dynamic, and heterogeneous.

The education sector is a significant generator, consumer and depository for educational content. These aspects position the current educational environment in a relevant sector in the development and integration of emerging solutions to management processes for the location and distribution of digital content. This openness in communication and the urgent need to achieve true interoperability between educational application environments have created the need for research in the search, retrieval and integration of heterogeneous educational content. The problems arising from the integration of educational content are usually caused by the multiple characterizations of the content, the vast amount of educational content distributed, and the access to them required by different users.

This paper proposes a model for the development of digital content retrieval based on the paradigm of virtual organizations of agents. The model allows the development of an open and flexible architecture called AIREH (Architecture for Intelligent Recovery of Educational content in Heterogeneous Environments) that supports the services necessary to conduct a personal search for distributed digital content dynamically.

The remainder of this paper is organized as follows: Section 2 describes relevant works related to Learning Object selection and recommendation, Section 3 introduces a new approach to applying CBR to the LO recommendation domain. Selected results of a comprehensive evaluation of the approach are presented in Section 4. The paper closes with relevant conclusions and an outlook to future work in section 5.

2 Related Work

With so many Learning Object Repositories, a major challenge is to efficiently find the most suitable LOs for the users. Researchers and developers of e-learning have begun to apply information retrieval techniques with technologies for recommendation, especially collaborative filtering [1], or web mining [2], for recommending educational content. A recent review of these applications can be seen in [3]. The features that handle these information filtering techniques in this context are the attribute information of education item (content-based approach) and the user context (collaborative approach).

One of the first works in this context was developed by Altered Vista: a system in which instructional techniques are evaluated based on collaborative filtering recommendation algorithms with close neighbors [4, 5]. These works explore how to collect user reviews of learning resources and propagate them in the form of word-of-mouth recommendations. RACOFI (Rule-Applying Collaborative Filtering) proposes a collaborative filtering by rules, with an architecture for the custom selection of educational content [5]. The author's recommendation combines both approaches to reduce recommendation by integrating a collaborative filtering algorithm that works with user ratings of a set of rules of inference, which creates an association between the content and rate of recommendation. McCalla [6] proposes an improvement to collaborative filtering called the ecological approach to designing e-learning systems. Key aspects of this proposal take into account the gradual accumulation of information and focus on end users. Manouselis *et al.* [7] have conducted a case study with data collected from the CELEBRATE portal users to determine an appropriate collaborative filtering algorithm.

Some solutions take a hybrid approach. [8, 9, 10] make use of algorithms based on reviews from other users according to interests which are extracted through nearest neighbor algorithms. However, all of the selected learning objects are treated equally without any distinction between them, which would allow more precise assessment criteria of the user, affecting the very pattern of preference of the user.

The works by [11, 12] suggest the need for selecting learning objects by taking into account the educational content described by their metadata, which falls in line with this work. They propose a mechanism called Contextualized Attention Metadata (CAM) to capture information about the actions along the life cycle of learning objects, including their creation, labeling, supply, selection, use and maintenance. Others are based on semantic aspects by considering contextual information automatically from the student's cognitive activities and the LO content structure [13, 14].

There are other approaches that require direct human intervention in their assessment such a LORI (Learning Object Review Instrument) tool, based on the advice of [15].

All the evaluated proposals concluded that the incorporation of mechanisms to assess attributes related to the educational content as well as aspects of user context and their interaction with the content, create effective recommendation mechanisms. There are a growing number of papers proposing systems to recommend learning resources, evidenced by the lack of operational solutions as confirmed by recent work [16].

The architecture proposed in this paper provides multiple perspectives to assess the recovery of educational content from a real, open and scalable environment, and will also will be a support mechanism to implement the recommendation or ranking for the recovered LOs.

3 The AIREH Proposal

The cornerstone of this work is the recovery of learning objects in a real environment using federated search in different Learning Objects repositories. It is necessary to provide the user with a framework that unifies the search and retrieval of objects, thus facilitating the learning process that filters and properly classifies learning objects retrieved according to some rules. The generation of such rules for the organization of the items recovered will be based on educational metadata and will provide useful content to end users.

The structural diagram, see Figure 1, adapted according to the pattern of congregation AIREH unit, contains many features (and products). In addition, supplier and customer roles are refined into these new units to specialize in the functionality or the utilization of specific services associated with these types of products.

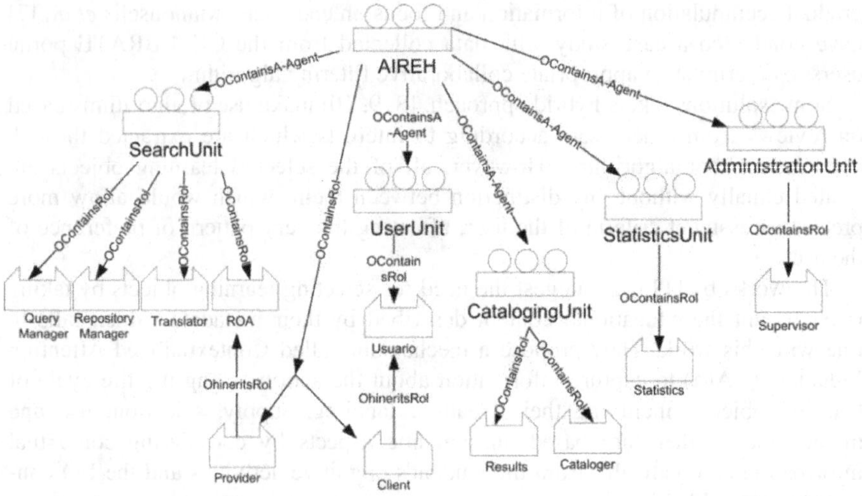

Fig. 1 Functional View

The federated search mechanism has been addressed as the resolution of the issue of content retrieval by solving three phases: (1) the selection of repositories, (2) the recovery of content, and finally (3) the merger of results. The first phase of the proposed federated search works by selecting the most appropriate ROA according to statistical parameters and performance during the search and retrieval session of content, and is based on user research through key terms. A second stage of filtering incorporates aspects of quality of the retrieved objects according to two criteria (1) the assessment of the quality of metadata retrieved and (2) the assessment of LO estimated by users through collaborative techniques. The sum of these criteria on the ranking of retrieved objects in the system provides a hybrid recommendation that begins with a refined content-based recommendation and collaborative features. This stage ends with the third phase of the federated search, and includes recommendations on merging the content, which improves the quality of the retrieved content for the user that generated the query.

3.1 Proposed Recommendation Strategy

The recommendations of real world sales are always based on knowledge about the items as well as about user tastes, preferences or interest determines the seller in the context. AI research has been focused the so-called recommender systems to model this tasks.

The Case Base Reasoning (CBR) is a particular search technique widely used in nearest neighbor recommender systems. Recovery techniques and their adaptation to CBR techniques have become effective for the development of recommender systems [17].

The development of a single ordered list of Learning Objects that incorporates user's relevance criteria in this work is one of the tasks that the agent model AIREH implements with a CBR reasoning model. For this initial problem is defined based on the elements of the context, as shown in Figure 2. In this regard, CBR-BDI agents use a system of case-based reasoning.

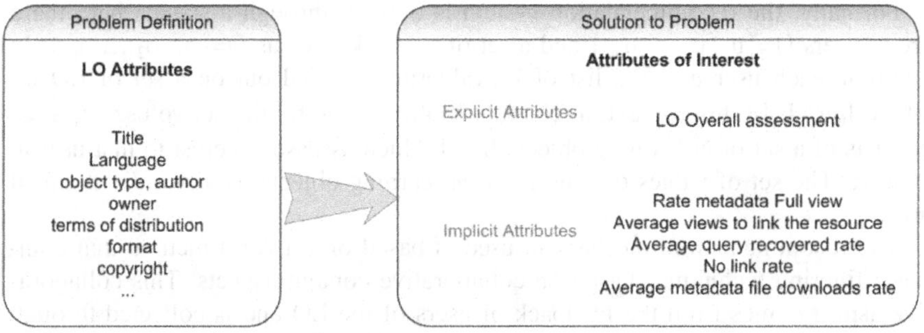

Fig. 2 Example of representation of cases in the LO domain

Given a set of educational content $p=\{p_1, p_2,...p_i\}$, each of them, p_i, is characterized by a set of attributes from the set goals total T of possible attributes. So each unit of educational content is represented by a vector $p_i=\{p_{i1}, p_{i2},...p_{in}\}$. In the specific case of learning objects the total set of attributes defined, T = {*title, language, keywords, format,...}*, is extracted mainly from the information in the markup language in accordance with standard tagged used (LOM, DC, etc.) While attributes are introduced that reflect other aspects of the environment such as the repository to which it belongs.

Field Case	Element
USER	UserProfile
QUERY	Query
PREF	User Preference
STIME	Time Stamp

Fig. 3 Case Attributes

Figure 3 shows a description of the problem formalized by case attributes included to estimate other objective property but removing it from the set of objective parameters favoring a modular design ranking algorithm allows flexibility in implementation. Using these attributes, the system is able to get the description of the problem to solve.

Each user in the process has a different level of interest in each of the items. These interests are expressed by the user, as part of the so-called explicit or attributes captured automatically by the system due to the interactivity of the user, generating the so-called implicit attributes. In this work it distinguishes the two types of parameters. The explicit parameter values user vote on any particular LO, while the implicit data is obtained through user interaction with the interface.

Each user has experience in several items. As each case represents the user experience on a particular article, the complete case base for a user is the user profile representation in the context modeling. Thus, the recommendation system maintains a database of cases representing each user profile.

Formally, the recommendation system is defined through a stage where there are m users U={u_1, u_2 … u_m } and a set of n LO, known as O={o_1, o_2 … o_n }. In addition, each user u_i has a list of k evaluations carried out on a set of LO I_{ui}, where $I_{ui} \subseteq I$. In this context, a recommendation made by the active user $u_a \in U$, consists of a set of N learning objects $I_r \subset I$ which assesses interest to that user in context. The set of values of *n* users on *m*-learning objects, is a two dimensional array.

The recommendation mechanism used is based on a hybrid method that combines filtering techniques based on collaborative content aspects. This collaborative aspect comes from the feedback of users of the LO and is collected through the interface, and becomes part of the attributes explicit in the case base described in this section.

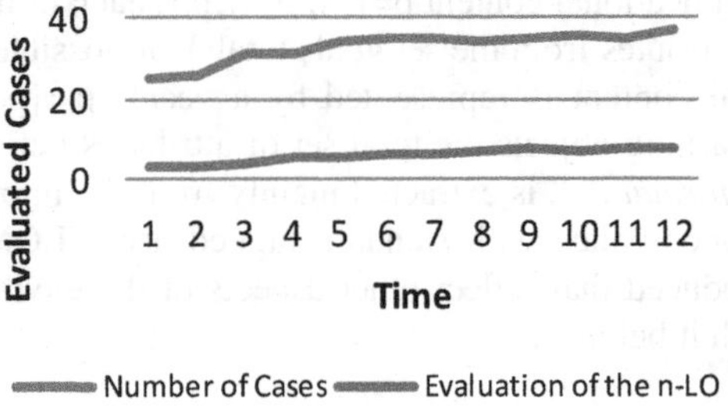

Fig. 4 Evaluation of the recommendations of the CBR

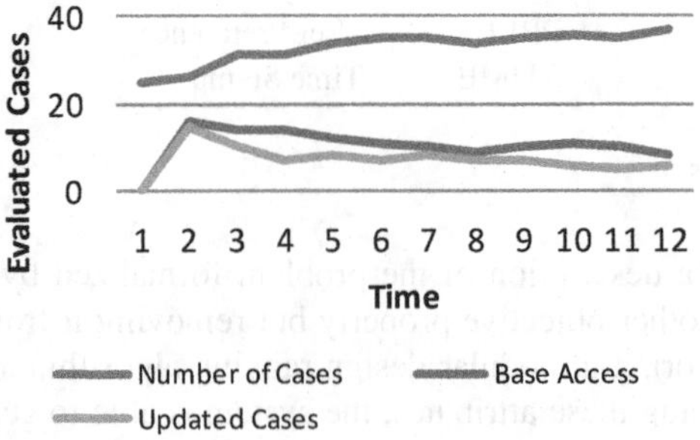

Fig. 5 Evolution of the CBR

4 Experimental Results

The recommendation is made according to the group of recovered cases. To validate the application, we compared the results obtained by the following evaluations of AIREH assessment by 6 months with 10 users. They perform a battery of queries from a selection of keywords from the computer science ground extracted from UNESCO codes.

The evolution of the number of cases in the case base allows for greater knowledge and appreciation of potential LO, as well as users. The Figure 4 shows how the user will increase the value of the first n recommended LO (n = 5) as the system solves new cases. Success in the system is evaluated through user interaction with the recommended LO, as well as the assessment it makes of each. This improvement is due to the system's ability to learn and adapt to lessons learned. Likewise, the experiences allow a better adaptation to the user profile.

Figure 5 shows that the number of updated cases decreases as the system acquires experiences (the x axis represents the number evaluations during the time period and the y axis quantifies the number of cases concerning the aspects evaluated (number of cases, access to the base and updates of cases). It is logical, since by increasing the variability of the cases captured and the number, the ability to find cases similar to the query that the user requires increases and may validate the recovered LO criteria defining user tastes and/or needs.

5 Conclusions and Further Work

A new approach for intelligent search of educational content was introduced by proposing an architecture called AIREH. This architecture is based on the application of virtual organizations of multi-agent systems and it has been used in a particular case study for federated search in repositories of learning objects and the subsequent recommendation of results.

This work demonstrate the possibility of modeling an efficient system for managing open systems from a model of an adaptive organization that provides personal recovery based on a CBR while is flexible and dynamic. The model and the technologies presented in this paper are an example of the potential for developing recovery systems for digital content based on the paradigm of virtual organizations of agents. The advantages of the proposed architecture are its flexibility, customization, integrative solution and efficiency.

Currently we have a prototype with some of the features implemented There are still many different perform evaluations to be covered for enhancing the performance of searching and retrieval educational content with personalization aspects. However, the research presented in this paper has provided suitable strategies for the future research towards enhanced content based retrieval systems.

Acknowledgements. This work has been partially supported by the project MICINN TIN 2009-13839-C03-03 and JCyL project SA225A11-2.

References

1. Bobadilla, J., Serradilla, F., Hernando, A., MovieLens: Collaborative filtering adapted to recommender systems of e-learning. Knowledge-Based Systems (2009), doi:10.1016/j.knosys.2009.01.008
2. Khribi, M.K., Jemni, M., Nasraoui, O.: Automatic recommendations for e-learning personalization based on web usage mining techniques and information retrieval. Educational Technology & Society 12(4), 30–42 (2009)
3. Manouselis, N., Vuorikari, R., Van Assche, F.: Collaborative recommendation of e-learning resources: an experimental investigation. Journal of Computer Assisted Learning 26, 227–242 (2010)
4. Recker, M., Walker, A., Lawless, K.: What do you recommend? Implementation and analyses of collaborative information filtering of web resources for education. Instructional Science 31(4-5), 299–316 (2003)
5. Lemire, D., Boley, H., McGrath, S., Ball, M.: Collaborative Filtering and Inference Rules for Context-Aware Learning Object Recommendation. Technolodgy and Smart Education 2(3), 179–188 (2005)
6. McCalla, G.: The Ecological Approach to the Design of E-Learning Environments: Purpose-based Capture and Use of Information about Learners. Journal of Interactive Media in Education, Special Issue on the Educational Semantic Web 1(7), 18 (2004)
7. Manouselis, N., Vuorikari, R., Van Assche, F.: Simulated Analysis of Collaborative Filtering for Learning Object Recommendation. In: SIRTEL Workshop, EC-TEL (2007)
8. Aijuan, D., Baoying, W.: Domain-based recommendation and retrieval of relevant materials in e-learning. In: IEEE International Workshop on Semantic Computing and Applications 2008 (IWSCA 2008), pp. 103–108 (2008)
9. Ghauth, K., Abdullah, N.: Learning materials recommendation using good learners' ratings and content-based filtering. In: Educational Technology Research and Development. Springer, Boston (2010), `http://dx.doi.org/10.1007/s11423-010-9155-4`, ISSN 1042-1629
10. Wang, T.I., Tsai, K.H., Lee, M.C., Chiu, T.K.: Personalized Learning Objects Recommendation based on the Semantic Aware Discovery and the Learner Preference Pattern. Educational Technology and Society 10(3), 84–105 (2007)
11. Ochoa, X., Duval, E.: Use of Contextualized Attention Metadata for Ranking and Recommending Learning Objects. In: Proceedings of 1st International Workshop on Contextualized Attention Metadata: Collecting, Managing and Exploiting of Rich Usage Information, pp. 9–16 (2006)
12. Wolpers, M., Najjar, J., Duval, E.: Tracking Actual Usage: the Attention Metadata Approach. Educational Technology & Society 10(3), 106–121 (2007)
13. Han, Q., Gao, F., Wang, H.: Ontology-based learning object recommendation for cognitive considerations. In: 8th World Congress on Intelligent Control and Automation (WCICA), July 7-9, pp. 2746–2750 (2010)
14. Ruiz-Iniesta, A., Jiménez-Díaz, G., Gómez-Albarrán, M.: Personalización en Recomendadores Basados en Contenido y su Aplicación a Repositorios de Objetos de Aprendizaje. IEEE-RITA 5(1), 31–38 (2010)

15. Vargo, J., Nesbit, J.C., Belfer, K., Archambault, A.: Learning object evaluation: Computer mediated collaboration and inter-rater reliability. International Journal of Computers and Applications 25(3), 198–205 (2003)
16. Manouselis, N., Drachsler, H., Vuorikari, R., Hummel, H., Koper, R.: Recommender Systems in Technology Enhanced Learning. In: Recommender Systems Handbook, pp. 387–415. Springer (2011)
17. Montaner, M., López, B., de la Rosa, J.L.: Opinion-Based Filtering Through Trust. In: Klusch, M., Ossowski, S., Shehory, O. (eds.) CIA 2002. LNCS (LNAI), vol. 2446, pp. 164–178. Springer, Heidelberg (2002)

Adaptation Model for PCMAT – Mathematics Collaborative Learning Platform

Marta Fernandes, Constantino Martins, Luiz Faria, Paulo Couto, Cristiano Valente, Cristina Bastos, Fátima Costa, and Eurico Carrapatoso

Abstract. The aim of this paper is to present an adaptation model for an Adaptive Educational Hypermedia System, PCMAT. The adaptation of the application is based on progressive self-assessment (exercises, tasks, and so on) and applies the constructivist learning theory and the learning styles theory. Our objective is the creation of a better, more adequate adaptation model that takes into account the complexities of different users.

Keywords: Adaptive Educational Hypermedia, Adaptation Model, Learning Objects.

1 Introduction

The main purpose of Adaptive Hypermedia Systems is adapting interface, content presentation, link navigation, and so on, to the specific objectives and characteristics of individual users [5, 8]. With Adaptive Educational Hypermedia Systems the focus is placed on helping the user achieve his learning goals, thus characteristics such as the user's knowledge and learning style are particularly important [6, 8].

Marta Fernandes · Constantino Martins · Luiz Faria, Paulo Couto · Cristiano Valente · Cristina Bastos · Fátima Costa
GECAD – Knowledge Engineering and Decision Support Group
Institute of Engineering – Polytechnic of Porto,
Rua Dr. António Bernardino de Almeida, 431
4200-072 Porto, Portugal
e-mail: {mmaf,acm,lef}@isep.ipp.pt

Eurico Carrapatoso
Faculty of Engineering – University of Porto, Rua Dr. Roberto Frias,
s/n 4200-465 Porto Portugal
e-mail: emc@fe.up.pt

J.B. Pérez et al. (Eds.): Highlights on PAAMS, AISC 156, pp. 95–102.
springerlink.com

Numerous such systems have been developed with good results; however, very little research has been made into adaptation techniques, which are still incipient and somewhat ineffective. More work is therefore required to provide users with proper adaptation [14] and is one of the purposes of this project.

According to the OECD PISA 2009 study, Portugal has made progress in mathematics performance, with most improvements occurring between 2006 and 2009 [19]. However, in spite of these good results Portugal is still significantly below the OECD average in mathematics performance. Portugal's lower rank is 29 and upper rank is 22 out of 34 OECD countries featured in the study. There's clearly much room for improvement and Portugal still has a long way to go before we can consider ourselves satisfied. That is the main drive behind the development of this project.

Learning styles represent models of how a person learns. Students learn in different ways and depend upon many different and personal factors [21]. Initially, the idea was that each individual had a single learning style. However, more recent studies have shown the majority of people are actually multimodal, meaning they have more than one learning style [10, 16]. Learning styles have many proponents and several studies support their use [12, 17, 20], but they have also been criticized by many [3, 22, 11]. There doesn't seem to be, however, evidence suggesting the use of learning styles is detrimental. The mathematics teachers working on this project have been teaching for many years and their personal opinion, based on what they have observed, is that learning styles might indeed be useful. For that reason and because learning styles don't appear to be harmful to the learner and there isn't a consensus on whether they are useful or not, we have decided to apply this theory when developing the Mathematics Collaborative Learning Platform (PCMAT). One of our objectives is putting this theory to the test, hopefully with positive results.

In this paper we introduce the adaptation model of PCMAT [15], an online collaborative learning platform with a constructivist approach, which assesses the user's knowledge and presents contents and activities adapted to the characteristics and learning style of students of mathematics in basic schools.

In section 2 we make a brief description of adaptive educational hypermedia. In section 3 we describe in detail the platform's adaptation model and in section 4 we take some conclusions and talk about future work.

2 Adaptive Educational Hypermedia

Unlike conventional hypermedia systems, Adaptive Hypermedia Systems (AHS) present users with content and/or navigation that is tailored to their specific characteristics, needs and interests. Brusilovsky [4] referred to it as the crossroads of hypermedia and user modeling. The purpose of such systems is to direct the user to the most relevant content while keeping him away from content that isn't interesting or suitable. As a user's needs, goals and characteristics evolve so does the content presented to him by the system.

A usual application of Adaptive Hypermedia is Adaptive Educational Hypermedia (AEH). In AEH the construction of the user model must take into account features such as the user's goals, abilities, needs, and interests but also the user's knowledge of the subject.

A common framework for AHS is the one proposed by Benyon [1] and De Bra [9] which consists of three interdependent modules:

- The User Model may contain personal information about the user, as well as data concerning his interests, goals, et cetera. In the case of AEH the Student Model must consider domain dependent data and domain independent data.
- The Domain Model consists of a semantic network of domain concepts. Its main purpose is providing a representation structure for the user's domain knowledge.
- The Adaptation model represents and defines the interaction between the user and the application. It displays information to the user based on the user model and updates its information according to the user's actions.

By relating the user model to the domain model the system can, through the adaptation model, adapt its content, navigation and interface to each user's specific needs. Triantafillou [24] underlined the importance of adaptivity in Educational Hypermedia, considering these systems are meant to be used by several learners.

3 Adaptation Model Development

The development of PCMAT took into account the constructivist learning theory. The system assesses the user's previous knowledge and based on that provides a path into the subject. It also presents the user with learning objects, such as content and activities, which are adapted to the student's characteristics and performance. In addition, with the purpose of consolidating the user's knowledge the system is able to make permanent automatic feedback and support, through instructional methodologies and educational activities explored in a constructivist manner.

PCMAT is based on AHA! (Adaptive Hypermedia Architecture) [25]. For that reason the adaptation model of the platform can be divided into two, intrinsically connected, parts: one that's dependent on AHA's adaptation engine and another that has been developed to address our platform's specific needs.

The AHA! runtime environment consists of Java Servlets and uses a XML database representation. We have chosen to develop the PCMAT adaptation engine using the same technologies. The system also works with freely available tools like the Tomcat Web server and the mySQL database system.

3.1 Adaptation Using AHA!

AHA's adaptation model uses the elements in the user model to define a specific domain concept graph, adapted from the domain model, which is used to address the user's needs. The scheme is set by the teacher, but each student's path in the graph is determined by the interaction with the system using progressive assessment, the student's knowledge representation and the user's characteristics in the user model.

AHA's adaptation is done through changes in content presentation, in the structure of links and in the links annotation. Every activity presented by the system corresponds to a set of concepts in the Domain Model and in the User Model (implemented through an overlay model). Each concept has a set of attributes. The *knowledge* attribute, for example, represents an estimate of the knowledge a student has of a particular concept and is expressed as an integer between 0 and 100. The *suitability* attribute indicates whether a given concept is suitable to the learner or not. The adaptation of content presentation is achieved by defining rules related to those elements. Each rule contains a boolean expression involving attributes of concepts or attributes of the User Model (learning style, for example). This expression must be true for the resulting sequencing action to be triggered, which will then update the values of some attributes of concepts or attributes of the User Model. Each page in a course may be composed by multiple fragments of content. These rules define which fragments are shown to a user and which fragments are omitted. Changes in the structure of links, also known as link adaptation, consist in hiding, disabling and removing certain links with the purpose of guiding the student through the course. The main objective is keeping the student away from information that is irrelevant or not yet appropriate, and directing him towards the most relevant information.

3.2 PCMAT's Adaptation Engine

The student's progression through the course is dependent on self-assessment exercises and tasks. We developed an adaptation engine in order to present the student with tasks that are adapted to his level of knowledge, competences, abilities, learning style and learning path, as well as to update the user model according to the student's responses.

The exercises shown to each student are kept in a repository of learning objects. Each exercise is related to a set of domain concepts and a specific learning style. Choosing an exercise appropriate to the student's current position in the learning path depends on several restrictions. Exercises are presented within the context of one or more domain concepts and must therefore be in accordance with those concepts. Exercises must also conform to the student's current dominant learning style. To guarantee such restrictions are met and only the appropriate exercises are chosen, the adaptation engine obtains that information from AHA's adaptation model. In that way, the exercises presented to the student are completely individualized to his personal characteristics and position in the learning path. In addition, the system maintains a record of which exercises the student has already performed and uses it to give priority to exercises that haven't been performed yet. This ensures the student won't be shown the same exercise twice, for as long as there are exercises in the repository he hasn't seen yet. In case a student has already performed all the exercises in the repository, which obey the concept and learning style restrictions, the system consults the student's record and chooses the exercise with the oldest timestamp. Once the exercise has been completed the system updates the student's record by adding a new timestamp to that exercise.

The exercises in the repository may be of two types: simple or parameterized. Exercises of the first type are made up of a single body of text with all the parameters in the text already defined. On the other hand, the body of text of parameterized exercises doesn't have defined parameters. The author of an exercise substitutes all or some of the possible parameters by variables and defines several sets of parameters. If the system chooses an exercise which is parameterized it then retrieves all possible sets of parameters for that exercise and randomly chooses one. The system uses the parameters in the set to instantiate each of the variables in the body of text and shows the user an exercise with defined parameters. This functionality has the advantage of allowing the creation of various exercises based on a common structure. Instead of manually creating several different exercises, the user only has to define the body of the exercise, which will be common to all exercises, and indicate the parameters that will be different in each exercise. This feature was implemented after observing that mathematics exercises about the same subject are often very similar, with only different parameters. Since a parameterized exercise might have various sets of parameters, special care was taken when referencing this type of exercise in the student's record of exercises already solved. Aside from a reference to the exercise, a reference to the specific set of parameters used is also added to the record. In this way, the same exercise might once again be chosen by the system, but it will be instantiated with a different set of parameters, thus presenting the student with an exercise different from the one previously shown.

The progressive assessment of the student contributes to the definition of his learning path. Therefore, evaluating the student's answers to the exercises shown is an important task of the adaptation engine. Exercises may be of multiple choice, true or false, or open-ended. Assessing multiple choice exercises and true or false exercises is a straightforward task. However, during the development of the adaptation engine it became clear assessing open-ended questions would be more challenging than initially expected. The correct answer to these exercises can be given in several different ways, that is to say its syntax may differ. To address that issue a probabilistic natural language parser, the Stanford Parser [23], is used in conjunction with a Portuguese language grammar [2, 13]. Processing an answer thus requires it to be previously tagged and tokenized. However, the Stanford Parser isn't capable of tokenizing text written in Portuguese, therefore a tokenizer and a part-of-speech tagger have also been developed. The Stanford Parser and the Portuguese language grammar allow the system to evaluate answers written in a correct manner. In practice, however, this doesn't always occur. Answers given by children in particular are often written in an incomplete or incorrect form. To address this issue the system uses a set of dependencies between words provided by the Parser, together with a list of object/attribute/value triples. This enables the system to validate answers it wouldn't be able to validate by using the Parser and the Portuguese grammar alone. The need for object/attribute/value triples means users/teachers must provide this information when creating open-ended questions.

Once the system has processed an answer, it checks whether it matches the respective exercise's correct answer, which is stored in the repository. If the student was successful he's given the possibility of advancing to the next concept in the course. If the student failed the exercise he may choose to go back and review previous concepts.

The evaluation of a student's answers results in changes in the student model. If an answer is correct the system accesses the student model and updates the value of each of the related domain concept's knowledge using the following mechanism:

```
Let A1,A2,A3,...,An be the set of concepts associated with the activity
For each i in {1,2,3,...,n}
     Ai.knowledge = min(Ai.knowledge+Ai.knowledge*0.25,100)
```

A similar update is performed for the concepts that, in the concept graph, precede the concepts to which the exercise is related. The following example shows how the learning style attributes are updated when the prominent learning style attribute associated with the exercise is personal.lst (learn by reading and hearing):

```
personal.lst = min(personal.lst + 1,10)

If personal.lsv >= personal.lsp then personal.lsv= max(personal.lsv-1,0)

If personal.lsv < personal.lsp then personal.lsp = max(personal.lsp-1,0)
```

These changes in a student's learning styles' values, according to his performance, agree with the idea that students may have multiple context dependent learning styles [10, 16]. By taking that into account the platform was developed to adapt to the student's changes in dominant learning style, thus providing him with appropriate learning objects at all times.

4 Conclusion and Future Work

With this project we have tried to contribute to the progress of AHS, in particular adaptation techniques, and have developed an adaptation model that not only provides adaptation, such as content adaptation and link adaptation, but does so by taking into account the constructivist learning theory and the learning styles theory. The system continuously adapts to the student's learning style in an attempt to achieve the best possible results. As the student's learning style changes, so does the content proposed by the system. Adaptation is achieved by considering the student's level of knowledge, as well as previous knowledge, guiding him towards appropriate content and helping him integrate and assimilate newly acquired knowledge.

A first version of the platform was already tested in the learning of mathematics in two basic schools and achieved good results. For example, the average of student scores from both schools in the experimental group was higher than the average of student scores in the control group, $\mu = 61.7$ ($\sigma = 19.9$) vs. $\mu = 54.4$ ($\sigma = 14.3$). Although the observed differences are not statistically significant ($p = 0.073$), the difference between groups seems clear. The two groups were statistically compared using a two sided, independent samples t test with a 0.05 (5%) critical level of significance ($t = 1.82$, degrees of freedom = 74). Students also perceived this tool as very relevant for their learning, as a self-operating application to be integrated in a more global learning strategy that also includes tutoring (direct contact with the teacher) and peer learning. Teachers agreed with these definitions of the platform as well.

The present version is now entering a new testing phase, with an increased sample size. We hope to obtain additional results that will allow us to conclude about the adequate features and true effectiveness of the PCMAT system. We also aim to assess the usefulness of learning styles.

Work on the adaptation model will continue as we feel other areas, such as the difficulty level of content and exercises, need to be addressed in order to provide better adaptation and help students learn mathematics and improve their results.

Acknowledgments. The authors would like to acknowledge FCT, FEDER, POCTI, POSI, POCI and POSC for their support to GECAD unit, and the project PCMAT (PTDS/CED/ 108339/2008).

References

1. Benyon, D.: Adaptive systems: A solution to Usability Problems. Journal of User Modeling and User Adapted Interaction 3(1), 1–22 (1993)
2. Branco, A., Silva, J.: Evaluating Solutions for the Rapid Development of State-of-the-Art POS Taggers for Portuguese. In: Proceedings of the 4th International Conference on Language Resources and Evaluation (LREC 2004), pp. 507–510. ELRA, Paris (2004)
3. Brown, E., Fisher, T., Brailsford, T.: Real Users, Real Results: Examining the Limitations of Learning Styles within AEH. In: Proc. 18th Conf. Hypertext and Hypermedia, pp. 57–66 (2007)
4. Brusilovsky, P.: Adaptive hypermedia. User Modeling and User Adapted Interaction. Ten Year Anniversary Issue 11(1/2), 87–110 (2001); Kobsa, A. (ed.)
5. Brusilovsky, P., Millán, E.: User Models for Adaptive Hypermedia and Adaptive Educational Systems. In: Brusilovsky, P., Kobsa, A., Nejdl, W. (eds.) Adaptive Web 2007. LNCS, vol. 4321, pp. 3–53. Springer, Heidelberg (2007)
6. Chepegin, V., Aroyo, L., De Bra, P., Heckman, D.: User Modeling for Modular Adaptive Hypermedia. In: SWEL 2004 Workshop at the AH 2004 Conference. TU/e Computing Science Report 04-19, pp. 366–371 (2004)
7. Couto, P., Martins, C., Faria, L., Fernandes, M., Carrapatoso, E.: PCMAT Metadata Authoring Tool. In: International Symposium on Computational Intelligence for Engineering Systems, ISCIES 2011 (2011)
8. De Bra, P.: Web-based educational hypermedia. In: Romero, C., Ventura, S. (eds.) Data Mining in E-Learning, pp. 3–17. Universidad de Cordoba, WIT Press, Spain (2006) ISBN 1-84564-152-3
9. De Bra, P., Aroyo, L., Chepegin, V.: The Next Big Thing: Adaptive Web-Based Systems. Journal of Digital Information 5(1), Article No. 247 (2004)

10. Fleming, N.D.: VARK 'A review of those who are multimodal' (2007), http://www.vark-learn.com/english/page_content/multimodality.html (accessed on October 1, 2011)
11. Hargreaves, D., et al.: About learning: Report of the Learning Working Group. Demos (2005)
12. Kolb, A., et al.: Learning styles and learning spaces: Enhancing experiential learning in higher education. Academy of Management Learning and Education 4(2), 193–212 (2005)
13. Language Resources and Technology for Portuguese, http://lxcenter.di.fc.ul.pt/
14. Martins, C., Faria, L., Carrapatoso, E.: Constructivist Approach for an Educational Adaptive Hypermedia Tool. In: The 8th IEEE International Conference on Advanced Learning Technologies (ICALT 2008). University of Cantabria, Santander (2008)
15. Martins, C., Couto, P., Fernandes, M., Bastos, C., Lobo, C., Faria, L., Carrapatoso, E.: PCMAT – Mathematics Collaborative Learning Platform. In: Pérez, J.B., Corchado, J.M., Moreno, M.N., Julián, V., Mathieu, P., Canada-Bago, J., Ortega, A., Caballero, A.F. (eds.) Highlights in PAAMS. AISC, vol. 89, pp. 93–100. Springer, Heidelberg (2011)
16. Miller, P.: Learning styles: the multimedia of the mind. Educational Resources Information Center ED 451 140 (2001)
17. Montgomery, S., Groat, L.N.: Student Learning Styles and Their Implications for Teaching. Occasional Paper, Center for Research on Learning and Teaching. University of Michigan (1998), http://www.crlt.umich.edu/publinks/CRLT_no10.pdf (accessed September 26, 2011)
18. OECD: PISA 2009 Results: What Students Know and Can Do – Student Performance in Reading. Mathematics and Science, vol. I (2010)
19. OECD: PISA 2009 Results: Learning Trends: Changes in Student Performance Since 2000, vol. V (2009)
20. Richmond, A.S., Cummings, R.: Implementing Kolb's learning styles into online distance education. International Journal of Technology in Teaching and Learning 1(1), 45–54 (2005)
21. Ritu, D., Sugata, M.: Learning Styles and Perceptions of Self. International Education Journal 1(1) (1999)
22. Stahl, S.A.: Different strokes for different folks? In: Abbeduto, L. (ed.) Taking Sides: Clashing on Controversial Issues in Educational Psychology, pp. 98–107. McGraw-Hill, Guilford (2002)
23. The Stanford Natural Language Processing Group, http://nlp.stanford.edu/software/lex-parser.shtml
24. Triantafillou, E., Pomportsis, A., Demetriadis, S.: The design and the formative evaluation of an adaptive educational system based on cognitive styles. Computers &Education 41, 87–103 (2003)
25. Wu, H., Houben, G.J., De Bra, P.: User Modeling in Adaptive Hypermedia Applications. In: Proceedings of the "Interdisciplinary Conferentie Informatiewetenschap", Amsterdam, pp. 10–21 (1999)

Distributed Active-Camera Control Architecture Based on Multi-Agent Systems

Alvaro Luis Bustamante, José M. Molina, and Miguel A. Patricio

Abstract. In this contribution a Multi-Agent System architecture is proposed to deal with the management of spatially distributed heterogeneous nets of sensors, specially is described the problem of Pan-Tilt-Zoom or active cameras. The design of surveillance multi-sensor systems implies undertaking to solve two related problems: data fusion and coordinated sensor-task management. Generally, proposed architectures for the coordinated operation of multiple sensors are based on centralization of management decisions at the fusion center. However, the existence of intelligent sensors capable of taking decisions brings the possibility of conceiving alternative decentralized architectures. This problem could be approached by means of a Multi-Agent System (MAS). In specific, this paper proposes a MAS architecture for automatically control sensors in video surveillance environments.

1 Introduction

Nowadays, video surveillance systems are evolving towards complex information systems, being capable of providing the operator with a great amount of data obtained through spatially distributed sensor networks. The advances of the underlying technologies like digital communication, video coding and transmission, and specially, wireless sensors and actuator networks (WSANs) [1] have lead an easy deployment of new video surveillance sensors for its use in environmental monitoring, health-care systems, homeland security, public safety, and in general critical environments.

There is many scientific research done to take advantage of the visual sensor networks, and also commercial systems, to provide rich features in video surveillance like object detection and tracking, object and color classification, activity recognition, alert definition and detection, database event indexing, and so on [7].

Alvaro Luis Bustamante · José M. Molina · Miguel A. Patricio
Universidad Carlos III de Madrid, Avda. de la Universidad Carlos III,
22. 28270 Colmenarejo, Madrid
e-mail: {aluis,mpatrici}@inf.uc3m.es, molina@ia.uc3m.es

J.B. Pérez et al. (Eds.): Highlights on PAAMS, AISC 156, pp. 103–112.
springerlink.com

But considering an environment plenty of Pan-Tilt-Zoom (PTZ) cameras with some existing overlapped fields of views, it makes sense to apply some kind of coordination, competition, and collaboration among PTZ cameras to satisfy or improve some established goals. The benefits of active cooperative sensing as compared to non-cooperative sensing has been well established in literature [5]. There are some interesting dual-camera frameworks proposed in which a master camera takes wide panoramic images and the slave zooms into the targets to get more accurate images, like those described in [11, 5]. More recent work like in [14] describes a method for the dynamic target assignment according to the availability and accuracy of each camera. In [8] is addressed the generic problem of collaboratively controlling a limited number of PTZ cameras to capture an observed number of subjects in an optimal fashion. However, the problem of these approaches is the camera control centralization in a single node, leading the architectures to fail in scalability, performance and fault-tolerance.

The design of this kind of visual surveillance systems implies undertaking to solve two main problems [10]. The first one is the data fusion, which is related with the combination of data from different sources in an optimal way [9]. The second one is the multi-sensor management. It assumes that the previous problem was solved, and it is in charge of optimizing the global management of the joint system through the application of individual operations in each sensor [12].

There are two main approaches to solve this problem, the centralized and the decentralized way. A centralized architecture uses to be based on a data fusion center or node of the network combining the whole system information to planify and execute actions over each sensor. Thus, a centralized architecture is affordable to build a prototype but have many problems related with scalability, and deployment when the sensors are highly distributed [12].

This research is focused on solving the two main problems described before, multi-sensor management and data fusion for PTZ cameras in a distributed way, since we assume that sensors may have a high degree of autonomy, so that the last decisions about task to be executed in the sensor is taken in their own management system [15]. In this way, this problem could be approached by means of a Multi-Agent architecture, as proposed in this paper.

The study of multi-agent systems (MAS) focuses on systems in which many intelligent agents interact with each other. So, this work proposes a MAS architecture for support advanced controls for active camera sensors in the video surveillance environment. There are some previous works done by our research group also facing the problem of still/active cameras management via MAS, like the proposed in [6, 4, 13]. However this previous works do not contemplate the data fusion as an integral part of the architecture itself and its management with MAS. This was presented in [3] but for different purposes, not for sensor management. In this paper we try to integrate these researches to build a common architecture dealing with data fusion and sensor management providing also an application example.

2 Multi-Agent System Architecture

In this section is described an overview of the architecture and the different underlying agents involved (see Figure 1) to solve the problem of sensor management and data fusion. Also is proposed a third layer related with the user interface, that lets an eventual operator to monitor and control the state of the environment.

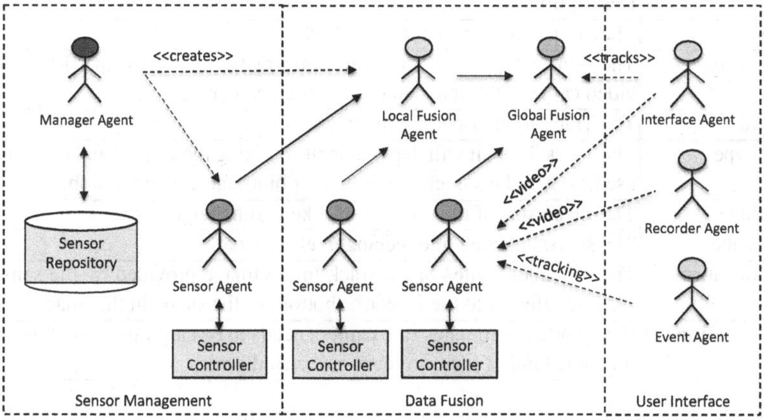

Fig. 1 Different types of agents in the architecture for solve problems of Sensor Management and Sensor Data fusion

In the sensor management side we have the Manager Agent and the Sensor Agent, that in general terms provides the management and control for the sensors presented in the system. There is only one Manager Agent in the environment, and one Sensor Agent for each camera. In the data fusion side, we have the Local fusion Agent and the Global Fusion Agent that will allow the system obtain non-redundant tracks of the different targets in the environment. All the agents are explained in more detail in the following subsections.

2.1 Sensor Agent

This agent is related with the sensor itself, and it basically provides an interface to other agents and systems to perform actions or get data from the sensor. This agent mainly depends on a software component called Sensor Controller, which is a middleware between the agent and the physical sensor. This middleware is actually developed and described in [2] and it currently provides three different interfaces: video, PTZ control, and a tracking interface which provides real-time tracking information. Thereby any Sensor Agent regarding to control a video sensor will only have to handle a common interface provided by the Sensor Controller and do not take care about the physical interfaces or protocols involved in the sensor. Depending on the underlying sensor, this agent will have different capabilities, like provide color video sequences, thermal information, PTZ control, tracks information, etc.

In the Sensor Controller middleware can be several tracking modes working at the same time depending on the features that other agents want to extract, like blob, color, and thermal tracking or face recognition. But the basic tracking information provided by the middleware is at least the presented in Table 1:

Table 1 Basic track information provided by the middleware

Value	Description
Sensor Id	The unique Sensor Id in the network
Sensor Type	The Sensor type used to get the information of the shape, like color video camera, thermal camera, infrared camera, etc.
Track Id	The Track Id provided by the tracker.
Track Type	The Track Type. It will depend on the tracking mode used and the sensor used, so may be something like color blob, face, thermal blob, etc.
Track Size	The shape size of the track itself, like width, height, etc.
Tack Shape	The shape detected like rectangle, ellipse, etc.
Track Location	The x, y coordinates of the track in the image provided by the sensor. May be referred to the center or bottom of the shape in the image.
Track Location (2)	Real world coordinates (the camera needs to be calibrated). Also is useful some kind of measurement-associated error.

2.2 *Manager Agent*

Manager Agent is the responsible of the MAS initialization in an easy way, so start the MAS architecture do not become a tedious task every time you want to achieve video surveillance. In this way, this agent relies on a Sensor Repository that contains information about all the sensors deployed. The information for each sensor is something like sensor id, sensor type, real-world sensor placement, calibration data, and the information about the Sensor Controller to employ. But not only contains information about single sensors, it also keeps information about the relationship between sensors, that is, when there exist sensors with overlapped fields of view. This information should be introduced manually as the initial setup of the architecture, thereby it can be reused in further launches of the system.

So the initial functioning of the Manager Agent can be divided in two steps. The first one is related with the dynamically creation/initialization of Sensor Agents. For each sensor present in the environment, the Manager Agent creates one Sensor Agent with the specified sensor information, so the Sensor Agent initializes the sensor and become available for other agents in the architecture. The second one relies on create static coalitions for data fusion between sensors. In this way, for each set of sensors with overlapped fields of view, it is created a Local Fusion Agent regarding tracks of the involved sensors.

2.3 *Local Fusion Agent*

This agent is dinamically created by the Manager Agent at the system startup. Its concern is to communicate with the assigned Sensor Agents to request them the tracking feature in order to achieve a local data fusion, which is a process where all the possible redundant tracks generated by overlapped fields of views can be fused to provide an unique representation of each track. This will provide to the system using this architecture a set of non-redundant tracks with all the information provided by different sensors. In addition it is useful, as is shown in the following sections, for the coordination task itself.

The proposed Local Fusion Agent will be able to perform the data fusion using the tracking information provided by each Sensor Agent, like the shape size, shape location, etc., presented in Table 1. Depending of the tracking capabilities of the Sensor Agents present in the coalition, this agent may try to use different fusion approaches. As a first step we will suppose that all the Sensor Agents are able to provide tracks with their corresponding locations in the real world, in this way, we can achieve a location-based fusion. In this way, the information that may provide the Agent as a result of the coalition is the one presented in Table 2.

Table 2 Track information as a result of a fusion

Value	Description
Track Id	An unique Track Id for the local coalition
Sensor-Track Mapping	This field will contain the relation of the underlying sensors and tracks that are contributing to generate this track, like: S1 Track 1 S2 Track 3 S3 Track 2 Also a track in the coalition may be obtained only by one single track, like: S1 Track 2
Track Location	Real world coordinates of the fused track. Also is useful some kind of measurement-associated error.
Additional Track information	As the fusion may be obtained from different kind of tracks, the idea is to provide here a general description of the combined tracks. Also it may be obtained with the Sensor-Track Mapping field.

2.4 *Global Fusion Agent*

Global Fusion Agent is the responsible of request and receive tracks both from Local Fusion Agents, and Sensor Agents that are not in a coalition with a Local Fusion Agent. All this information is integrated in order to provide a global view of the environment being monitored by all the Sensor Agents. It also allows obtaining a continuity of targets across the whole area covered by the sensor network. The information provided is quite similar to this one presented in Table 2, changing only the Track Id, as it will be an unique Id in the whole system.

2.5 *User Interface Agents*

In this group we may find different interesting agents supporting the communication with the operator to present the data and allow the management of the different surveillance agents. For example there is defined an Interface Agent that will be able to present the different video feeds and also a global vision of the tracks presented in the system through the Global Fusion Agent. Other Agent may be the Recorder Agent that can be used for retrieving video sequences of the different cameras for its recording. And finally the Event Agent that can be used along with Sensor Agents for receiving tracks in order to provide events to the operator, for example if a track enters in a field of view, or in a specific zone of the environment.

3 Application Example

After the general architecture functioning has been described, this section wants to provide an example of how this architecture can be used to control active or pan-tilt-zoom cameras. In this case is defined a PTZ Agent interacting with the architecture. So, suppose that we have a set of cameras in which three of them have an overlapped field of view, meanwhile there is another one working alone. With this configuration we can start thinking about different utilities for the PTZ Agent, both in selfish or coordinated operation modes.

In this way, when a new PTZ Agent is deployed in the architecture, it should start dealing with the Sensor Agent assigned. The PTZ Agent should check that the capabilities of the Sensor Agent allow pan-tilt-zoom controls in order to satisfy its own goals. The goals of the agent may vary depending on the operator preferences, in this work we propose different examples that can be solved with the use of the architecture.

3.1 Selfish Operation Mode

At first we should consider different kinds of PTZ controls, for example, some goals may depend on using some tracking features of the Sensor Agent, and some others may be achieved only with the PTZ feature.

- Automatic movements: the agent will turn the camera orientation automatically to monitor different areas. Can be movements like continuous or aleatory scanning, swapping between predefined areas, or follow simple predefined paths in the environment. This is the most basic and simple feature that this agent will provide using only the PTZ control.
- Target tracking: In this case this agent may request some of the tracking features available in the Sensor Agent, like simple blob tracking, face detection, or color tracking. So, with a set of tracks and the PTZ control, it can start to perform different target tracking, like maximizing the number of targets in the image, or follow the most relevant track (depending on the user preferences and the track features). Also is possible to request different tracking modes at the same time to the Sensor Agent like blob tracking and face detection and follow the tracks depending on the information given. For example, if there are only simple blobs, track them, but if a face is detected in a given moment the agent can switch to start its tracking. Also this operation mode can be combined with the previous

one, i.e, if there is no tracks in the scene, it can start looking different areas of the environment to detect them.

3.2 Coordinated Operation Mode

The previous defined examples may be useful for cameras that do not have overlapped fields of view and do not have to achieve a priori any kind of coordination. However, when we have some cameras covering the same area, we can obtain a better surveillance performance by letting the cameras get coordinated to satisfy some goal. For example, ensure that every target in the shared environment is being monitored (when possible), or each camera is tracking different targets, or many cameras collaborates to acquire as much information as possible about one single target selected by the operator, or also, not in the tracking field, all the cameras get synchronized to perform scanning of the environment.

In this way is necessary to define how to start the collaboration mechanisms and how to achieve them using the proposed architecture. In the following is described how the system may work:

- Local Fusion Agent Events: Each new PTZ Agent included in the system to control a Sensor Agent, can be joined to the Local Fusion Agent coalition in order to retrieve information about the presence of new tracks and fusions on the shared area. These events will let the agents listening to determine if there is necessary create new dynamic coalitions with other PTZ Agents to achieve collaboration. The information we propose to send in general terms is the described in table 3.
- PTZ Agents reactions: Each PTZ Agent can react in different ways to the events received, and it may depend on its current state or the personal goals to satisfy. But it is supposed that if a PTZ Agent has been registered for receive events, is

Table 3 Basic track information provided by the middleware

Event Name	Description
Track Creation	When a new track is detected in the Local Fusion (can be from any camera in the coalition), a message is sent to all the PTZ Agents listening with information presented in Table 1 to identify the track.
Track Deletion	This message is sent when some track no longer exists in the Local Fusion coalition. This message contains the information presented in Table 1.
Fusion Creation	This message is sent when a new fusion has been created in the Local Fusion coalition, that is, at least two tracks of different source sensors has been combined into one single track. Also may be sent when a new track joins to a previous existing fusion. The information sent is the information provided in Table 2.
Fusion Deletion	AThis message is sent when a fusion has been destroyed, that is, a previous existing combination of tracks no longer exists in the Local Fusion coaliton. Also may be sent when a track no longer belongs to an existing fusion. The information sent is the Fusion information presented in Table 2.

for collaborate with other agents present in the architecture. Thus we can suppose that in general the PTZ Agents can achieve the following reactions:

- Track Creation: This event may be omitted by many PTZ Agents but may be useful for alert it about the presence of new targets, and if necessary, switch the camera orientation to start monitoring them.
- Track Deletion: Similar to the track creation, this event may be used for alert that an agent has ceased to see some target, thereby any other agent in the environment can switch to the latest known target location to see if the target is available from its field of view.
- Fusion Creation: This event will notify that at least two cameras have in their fields of view at least one shared target. It may be used for different purposes, but we think that the interesting one is the creation of dynamic coalitions between the involved PTZ Agents in order to achieve coordination about the shared information. The purposes of dynamic coalitions are discussed in the following point.
- Fusion Deletion: Similar to the fusion creation, this event may induce the elimination of a dynamic coalitions between PTZ Agents or also get out some specific PTZ Agent from one coalition.

• Dynamic Coalition of PTZ Agents: The purpose of a dynamic coalition is to let the involved PTZ Agents (at least two) to collaborate in a common pursued goal. Suppose that a PTZ Agent (P1) listening in the Fusion Agent coalition receives a notification about a new fusion done with other PTZ Agent (P2). Depending on P1 goals, it may suggest to P2 create a new coalition in order to start a collaboration to satisfy a collaborative goal (suppose for this case obtain as much information about the fused target). If P2 agent accepts the coalition creation, then it is started a new temporary agent called Coalition Agent, which will be able to coordinate P1, and P2. This Coalition Agent will start to receive the tracks data from P1 and P2 from the Local Fusion Agent in order to decide or planify the tasks that should be achieved by P1 and P2 (in this case specify P1 and P2 to track the same fused target). This interaction for the dynamic coalition creation is presented in Figure 2.

In this way, different coalitions may response to different behaviors, depending on the coalitions goals. For example suppose that we have a Coalition Agent specialized in track different targets for each camera, or another one that priorices the targets by some rule or inference and select what target should be tracked by each PTZ agent (maybe the same for all if its is visible), or also another one that avoid overlapped fields of view depending on the fusion information. Thus, there can be many examples that can be pursued using this architecture.

This kind of dynamic coalitions that allows coordination between cameras may be started by agents (as shown in Figure 2) depending on the fusion events, or also by an eventual operator using the system, so rich coalitions with different goals can be established to improve the survelliace system. In general terms, we consider that if a PTZ Agent is actually working in a coalition, it cannot enter to collaborate to another coalition.

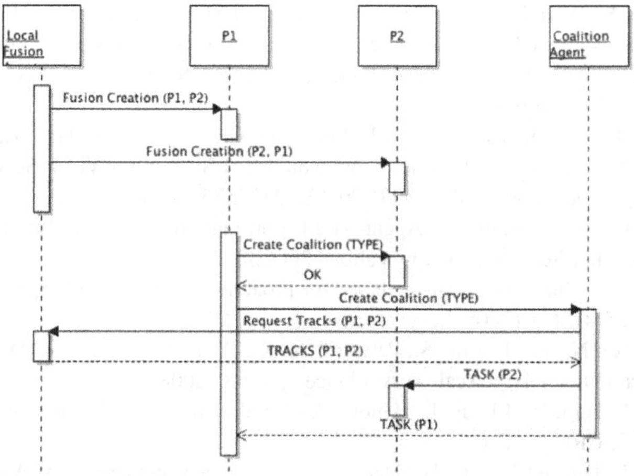

Fig. 2 Dynamic coalition formation of two PTZ agents, P1 and P2

4 Conclusions

In this work[1] we have defined a MAS architecture in order to support the distributed management of a visual sensor network, specially for PTZ devices. It supposed to deal with the common problems of data fusion and sensor management. In this way several agents where defined to accomplish different tasks. Also we define the data fusion as an integral part of the architecture that can provide mechanisms for coordination and competition between different agents. We have also presented an application example based on the architecture to understand how it can be used for different purposes. In this way dynamic coalition between PTZ Agents has emerged as a natural way of coordination.

References

1. Akyildiz, I., Kasimoglu, I.: Wireless sensor and actor networks: research challenges. Ad Hoc Networks 22(4), 351–367 (2004)
2. Bustamante, A.L., Molina, J.M., Patricio, M.A.: Multi-camera Control and Video Transmission Architecture for Distributed Systems. In: Molina, J.M., Corredera, J.R.C., Pérez, M.F.C., Ortega-García, J., Barbolla, A.M.B. (eds.) User-Centric Technologies and Applications. AISC, vol. 94, pp. 37–45. Springer, Heidelberg (2011)
3. Castanedo, F., García, J., Patricio, M.A., Molina, J.M.: Data fusion to improve trajectory tracking in a Cooperative Surveillance Multi-Agent Architecture. Information Fusion 11(3), 243–255 (2010)

[1] This work was supported in part by Projects CICYT TIN2008-06742-C02-02/TSI, CICYT TEC2008-06732-C02-02/TEC, CAM CONTEXTS (S2009/ TIC-1485) and DPS2008-07029-C02-02.

4. Castanedo, F., Patricio, M.A., García, J., Molina, J.M.: Extending surveillance systems capabilities using BDI cooperative sensor agents. In: Proceedings of the 4th ACM International Workshop on Video Surveillance and Sensor Networks - VSSN 2006, p. 131. ACM Press, New York (2006)
5. Collins, R.T., Lipton, A.J., Kanade, T., Fujiyoshi, H., Duggins, D., Tsin, Y., Tolliver, D., Enomoto, N., Hasegawa, O., Burt, P., Wixson, L.: A System for Video Surveillance and Monitoring. System 69(CMU-RI-TR-00-12), 573–575 (2011)
6. Garcia, J., Carbo, J., Molina, J.: Agent-based coordination of cameras. International Journal of Computer Science and Applications 2(1), 33–37 (2005)
7. Hampapur, A.: Smart video surveillance for proactive security. IEEE Signal Processing Magazine 25(4), 136 (2008)
8. Krahnstoever, N., Yu, T., Lim, S., Patwardhan, K., Tu, P.: Collaborative real-time control of active cameras in large scale surveillance systems (2008)
9. Liggins, M., Hall, D., Llinas, J.: Handbook of multisensor data fusion: theory and practice, vol. 22. CRC (2008)
10. Manyika, J., Durrant-Whyte, H.: Data fusion and sensor management: A decentralized information-theoretic approach (1994)
11. Marchesotti, L., Messina, A., Marcenaro, L., Regazzoni, C.: A cooperative multisensor system for face detection in video surveillance applications. Acta Automatica Sinica 29(3), 423–433 (2003)
12. Molina López, J.M., García Herrero, J., Jiménez Rodríguez, F.J., Casar Corredera, J.R.: Cooperative management of a net of intelligent surveillance agent sensors. International Journal of Intelligent Systems 18(3), 279–307 (2003)
13. Patricio, M.A., Carbó, J., Pérez, O., García, J., Molina, J.M.: Multi-Agent Framework in Visual Sensor Networks. EURASIP Journal on Advances in Signal Processing 2007, 1–22 (2007)
14. Singh, V.K., Atrey, P.K., Kankanhalli, M.S.: Coopetitive multi-camera surveillance using model predictive control. Machine Vision and Applications 19(5-6), 375–393 (2007)
15. Wesson, R., Hayes-Roth, F., Burge, J.W., Stasz, C., Sunshine, C.A.: Network structures for distributed situation assessment. IEEE Transactions on Systems, Man and Cybernetics 11(1), 5–23 (1981)

A Meta-model-Based Tool for Developing Monitoring and Activity Interpretation Systems

José Carlos Castillo, José Manuel Gascueña, Elena Navarro, and Antonio Fernández-Caballero

Abstract. Monitoring systems are often modeled through agent-based technologies. But current approaches to monitoring and activity interpretation suffer from a lack of consensus in terms of operation levels. Most approaches are standalone, which makes difficult to reuse and integrate code. This paper presents a generic tool that enables users to design and implement monitoring systems that meet their needs, regarding to the essential levels of their architecture. The specification of a meta-model, to reflect the vocabulary (concepts and relations between them) used in the specific domain of monitoring systems, guides the development of the proposed tool. The components created in the tool are implemented as intelligent agents.

Keywords: Monitoring, Activity interpretation, Meta-models, Model-driven architectures.

1 Introduction

Monitoring and activity interpretation is for sure an excellent application area for intelligent agents and multiagent systems [8]. Nowadays, a broad range of systems for monitoring and activity interpretation have arisen in academia [9, 12, 21, 6, 1, 7, 19, 4] as well as in the commercial field [14, 5, 13]. These systems operate at several processing levels. Notice that when comparing two commercial systems devoted to activity detection, strong differences can be found. For instance, the *Detect* [5] surveillance system uses cameras to track objects as well as to detect simple activities. On the other hand, the *Nicta Open Sensorweb Architecture* (*NOSA*) [13] multisensory monitoring system is proposed for the detection of activities in several domains that can range from detecting tsunamis to monitoring roads and means of transport. Despite the main goal of both systems is the detection of activities,

José C. Castillo · José M. Gascueña · Elena Navarro · Antonio Fernández-Caballero
Universidad de Castilla-La Mancha, Departamento de Sistemas Informáticos & Instituto de Investigación en Informática de Albacete, 02071-Albacete, Spain
e-mail: JoseCarlos.Castillo@uclm.es

J.B. Pérez et al. (Eds.): Highlights on PAAMS, AISC 156, pp. 113–120.
springerlink.com

NOSA allows users to work at higher abstraction levels than *Detect*. Indeed, *NOSA* performs the fusion of the information coming from the sensors, the detection of objects, their tracking and classification and, finally, the detection of activities. On the contrary, *Detect* only processes the detection of objects and activities. In the academic field different processing levels are also proposed to achieve similar goals [18, 3]. The former proposes a system for surveillance applications that can also be applied to Human-Computer Interaction or video content management. This vision-based system performs object detection, tracking and classification prior to the detection of activities. The latter consists in an intruder detection system that performs intruder detection and tracking to detect activities without a previous classification step. In short, there is no consensus about the operation levels needed to satisfy the monitoring functionalities.

Now, notice that most monitoring systems are standalone approaches devoted to a specific purpose. They select a set of operation levels customized according to their requirements, as previously described. Thus, our goal is to provide a generic tool that allows users to design and implement monitoring systems according to the levels they require to be present in the system architecture in order to meet their needs. Specifically, the proposed specification of a meta-model [10], to reflect the vocabulary (concepts and relations between them) used in the monitoring systems domain, guides us through the development of this tool. The components specified with the tool proposed to model monitoring and activity interpretation systems will be implemented following the agent-based paradigm. The rest of this paper is organized as follows. In section 2, the meta-model for monitoring and interpretation systems is described. In section 3, a case of study to illustrate the use of the tool developed using the meta-model is presented. And, finally, in section 4, some of the drawn conclusions are described.

2 Meta-model of Monitoring and Interpretation Systems

The Model-Driven Architecture (MDA), a four-layer meta-model hierarchy proposed by the Object Management Group [15], is used as guideline for developing a visual tool to create models of monitoring and interpretation systems. This is performed according to the meta-model described in this section. In the proposed hierarchy, at the meta-meta-modeling layer (M3) a collection of primitives is offered to define meta-models at the M2 layer. That is, the meta-meta-model describes the properties of the meta-models. At the meta-modeling layer (M2), the defined meta-elements are used to instantiate the elements that make up models at M1 layer. At the modeling layer (M1) the application model is specified. Finally, it is at the M0 layer where instances of M1 models are specified. Indeed, as illustrated in Fig. 1, a direct relation can be established between our tool and the four-layer architecture. In this work, UML [17, 16] (M3 layer) is used to specify the meta-model of monitoring and interpretation systems (M2). At M1 layer, models of monitoring and interpretation systems are specified according to the meta-model described at M2 layer. Finally, at M0 layer the instances of the models at M1 layer are instantiated.

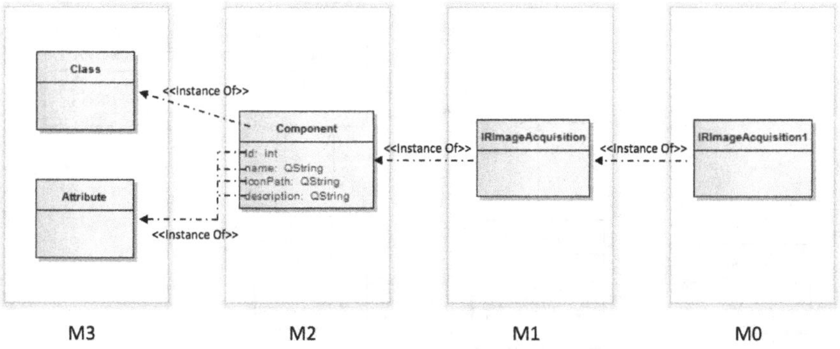

Fig. 1 MDA four-layer meta-model hierarchy

The basic idea of a meta-model is to identify the main concepts and their relations of a given problem domain used to describe the models of that domain [11]. The two following subsections describe the fragments of the meta-model for defining components and concrete configurations, respectively, of monitoring and interpretation systems.

2.1 Meta-model for Defining a Repository of Components

The meta-model fragment structure to manage the definition of a repository of components for monitoring and interpretation systems (see Fig. 2) is summarized as follows:

- A repository is made up of a collection of components. The multiplicity of composition relationship is set to 0..* to denote that "Repository" instances are related to zero or more instances of "Component".
- Component is a class with four attributes to specify the identifier, the name, the path to a picture for a visual representation and a brief description, respectively, of an entity of the repository.
- Each component belongs to one level according to the operation it performs. This is denoted by the "hasLevel" relationship. It can have one or more attributes, such as described by the multiplicity of the "hasAttribute" relationship (1..*).
- An attribute is defined as an input or output parameter, as it can be deduced from the multiplicity indicated in the "hasDataLevel" relationship defined between "Attribute" and "DataLevel". The "inout" attribute specified in class "DataLevel" enables one to specify whether the attribute is used as an input (value in) or output (value out) to/ from the component.
- Every attribute has a certain data type defined in the repository (see the connection Attribute → DataLevel → DataType in Fig. 2. It should be noted that the same data type can be used at different levels - notice that a multiplicity one or many (1 ..*) is established in the relationship "belongsDataLevel" established between "DataType" and "DataLevel".

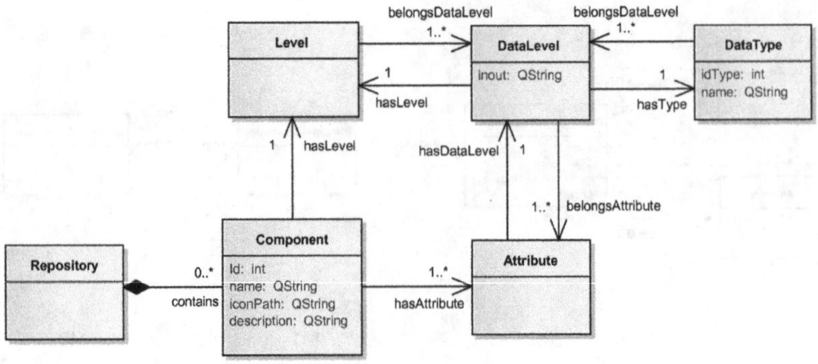

Fig. 2 Fragment of the meta-model for managing components

2.2 *Meta-model for Defining Configurations*

Establishing the proper relations between the instances of the selected components specifies the *configuration* of the monitoring and interpretation system. The meta-model fragment (see Fig. 3) to define these configurations can be summarized as follows:

- A configuration is made up of a collection of component instances. In this case, the multiplicity of the composition relationship is set to 1..* to denote that at least one instance of "ComponentInstance" belonging to an instantiated configuration must exist.
- Each component instance belongs to a type of component defined in the repository (see relationship "isTypeOf").
- Regarding to the connection of component instances, it should be noted that the meta-model is generic as it allows one to specify models of configurations in which component instances are connected in consecutive levels or in non-adjacent levels. The relationships "connectLowerLevel" and "connectUpperLevel" allow one to specify to which component instances (defined in the lower and upper levels, respectively) a given component instance is connected.
- The components defined at bottom/top level have no inputs/outputs, respectively. This is why 0 is specified as the multiplicity lower limit of the relationships "connectLowerLevel" and "connectUpperLevel".
- Finally, the tool verifies that the user has created a correct configuration by using the information of the attributes of the components instantiated when the configuration is specified. Specifically, it is checked that the data types of the input and output attributes of the types of components involved in each connection are compatible. This checking is done by looking for information of the type of parameter ("DataLevel") and the type of data ("DataType") of each attribute.

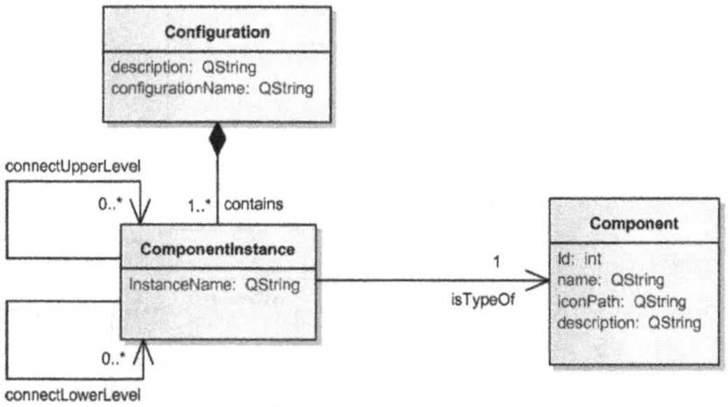

Fig. 3 Fragment of the meta-model for managing configurations

3 Case of Study

This section introduces a case of study that shows the definition of a repository of components and the configuration for a monitoring and fall detection system currently being developed by our research team. Specifically, the addition of a specific component for the detection of blobs from infrared images (see Fig. 4) is illustrated. According to the definition accepted by the image processing community, a blob is a set of connected pixels of an image. Blob detection aims at highlighting the presence of objects of interest in the analyzed images. In Fig. 4 several relations are described. Firstly, it can be noticed how the "IRBlobDetection" component is placed through the relation "hasLevel" in the detection level, which is described by means of the "Detection" class shown in Fig. 4. This is an intermediate level of the processing stack associated to monitoring and interpretation systems. The component holds two attributes: Image and Blob. Image is described as an input attribute through its relation with the "DataLevel2" class. Besides, the type of this attribute is described by means of the "DataType2" class as an "IplImage" pointer. This is an image data type belonging to the OpenCV image processing library, which is used to implement the monitoring and fall detection system. The second attribute, "Blob", is described by means of "DataLevel1" class as an output attribute. As described in the "DataType1" class, each blob is defined to have a "Blob" type, which is a specifically created data type for the application described next. The rest of components are added to the repository in a similar way.

Once the components have been specified in the repository, the proposed tool is used to define the configuration of the system being developed. For example, Fig. 5 depicts a configuration of a monitoring and fall detection system. "FallDetectionConfiguration" class contains all the instances of the components that belong to a given configuration. As the processing in monitoring systems is sequential, component instances are just connected to the components placed at an upper and lower level. Obviously, component instances at the bottom and top of the processing stack have only one link to another one through "connectUpperLevel" and "connectLowerLevel" relations, respectively. The types of the specified instances are the usual of the stack of traditional monitoring and activity interpretation systems, namely, image acquisition, image segmentation, blob

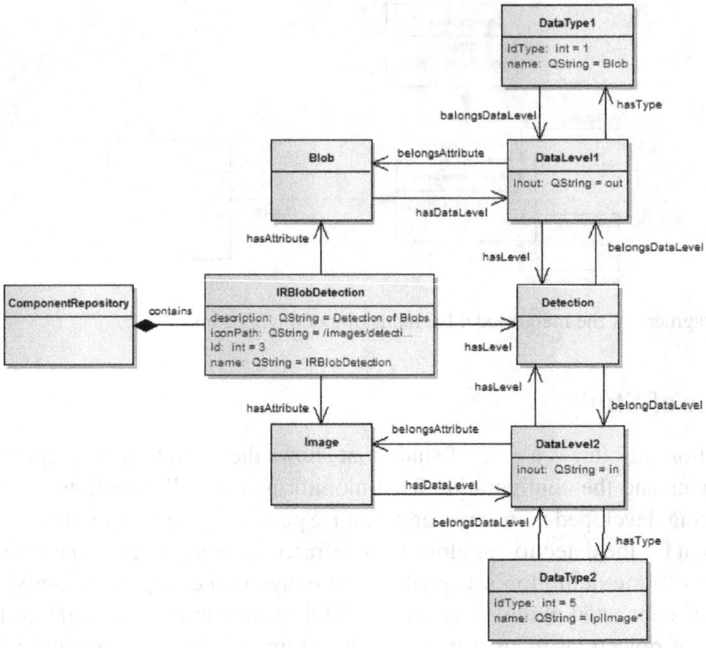

Fig. 4 "IRBlobDetection" component

detection, object identification, object tracking, and fall detection (corresponding to "IRImageAcquisiton", "IRImageSegmentation", "BlobDetection", "ObjectIdentification", "ObjectTracking" and "FallingDetection" classes, respectively). Along the processing flow, the information grows in its abstraction level as the different component instances process it. The "isTypeOf" relation relates each component with its type. As previously shown at the beginning of this section, the components have a set of attributes included in their definition.

4 Conclusions

After analyzing in the literature several approaches to the development of monitoring and interpretation systems, it has been observed that there is a generalized lack of consensus to carry out the design of components for surveillance. For this reason, it can be a very valuable asset for the activity interpretation community to have on hand a meta-model-based tool that provides support for the specification of concepts used in the development of monitoring systems. This paper has proposed a preliminary unified meta-model to generate generic monitoring and activity interpretation systems, independently of the final application field. The two fragments of the meta-model shown in Fig. 2 and Fig. 3 constitute this meta-model of monitoring interpretation systems. Agent technology is used to implement most components of the monitoring and activity interpretation systems, mainly in the higher levels [20].

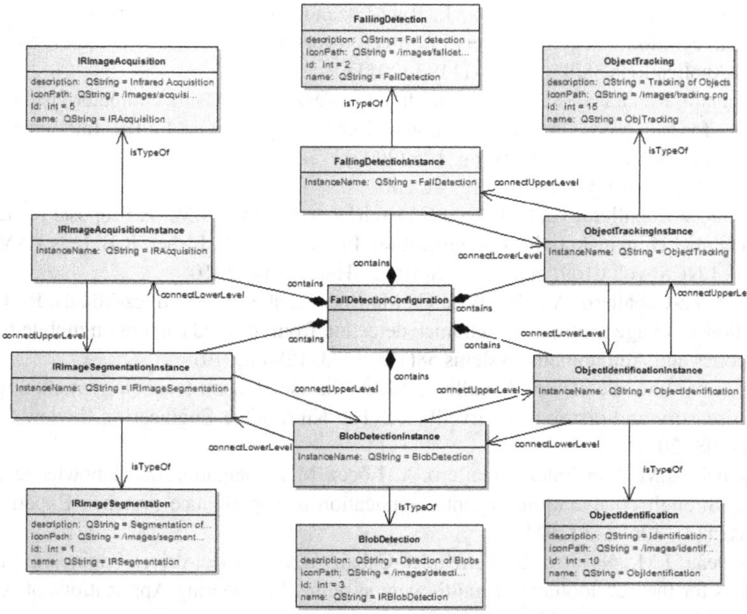

Fig. 5 Component configuration for fall detection system

The aim is to use agents to efficiently manage the great amount of data captured by the environmental sensors for the recognition of the activities carried out by the objects in the scenario.

Additional work is being developed. For example, alternatives followed to specify which node deploys each component instance are being analyzed. Another ongoing work is related to the generation of code in an automatic way. In particular, we are trying to use the described tool to improve the development of applications using our own multisensory monitoring and interpretation framework [2].

Acknowledgements. This work was partially supported by Spanish Ministerio de Ciencia e Innovación /FEDER under TIN2010-20845-C03-01 and CENIT A-78423480 grants, and by Junta de Comunidades de Castilla-La Mancha / FEDER under PII2I09-0069-0994 and PEII09-0054-9581 grants.

References

1. Castanedo, F., García, J., Patricio, M.A., Molina, J.M.: A multi-agent architecture based on the BDI model for data fusion in visual sensor network. Journal of Intelligent and Robotic Systems 62(3), 299–328 (2010)
2. Castillo, J.C., Rivas-Casado, A., Fernández-Caballero, A., López, M.T., Martínez-Tomás, R.: A Multisensory Monitoring and Interpretation Framework Based on the Model–View–Controller Paradigm. In: Ferrández, J.M., Álvarez Sánchez, J.R., de la Paz, F., Toledo, F.J. (eds.) IWINAC 2011, Part I. LNCS, vol. 6686, pp. 441–450. Springer, Heidelberg (2011)

3. Castro, J., Delgado, M., Medina, J., Ruiz-Lozano, M.: Intelligent surveillance system with integration of heterogeneous information for intrusion detection. Expert Systems with Applications 38(9), 11182–11192 (2011)
4. Cucchiara, R., Grana, C., Prati, A., Tardini, G., Vezzani, R.: Using computer vision techniques for dangerous situation detection in domotic applications. In: IEE Intelligent Distributed Surveilliance Systems, pp. 1–5 (2004)
5. Detec (2011), `http://www.detec.no`
6. Fernández-Caballero, A., Castillo, J.C., Rodríguez-Sánchez, J.M.: A Proposal for Local and Global Human Activities Identification. In: Perales, F.J., Fisher, R.B. (eds.) AMDO 2010. LNCS, vol. 6169, pp. 78–87. Springer, Heidelberg (2010)
7. Fernández-Caballero, A., Castillo, J.C., Martínez-Cantos, J., Martínez-Tomás, R.: Optical flow or image subtraction in human detection from infrared camera on mobile robot. Robotics and Autonomous Systems 58(12), 1273–1280 (2010)
8. Gascueña, J.M., Fernández-Caballero, A.: On the use of agent technology in intelligent, multisensory and distributed surveillance. The Knowledge Engineering Review 26(2), 191–208 (2011)
9. Gascueña, J.M., Fernández-Caballero, A., López, M.T., Delgado, A.E.: Knowledge modeling through computational agents: application to surveillance systems. Expert Systems 28(4), 306–323 (2011)
10. Gascueña, J.M., Navarro, E., Fernández-Caballero, A.: Model-driven engineering techniques for the development of multi-agent systems. Engineering Applications of Artificial Intelligence 25(1), 159–173 (2012)
11. Gómez-Sanz, J., Pavón, J.: Meta-modelling in Agent Oriented Software Engineering. In: Garijo, F.J., Riquelme, J.-C., Toro, M. (eds.) IBERAMIA 2002. LNCS (LNAI), vol. 2527, pp. 606–615. Springer, Heidelberg (2002)
12. Kieran, D., Yan, W.: A framework for an event driven video surveillance system. Journal of Multimedia 6(1), 3–13 (2011)
13. Nicta Open Sensorweb Architecture, NOSA (2008), `http://www.nicta.com.au`
14. Object Video (2010), `http://www.objectvideo.com/`
15. Object Management Group, Meta object facility (mof) specification - version 1.4 (April 2002), `http://www.omg.org/spec/MOF/1.4/`
16. Object Management Group, OMG Unified Modeling Language (OMG UML), Superstructure, V2.1.2 (2007), `http://www.omg.org/spec/UML/2.1.2/Superstructure/PDF/`
17. Object Management Group, OMG Unified Modeling Language (OMG UML), Infrastructure, V2.1.2 (2007), `http://www.omg.org/spec/UML/2.1.2/Infrastructure/PDF/`
18. Onut, V., Aldridge, D., Mindel, M., Perelgut, S.: Smart surveillance system applications. In: Proceedings of the 2010 Conference of the Center for Advanced Studies on Collaborative Research, pp. 430–432 (2010)
19. Pavón, J., Gómez-Sanz, J., Fernández-Caballero, A., Valencia-Jiménez, J.J.: Development of intelligent multi-sensor surveillance systems with agents. Robotics and Autonomous Systems 55(12), 892–903 (2007)
20. Rivas-Casado, A., Martinez-Tomás, R., Fernández-Caballero, A.: Multiagent system for knowledge-based event recognition and composition. Expert Systems (2012), doi:10.1111/j.1468-0394.2010.00578.x
21. Vallejo, D., Albusac, J., Castro-Schez, J.J., Glez-Morcillo, C., Jiménez, L.: A multi-agent architecture for supporting distributed normality-based intelligent surveillance. Engineering Applications of Artificial Intelligence 24(2), 325–340 (2011)

Unconditionally Secure Protocols with Genetic Algorithms

Ignacio Hernández-Antón, Fernando Soler-Toscano,
and Hans van Ditmarsch

Abstract. This paper presents genetic algorithms as a tool for searching unconditionally secure protocols in card game scenarios. We model cards protocols with genetic algorithms and run an experiment to determine the influence of weighing differently the protocol requirements.

1 Introduction

We present an algorithmic approach to search unconditionally secure protocols of communicating agents within the well-known russian cards problem scenario. In public/private key approaches as AES (Advanced Encription Standard), RSA (Rivest, Shamir, Adleman technique), DSA (Digital Signature Algorithm) or the ECC (Elliptic Curve Cryptography), secret information if safeguarded because of the high complexity of computational operations to decrypt the message, for instance, RSA uses the IFD, the Integer Factorization Problem [7]. Instead, in this work, we choose an information-based approach to protocol design. We model the communicating agents as cards players and the communicating secret is the ownership of the cards in the game. In that approach, it is also possible to define good protocols regardless of the computational complexity of encryption [3, 6, 4].

We just study the security aspects of communication in order to avoid eavesdropping. We study how to guarantee the privacy of the message which should only be shared by those principal agents we legitimated. This will occur regardless of other agents listening passively to the information passed. There is a logical approach where this problem is formalized using dynamic epistemic logic [4, 3].

We employ genetic algorithms to search for card deal protocols [10, 8, 11]. Genetic algorithms are a bio-inspired family of computational techniques

Ignacio Hernández–Antón · Fernando Soler–Toscano · Hans van Ditmarsch
University of Sevilla, Spain
e-mail: {iha,fsoler,hvd}@us.es

J.B. Pérez et al. (Eds.): Highlights on PAAMS, AISC 156, pp. 121–128.
springerlink.com

which have natural evolution as a model for encoding some critical aspects of solutions as chromosomes-like data structures. An initial population is transformed by genetically inspired operations in order to produce new generations. The most important ones are **selection** (of the fittest), **crossing-over**, and **mutation**. We use JAVA to specify the russian cards problem and a genetic engine called *jgap* to search for protocols. For further details on the genetic engine see `http://jgap.sourceforge.net/`.

2 The Russian Cards Problem

From a pack of seven known cards (for instance 0-6) two players (a, b) each draw three cards and a third player (c) gets the remaining card. How can the two first players (those with three cards) openly (publicly) inform each other about their cards without cyphering the messages and without the third player learning from any of their cards who holds it?

Although this presentation of the problem has 7 cards in the stack and the deal distribution is 3.3.1 (a draws 3, b draws 3 and c draws 1), one may consider other scenarios, e.g., a 10-cards stack with a 4.4.2 deal. To become familiar with a basic game scenario, let us call agents a, b and c. The cards are named 0, ..., 6. Deals distributions (size) are noted as integer strings, for instance 3.3.1. Legitimated principals are a and b while c is the "intruder". We suppose the actual deal is 012.345.6. Communication is done by truthful and public announcements, see [9]. A public announcement for an agent a is a set of a's possible set of hands (we use a simplified notation to denote set os hands, e.g., $\{012, 125, 156\}$ instead of $\{\{0,1,2\},\{1,2,5\},\{1,5,6\}\}$).

A secure announcement in this scenario should keep c ignorant throughout the whole communication and guarantee the common knowledge of this agent's ignorance. According to that approach, a good protocol comprises an announcement sequence that verifies that:

- **Informativeness 1:** Principals a and b know each other cards.
- **Informativeness 2:** It is common knowledge, at least for the principals, that they do know each other's cards.
- **Security 1:** The intruder, c, remains ignorant.
- **Security 2:** It is common knowledge for all agents that the intruder remains ignorant.
- **Knowledge-based:** Protocol steps are modelled as public announcements.

The reason we split the informativeness and security requirements into two parts can be found in [4]. A protocol is then a finite sequence of instructions determining sequences of announcements. Each agent a chooses an announcement conditional on that agent's knowledge. The protocol is assumed common knowledge among all agents.

One knowledge-based protocol that consitutes a solution for the riddle is as follows. Suppose that the actual deal of cards is that agent a has $\{0,1,2\}$, b has $\{3,4,5\}$ and c has $\{6\}$.

- a says: My hand is one of $\{012, 046, 136, 145, 235\}$.
- Then, b says: c's card is 6.

After this, it is common knowledge to the three agents that a knows the hand of b, that b knows the hand of a, and that c is ignorant of the ownership of any card not held by itself. For further details on the notion of common knowledge see [5].

We can also see these two sequences as the execution of a knowledge-based protocol. Given a's hand of cards, there is a (non-deterministic) way to produce her announcement, to which b responds by announcing c's card. The protocol is knowledge-based, because the agents initially only know their own hand of cards, and have public knowledge of the deck of cards and how many cards each agent has drawn from the pack. It can be viewed as an *unconditionally secure* protocol, as c cannot learn any of the cards of a and b, no matter their computational resources. The security is therefore not conditional on the high complexity of some computation.

3 Modelling Cards Protocols with Genetic Algorithms

The final objective of modelling cards protocols with genetic algorithms is to find protocols for card deal sizes where an analytic solution is lacking. We reinvestigate protocols for 3.3.1 with genetic algorithms. The use of genetic algorithms requires to satisfy two conditions: possible solutions can be represented by chromosomes and an evaluation function can be defined in order to assign a value to each chromosome. Regarding this encoding representation we observe that the set of possible hands of an agent can be arranged in lexicographic order, e.g., for Russian Cards, the 35 hands are listed as 012, 013, ..., 456. Then we assign a binary gene to each possible hand. An announcement is then represented by a 35-bitstring. To illustrate a mapping like this see Figure 1 (right), we can encode several announcements into one chromosome as in Figure 1 (left) representing this way an entire protocol. To evolve the population a fitness function assigns to each protocol a value. As the protocols are knowledge-based, this function needs some epistemic aspects to be implemented. As we are looking for a two-step protocol [2], we only need one announcement. The fitness function evaluates possible announcements. Note that if D is the set of cards, these announcements are elements of $\mathcal{P}(\mathcal{P}(D))$ with certain properties.

The first epistemic function is $compCardsGivenAnnounce(Ann, Hand)$ which returns, for a given announcement Ann, the set of possible hands that an agent having $Hand$ considers for the agent making the announcement:

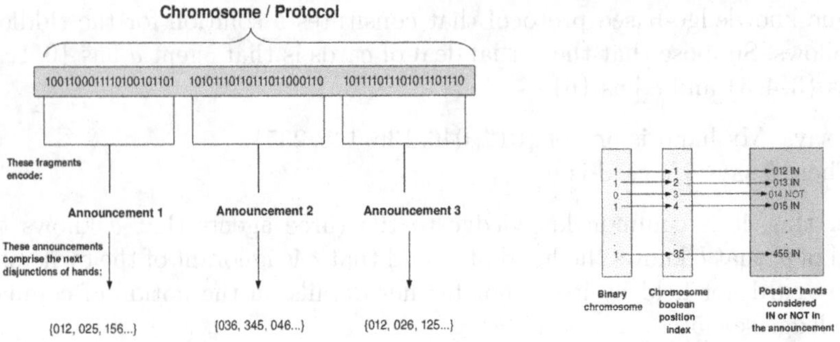

(a) Mapping announcements (b) Allele decoding

Fig. 1 Chromosome general structure

$$compCardsGivenAnnounce : \mathcal{P}(\mathcal{P}(D)) \times \mathcal{P}(D) \mapsto \mathcal{P}(\mathcal{P}(D))$$
$$compCardsGivenAnnounce(Ann, Hand) = \{h \in Ann \mid h \cap Hand = \emptyset\}$$

Now we define $whatAgentLearnsFromAnnounce(Ann, Hand)$ that returns the set of cards that an agent having $Hand$ learns from the agent. The function is defined as:

$$whatAgentLearnsFromAnnounce : \mathcal{P}(\mathcal{P}(D)) \times \mathcal{P}(D) \mapsto \mathcal{P}(D)$$
$$whatAgentLearnsFromAnnounce(Ann, Hand) = \bigcap compCardsGivenAnnounce(Ann, Hand)$$

As an example, consider that, in the 3.3.1 setting with deal 012.345.6, a announces that her hand is one of $\{012, 016, 234\}$. Then, a's compatible hands for b are $\{012, 016\}$ and b learns that a has $\{0, 1\}$. However an agent may learn cards not only from the agent making the announcement, but also from the remaining one. In our example, as c's hand is $\{6\}$, c can also apply the previous function to learn that a holds card 2. But not that a's compatible hands for c are $\{012, 234\}$, so c learns that b holds cards 5. So c learns two cards, one from a and the other from b. The following function calculates this:

$$howManyAgentLearnsFromDeal : \mathcal{P}(\mathcal{P}(D)) \times \mathcal{P}(D) \mapsto \mathbb{N}$$
$$howManyAgentLearnsFromDeal(Ann, Hand) = |whatAgentLearnsFromAnnounce(Ann, Hand)| + |D| - |Hand| - |\bigcup compCardsGivenAnnounce(Ann, Hand)|$$

In the fitness function we have to consider not only one deal (as that in our example) but all possible deals, in order to ensure that the announcement is unconditionally secure. The following function calculates, for a given announcement Ann and an agent having x cards, the minimum number of cards that the agent learns from the deal, by considering all possible agent's hands consistent with Ann:

$$minLearn : \mathcal{P}(\mathcal{P}(D)) \times \mathbb{N} \mapsto \mathbb{N}$$
$$minLearn(Ann, x) = min(\{howManyAgentLearnsFromDeal(Ann, H) \mid H \subseteq D, |H| = x, compCardsGivenAnnounce(Ann, H) \neq \emptyset\})$$

In the same way, we can define a function $maxLearn$ to calculate the maximum number of cards that an agent may learn. Then, the fitness function, given that the size of the deal is $n.m.k$, is defined as:

$$fitness : \mathcal{P}(\mathcal{P}(D)) \times \mathbb{N} \times \mathbb{N} \mapsto \mathbb{Z}$$
$$fitness(Ann, wI, wS) = wI \cdot minLearn(Ann, m) - wS \cdot maxLearn(Ann, k)$$

The two values wI and wS are two natural numbers which measure the relevance of b's knowledge and c's ignorance, respectively. Obviously, the maximum value of the fitness function is obtained when b learns $n + k$ cards (all a's and c's cards) and c learns nothing. Then, the value of the fitness function is $wI(n + k)$. The minimum value is $-wS(n + m)$.

In the Java implementation we work with JGAP's chromosomes, that are sequences of binary values. Prior to apply the fitness function we need to decode the binary sequence into an announcement, as explained above. As the fitness function is only allowed to return non-negative values, $S(n + m)$ is added to every result of our previous $fitness$ function.

4 Weighing Informativeness and Ignorance

It seems reasonable to suppose that there is a correspondence between weighing informativeness and security, on the one hand, and the efficiency in time of the search, on the other hand. In this section we demonstrate by statistical analysis that this is not the case.

Weighing the fitness function with wI and wS we can influence the search, prioritizing one aspect or the other. If we consider informativeness more important than security, the algorithm lets survise to the next generation announcements where c could learn some cards. But if we focus on security and wish to avoid that situation, we will prioritize the security weight in order to devaluate the announcement where c can learn.

Apart from the number of experiments, the size of the populations, and the number of generations set in the algorithm, the different weight combinations also allow us to create different scenarios and configurations. Those will be useful to collect data for statistical analysis. We wish to determine if there is a combination of weights that makes the search go faster.

The results of the algorithm search are stored and a statistic data analyzer goes through them in order to find relevant information. Figure 2 shows an 3.3.1-case sample summary of the output of this analysis.

The first line of Figure 2 represents the weights that have the best mean search time. Those weights could be good candidates to weigh other card deals (than 3.3.1), in order to investigate if there is a relation between weighing and search time. Expression `minTime sPR TimeOfSearching 465` denotes

```
Best weighing MEAN preformance: 7,1
minTime sPR TimeOfSearching 465,
announcement: [[0, 2, 5], [0, 4, 6], [1, 3, 4], [1, 5, 6], [2, 3, 6]],
wInformativeness: 9,
wSecurity: 9,
Deal: 3.3.1
maxTime sPR TimeOfSearching 48875,
announcement: [[0, 1, 2], [0, 3, 4], [0, 3, 6], [1, 2, 6], [1, 3, 5],
 [1, 3, 6], [2, 3, 4], [2, 3, 5], [4, 5, 6]],
wInformativeness: 6,
wSecurity: 7,
Deal: 3.3.1
(max-min) range time sPR 48410.0
Whole experiment mean timeOfSearching: 4031.433
Data analyzer timing: 915
```

Fig. 2 Stats analyzer output for 30-runs experiment $wI, wS = [1..10]$

the lowest time, in millis, (the fastest protocol found) in the experiment that is associated to the announcement `[[0, 2, 5], [0, 4, 6], [1, 3, 4], [1, 5, 6], [2, 3, 6]]` that has the weight $wI =$ `wInformativeness: 9`, $wS =$ `wSecurity: 9`. We also see similar parameters for the protocol with the highest search time. At the end, the search time range and mean of the entire experiment.

There are also other 7-cards distributions where the genetic approach can search for protocols. Considering the constraints among a, b, c cards presented in [1], we obtain just two 7-cards scenarios where there exist protocols to search, namely: 3.3.1 and 4.2.1. Another case is 2.4.1. Then, there is no protocol where first a makes the announcement. But swapping the roles of a and b we can apply a 4.2.1 protocol.

The first experiment we did was a 30-runs from 1 to 10 different combinations of weights. We can represent that as ($\texttt{SC}$ $30, 500, 200, wI, wS, 3, 3, 1$) where:

- SC stands for scenario configuration
- 500 is the maximun number of generations allowed to evolve
- 200 is the initial population size
- 30 means thirty runs of the algorithm
- wI is the weight for informativeness
- wS is the weight for security
- $3, 3, 1$ is the deal distribution

Notice that wI and wS will vary from 1 to 10, generating different scenarios configurations in order to search for weight with the fastest result. The first three results from a 30-run sample experiments with $wI = 4$ and $wS = 7$ are depicted in Figure 3.

Each two lines represent the execution of the serach for a 3.3.1 scenario. For example, considering the first protocol result, those figures mean:

```
[[0, 2, 5], [0, 3, 4], [1, 3, 5], [1, 4, 6], [2, 3, 6]]:
58.0, 58.0, 73, 6464, 4, 0, 500, 200
[[0, 1, 6], [0, 2, 3], [1, 2, 4], [2, 5, 6], [3, 4, 5]]:
58.0, 58.0, 24, 2587, 4, 0, 500, 200
[[0, 1, 5], [0, 3, 4], [1, 2, 6], [2, 4, 5], [3, 5, 6]]:
58.0, 58.0, 48, 4478, 4, 0, 500, 200
. . .
```

Fig. 3 `Partial output of SC` 30, 500, 200, 4, 7, 3, 3, 1

- `[[0, 2, 5], [0, 3, 4], [1, 3, 5], [1, 4, 6], [2, 3, 6]]` is the announcement sequence. So a annouces $\{025, 034, 135, 146, 236\}$.
- `58.0` means the value the fitness functions assigned to that protocol.
- `58.0` (the second occurrence) represents the maximum fitness value that can be reached for any protocol.
- `73` is the generation where this protocol was found.
- `6464` is the time of searching in milliseconds.
- `4` means the minimum number of cards b can learn using this protocol.
- `0` is the maximum number of cards c can learn.
- `500` is the maximun number of generations allowed to evolve.
- `200` is the initial population size.

We executed a 30-runs experiment varying both weights from 1 to 10 generating a total of $30 \times 10 \times 10 = 3000$ protocol results. Figure 4 depicts information about the runtime for different weights for 3.3.1: on the left, the relation between the differents weights assignments and the search mean time of those; on the right, the standard deviation extracted from the first. At first sight one can infer there is not a regular relation among those parameters. Hence we think there is no use in extracting the best weights for ignorance and informativeness from this experiment and scale them up to larger deals.

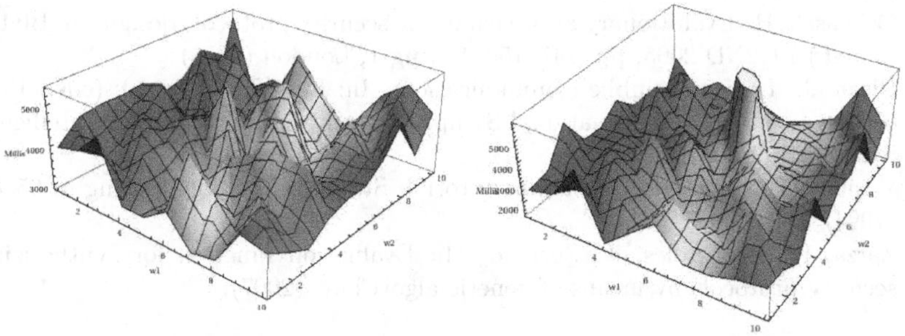

(a) weighing and mean time (b) weighing and deviation time

Fig. 4 3.3.1 weighing and mean stats graphics

5 Conclusions and Future Work

We confirmed the possibiliy of modelling unconditionally secure protocols search using Genetic Algorithms. The statistical analyzer showed the non-regular relation between weighing protocol (main) requirements and the mean time of searching. Althouth the genetic engine has been used for small deals, it has now been studied in order to improve the time/memory efficiency to use it for larger deals where several announcements comprise a protocol and a huge number of operations is presumed to be executed. We project new features regarding not just the statistical analysis over the protocols found but the search for symmetry properties in protocols that can give a clue about a possible analytic solution for larger sized deals.

References

1. Albert, M., Aldred, R., Atkinson, M., van Ditmarsch, H., Handley, C.: Safe communication for card players by combinatorial designs for two-step protocols. Australasian Journal of Combinatorics 33, 33–46 (2005)
2. Albert, M., Cordón-Franco, A., van Ditmarsch, H., Fernández-Duque, D., Joosten, J.J., Soler-Toscano, F.: Secure Communication of Local States in Interpreted Systems. In: Abraham, A., Corchado, J.M., González, S.R., De Paz Santana, J.F. (eds.) International Symposium on Distributed Computing and Artificial Intelligence. AISC, vol. 91, pp. 117–124. Springer, Heidelberg (2011)
3. van Ditmarsch, H.: The Russian cards problem. Studia Logica 75, 31–62 (2003)
4. van Ditmarsch, H., van der Hoek, W., Kooi, B.: Dynamic Epistemic Logic. Synthese Library, vol. 337. Springer, Heidelberg (2007)
5. Fagin, R., Halpern, J., Moses, Y., Vardi, M.: Reasoning about Knowledge. MIT Press, Cambridge (1995)
6. Fischer, M., Wright, R.: Bounds on secret key exchange using a random deal of cards. Journal of Cryptology 9(2), 71–99 (1996)
7. Kumar, A., Ghose, M.K.: Overview of information security using genetic algorithm and chaos. Information Security Journal: A Global Perspective 18(6), 306–315 (2009)
8. Ocenasek, P.: Evolutionary approach in the security protocols design. In: Blyth, A. (ed.) EC2ND 2005, pp. 147–156. Springer, London (2006)
9. Plaza, J.: Logics of public communications. In: Emrich, M., et al. (eds.) Proceedings of the 4th International Symposium on Methodologies for Intelligent Systems, pp. 201–216 (1989)
10. Whitley, D.: A genetic algorithm tutorial. Statistics and Computing 4, 65–85 (1994)
11. Zarza, L., Pegueroles, J., Soriano, M.: Evaluation function for synthesizing security protocols by means of genetic algorithms (2007)

Cooperative Sensor and Actor Networks in Distributed Surveillance Context

Alaa Khamis

Abstract. In this paper, we envision a cooperative sensor and actor network (C-SANET) that encompasses a set of heterogeneous sensing agents, acting agents, situation awareness agents, resource management agents and decision support/making agents. These agents are endowed with know-how capability for solving problems in an autonomous way and a know-how-to-cooperate capability by which the agents can share common interests and interact with each other. These spatially distributed agents, when properly managed, can sense collaboratively and continuously a volume of interest and physically manipulate and interact with it. The paper provides a comprehensive introduction to cooperative sensor and actor networks and discusses different forms of cooperation in C-SANETs in the context of distributed surveillance.

1 Introduction

In recent years, Sensor and Actor Network (SANET) research has received an increasing amount of attention from researchers in academia, government laboratories and industry. This research activity has borne some fruit in tackling some of the challenging problems of SANET that are still open.Among these problems is how to achieve effective and robust cooperation in SANETs.

Achieving cooperation in artificial systems is engineering and science inspired by different domains such as biology, artificial life, psychology and cognitive science. Cooperative Sensor and Actor Network (C-SAENT) is a distributed system that incorporates a set of heterogeneous sensing agents, acting agents, situation awareness agents, management agents and decision support/making agents with communication

Alaa Khamis
Robotics and Autonomous Systems Research Group,
Department of Mechatronics Engineering, German University in Cairo (GUC),
New Cairo City Egypt
e-mail: alaa.khamis@guc.edu.eg

J.B. Pérez et al. (Eds.): Highlights on PAAMS, AISC 156, pp. 129–138.
springerlink.com

modules. These spatially distributed agents, when properly managed, can sense collaboratively and continuously a volume of interest and physically manipulate and interact with it.

Effective and robust cooperation among these agents can synergistically improve the performance of sensor and actor networks and can endow them with higher-level faculties, such as dynamic task allocation, communication relaying, cooperative target detection and tracking and shared situation awareness. To achieve this effective cooperation in SANET, the agents must have know-how for solving problems in an autonomous way and a know-how-to-cooperate [1] by which agents can share common interests and interact with each other. Such cooperative sensor and actor networks can be applied to many pertinent areas of industrial and commercial importance such as distributed surveillance, perimeter security, reconnaissance, search and rescue, environment monitoring, disaster management, industrial process control, health care and home intelligence.

Among these applications is distributed surveillance. The objective of distributed surveillance systems is to provide systemic observation that includes the timely detection, localization, recognition and identification of objects and events, their relationships, activities, and plans, in a given volume of interest (VOI) in order to determine whether they are behaving normally or if there is any deviation from their expected behavior [2]. This paper provides a comprehensive introduction to cooperative sensor and actor networks and discusses the role of cooperation in such networks. It describes different problems that have been solved through achieving cooperation between the network agents in the context of distributed surveillance.

The remainder of the paper is structured as follows: Section 2 introduces cooperative sensor and actor networks. section 3 discusses the concept of cooperation in C-SANETs followed by addressing different forms of cooperation in section 4. Finally conclusions and future work are summarized in section 5.

2 Cooperative Sensor and Actor Network

Sensor and Actor Networks (SANETs) refer to a group of sensors and actors linked over a wireless medium to perform distributed sensing and acting tasks [3]. SANETs can be seen as an extension of sensor networks by embedded actuation facilities [4]. This combination enhances the capabilities of sensor networks and widens their field of application. We define cooperative sensor and actor networks (C-SANET) as a distributed system that incorporates a set of heterogeneous sensing agents, acting agents, situation awareness agents, management agents and decision support/making agents with know-how and know-how-to-cooperate capabilities as illustrated in Fig. 1. The following subsections describe the different agents that form a C-SANET.

2.1 Sensing Agents

Sensing agents represent a set of spatially distributed data sources that provide observations or measurements about the different activities in the VOI. These sensing agents can be classified based on the nature of the sensors into sensor-space (S-Space) sensing agents, human-space (H-Space) sensing agents and Internet-space (I-Space) sensing agents.

- S-Space Sensing Agents: Sensor Space or S-Space sensing agents refer to the physics-based sources of information. There is a huge array of sensors available for measuring any phenomenon. This space incorporates a vast array of spatially distributed static and mobile sensors of different modalities that can sense collaboratively and continuously the VOI. This array can include, but is not limited to, air quality sensors, temperature sensors, radiation sensors, range finding sensors, optical imagery systems formed by small but highly capable satellites that provide high resolution imagery to the decision makers, integrated multi-spectral imaging, Electronic Support Measures (ESM), Conventional Scanning Radars (CSR) and/or Electronically Scanned Arrays (ESA) to achieve day and night all-weather performance. The data coming from S-space sensing agents used to be called "hard data". This data is usually highly structured and quantitative in nature and thus has a straightforward mathematical representation [5].
- H-Space Sensing Agents: Human Space or H-Space sensing agents represent human-generated data sources. These dynamic observation resources allow humans to act as soft sensors providing information in form of a report. This type of data is called "soft data". For example, in urban surveillance mission or during mass emergencies and large-scale man-made or natural disasters, information coming from human observers is of crucial importance. Handling soft

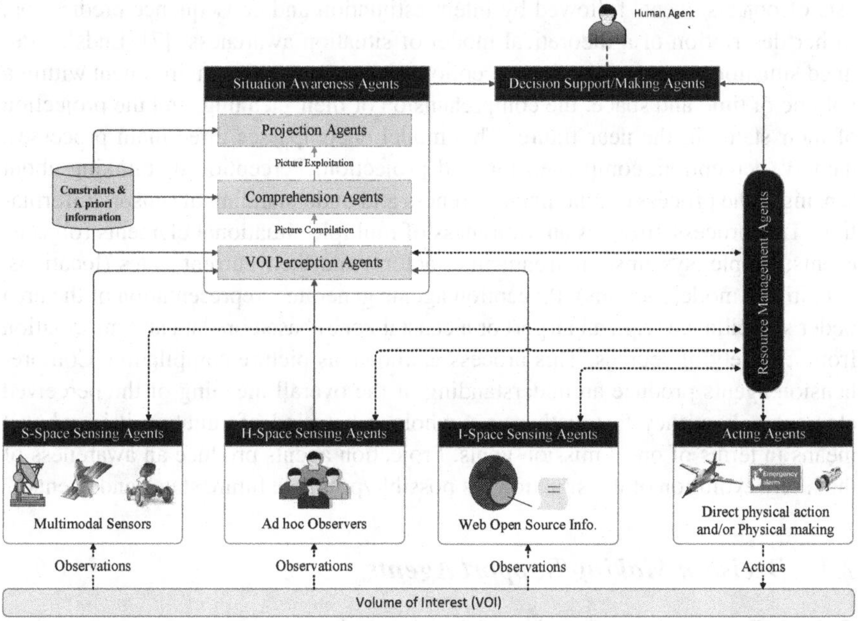

Fig. 1 Cooperative Sensor and Actor Network (C-SANET) in Disutrubted Surveillance

data is more challenging than hard data due to its qualitative nature, subjectivity and incompleteness. Human observations are based on inherently qualitative models used to perceive the environment resulting in vague measurements. Each human conveys information using their own context-dependent language(s) over bandwidth-limited communication channels and phenomenon observed by humans is usually complex and thus may not be fully described by a single person [5].

- I-Space Sensing Agents: Another source of soft data is the open source information on the Internet (e.g., Facebook, Twitter, MySpace, YouTube, eBay, Craigslist, Wikipedia, Blogger, Photobucket and Flikr) [6]. These Internet-based information resources are called Internet Space or I-Space sensing agents. The Web is considered as an open source of information that can be retrieved and mined in order to provide more observation about the VOI. This soft data source results in many opportunities as well as many challenges due to the extremely large-scale, dynamic, heterogeneous and hyperlinked nature of the Web and the unstructured nature of its content.

2.2 Situation Awareness Agents

A key requirement for successful monitoring using C-SANETs is situation awareness. In C-SANET situation awareness agents build collaboratively a complete picture of the situation by collecting the relevant data in order to identify the different entities and their relationships. Then, the agents perform a relational analysis of objects-events followed by intent estimation and consequence predication. In her description of a theoretical model of situation awareness [7], Endsley defined situation awareness as the perception of elements in the environment within a volume of time and space, the comprehension of their meaning, and the projection of their status in the near future. This model encompasses three main processes, namely, perception, comprehension and projection. Perception or thinking about sensing is the process of attaining awareness and understanding of sensory information. This process provides an awareness of multiple situational elements (objects, events, people, systems, environmental factors) and their current states (locations, conditions, modes, actions). Perception agents generate a representation of the area under surveillance (e.g., an airport or a coastal zone) based on data and information from a variety of sources. This process is known as picture compilation. Comprehension agents produce an understanding of the overall meaning of the perceived elements - how they fit together as a whole, what kind of situation it is, what it means in terms of one's mission goals. Projection agents produce an awareness of the likely evolution of the situation, its possible/probable future states and events.

2.3 Decision Making/Support Agents

C-SANETs encompass a number of decision making or support agents (based on whether the system is fully or semi-autonomous) that are responsible for making a decision in the absence of certainty. This uncertainty can result in wrong beliefs about system state and/or environment state. Von Neuman-Morgenstern utility model and Bayesian decision networks provide robust framework to handle sequential decision making under uncertainty. Following state-of-the-world-decision-making model, the decision making agents first choose an action from a set of available action. Then the state of the world is observed.

2.4 Resource Management Agents

Resource Management refers to the process that plans and controls the use of a set of distributed sensing and acting resources in a manner that synergistically improves the process of data fusion and ultimately enhances situation awareness and decision making. Managing scarce and heterogeneous sensors and actors involves making decisions and compromises regarding alternate sensing and acting strategies under time and resource availability constraints. Resource management agents can be modeled as utility maximizers who inhabit some kind of Markov decision process (MDP). MDPs provide a mathematical framework for modeling decision-making in situations where outcomes are partly random and partly under the control of a decision maker.

2.5 Acting Agents

Acting agents physically manipulate and interact with the VOI. Actuation may be a direct physical action upon the process, such as moving a mobile sensor to keep tracking certain feature; or a physical making of an electrical circuit, which in turn has a direct effect upon the process. An example would be an actuator (relay) that activates an alarm or deliver warning messages to the potentially affected locations to alert local and regional governmental agencies. The messages need to be reliable, synthetic and simple to be understood by authorities and public. In [8], Vuran et al explain the difference between actuator and actor. They claim that an actuator is a device such as flow-control valve, pump, positioning drive, motor, switch, relay or meter that converts an electrical control signal to a physical action. On the other hand, an actor, besides being able to act on the environment by means of one or several actuators, is also a single network entity that performs networking-related functionalities. The actor then is an agent with know-how and know-how-to-cooperate capabilities.

2.6 Communication Medium

C-SANETs represent a class of distributed system that emphasizes communication as well as coordination aspects. An effective C-SANET needs an effective communication system in order to timely react to sensor information with an effective action. C-SANET agents have various energy and computational constraints because of their inexpensive nature and ad-hoc method of deployment. Moreover, in challenging environments such as underwater and underground environments, the medium propagation characteristics require new communication methods and protocols explicitly designed for such environments. Considerable research has been focused at overcoming these deficiencies through more energy efficient routing, localization algorithms and system design [8].

If the C-SANET encompasses mobile agents such as mobile sensors or mobile actors, GSM/Wireless connectivity is the central means of carrying information because of the nature of the assets as moving platforms and field deployed basing. This communication medium used to be noisy, error prone, and time varying. The connectivity may be frequently disrupted because of channel fading or signal attenuation. This leads to the necessity for novel forms of cooperation between the distributed agents based on indirect communication concepts such as stigmergy (communication via environment) or communication via observation [9].

3 Cooperation in Sensor and Actor Network

Effective and robust cooperation among the artificial and human agents that form the C-SANET in distributed surveillance can synergistically improve its performance and can endow it with higher-level faculties such as the following merits: complex task allocation, communication relaying, cooperative target detection and tracking, sensemaking and context-awareness [2].

Achieving robust and productive cooperation between various system components is engineering and science inspired by different domains such as biology, artificial life, psychology and cognitive science in order to build artificially cooperative intelligent systems. Cooperation is a purposive positive interference of agents to further the achievement of a common goal or goals compatible with their own [2]. In cooperative sensor and actor networks (C-SANETs), cooperation is directed toward achieving intelligent connection between situation awareness and acting through joint gathering and sharing of information. In these systems, there are a set of cooperative artificial and human agents with know-how and know-how-to-cooperate capabilities.

3.1 Know-How Capability

The know-how gives a model of the individual activity of an agent. An agent cannot be guaranteed to succeed with its intentions if it lacks the know-how to achieve them [10]. The goal-directed agents build their know-how by performing four main activities, namely: information gathering/elaboration, diagnosis, intervention decision and action [1]. In [11], Belief-Desire-Intention (BDI) model is used to implement the agent's know-how. A BDI agent has beliefs about itself, other agents and its environment, desires about future states (i.e., goals) and intentions about its own future actions (i.e. plans).

3.2 Know-How-to-Cooperate Capability

The know-how-to-cooperate allows agents to manage interference between their goals, resources, etc., and allows agents to perform their own activities taking into account the activities of the other agents [1]. Each agent relies on its knowledge of other agents and the domain to achieve a high degree of efficiency in reaching both local and global goals. This interaction is characterized by mutual interest and only a partial knowledge of the reasoning steps of the other agents and of the environment. The goals might or might not be known to the agents explicitly, depending on whether or not they are goal-directed. Goal-directed agents may also change their goals to suit the needs of other agents in order to ensure cohesion and coordination. Different models have been proposed to achieve the know-how-to-cooperative such as Joint Responsibility model [12], Joint Intentions [13], TeamLog model [14], SOAR model and Satisfaction-Altruism model [16].

4 Forms of Cooperation in C-SANET

C-SANET incorporates a set of heterogeneous sensing agents, acting agents, situation awareness agents, management agents and decision support/making agents with know-how and know-how-to-cooperate capabilities. Three forms of cooperation can be found between these agents based on the factors that motivate this cooperation, namely, augmentative cooperation, integrative cooperation and debative cooperation [17]. The following subsections describe these forms and give examples for each in the context of distributed surveillance based on our previous work in this area. The details of these examples are not included in this paper due to lack of space.

4.1 Augmentative Cooperation

Augmentative cooperation occurs when agents have a similar know-how, but they must be multiplied to perform a task that is too demanding for only one agent. This task is then shared into similar sub-tasks. Examples of augmentative cooperation in distributed surveillance system include, but are not limited to, search and rescue [18], communication relaying [19], complex task allocation [20] and target cueing and hand-off. In distributed surveillance systems, C-SANET's sensing and acting agents can cooperatively detect and track multiple targets within the VOI. Such target search and tracking operations combine different forms of cooperation. For instance, cooperation can consist in dynamically tasking some sensors to fill the coverage gaps of other sensors, and therefore providing relevant observations in the areas of interest. In order to enable such a continuous spatial/temporal coverage of objects/events using multiple dispersed sensors and platforms, two cooperation mechanisms, namely target cueing and handoff, are often used. Cueing or slaving is the process of using data from sensor S1 to point sensor S2 towards the same target or event. S1 and S2 may be co-localized on-board the same platform P1 (intra-platform cueing), or distributed over two platforms P1 and P2 (inter-platform cueing). Cueing is done in order to alert, or prepare a sensor for the impending arrival of an object/event of interest and thus improve its response time/performance. Handoff occurs when sensor S1 cues sensor S2 and transfers to it the surveillance responsibility. Here also, S1 and S2 may be co-localized on-board the same platform P1 (inter-platform handoff), or distributed over two platforms P1 and P2 (inter-platform handoff). Sensor S2 must then search for the object of interest and verify that it has been acquired. Handoff is often used to ensure a continuity of the tracking process, when a tracked object exits the (spatial/temporal) coverage of one sensor to enter that of another. Both target cueing and handoff can be considered as augmentative forms of cooperation among set of sensors with similar know-hows. In [2], a case study of littoral region surveillance showed the importance of these augmentative cooperative behaviors among surveillance units to improve response time/performance and to ensure a maximum continuity of tracking of critical targets.

4.2 Integrative Cooperation

In integrative cooperation, agents have different complementary know-hows and it is necessary to integrate their contribution for achieving a task. Multimodal detection and tracking [21] and human-assisted tracking are examples of integrative cooperation in distributed surveillance systems. In human-assisted tracking, soft data obtained from a human agent is used to resolve the sensing inconsistency issue of conventional sensors, a problem common to asymmetric warfare situations. This soft data is used as a complementary source of information fused in concert with hard data using random finite set (RFS) theory in order to keep the continuity of target tracking [22]. This scenario represents an integrative form of cooperation where artificial agents (hard sensors) and human agents (soft sensors) have different complementary know-hows and it is necessary to integrate their contribution for target tracking. The results obtained in this work support the notion of RFS theoretic approach to fusion of soft/hard data. Compared to other alternative approaches of dealing with data uncertainty (imperfection), RFS theory appears to provide the highest level of flexibility in dealing with complex data while still operating within the popular and well-studied framework of Bayesian inference. Furthermore, achieving integrative cooperation through incorporation of soft data into fusion process was shown to enhance tracking performance and even allow the linear KEF to partially handle non-linearity due to maneuvering target [22].

4.3 Debative Cooperation

Debative cooperation occurs when agents have a similar know-how and are faced with a unique task, for which they seek the best solution by comparing their results. Multisensor single target cooperative tracking is an example of debative cooperation [23]. Target cueing (subsection 4.1) can also allow for debative cooperation, in situations where the tracking task is accomplished simultaneously by each sensor in order to identify the differences and to choose the sensor with the better performance. The debative form of cooperation during cueing is only possible if there is a minimum overlap among the sensing coverage of the sensors engaged in the cooperative activities. Without such an overlap, only augmentative cooperation is possible. In this case, continuity of tracking is not guaranteed.

5 Conclusion

This paper presented cooperative sensor and actor networks (C-SANETs) and discussed different forms of cooperation between the network agents in the context of distributed surveillance. In C-SANET, several agents with know-how and know-how-to-cooperate attempt, through their interaction, to jointly solve tasks or to maximize utility. The main challenging problem of C-SAENT is how to design low-level behaviors for a given desired collective behavior and how to combine actions of many agents to archive coordinated behavior on the global scale. The link between individual and collective behavior is still a challenging issue to be understood.

References

1. Pacaux-Lemoine, J., Debernard, S.: Common Work Space for Human-Machine Cooperation in Air Traffic Control. Control Engineering Practice 10(5), 571–576 (2002)
2. Benaskeur, A., Khamis, A., Irandoust, H.: Cooperation in Distributed Surveillance Systems for Dense Regions. International Journal of Intelligent Defence Support Systems 4(1), 20–49 (2011)
3. Akyildiz, I., Kasimoglu, I.: Wireless Sensor and Actor Networks. IEEE Communication Magazine 43(9), 23–30 (2004)
4. Dressler, F.: Self-orgnization in Sensor and Actor Networks. Wiley (2007)
5. Khaleghi, B., Khamis, A., Karray, F.: Multisensor Data Fusion: A Review of the State-of-the-art. Information Fusion (2011), doi:10.1016/j.inffus.2011.08.00
6. Hall, D., Llinas, J., Mullen, T., McNeese, M.: A Framework for Dynamic Hard/Soft Fusion. In: Int. Conf. on Information Fusion (2008)
7. Endsley, M.: Toward a Theory of Situation Awareness in Dynamic Systems. Human Factors 37(1), 32–64 (1995)
8. Vuran, M., Pompili, D., Melodia, T.: Future Trends in Wireless Sensor Networks. In: Zheng, J., Jamalipour, A. (eds.) Wireless Sensor Networks: A Networking Perspective. Wiley, New York (2009)
9. Khamis, A., Kamel, M., Slaichs, M.: Cooperation: Concepts and General Typology. In: 2006 IEEE International Conference on Systems, Man, and Cybernetics (2006)
10. Singh, M.: Multiagent Systems: A Theoretical Framework for Intentions, Know-How, and Communications. Springer, Heidelberg (1994)
11. Hilal, A., Khamis, A., Basir, O.: HASM: A Hybrid Architecture for Sensor Management in a Distributed Surveillance Context. In: 2011 IEEE International Conference on Networking, Sensing and Control (2011)
12. Jennings, N.: Towards a Cooperation Knowlege Level for Collaborative Problem Solving. In: Proc. 10th European Conference on Artificial Intelligence, Vienna, Austria, pp. 224–228 (1992)

13. Brazier, F., Jonker, C., Treur, J.: Formalization of a Cooperation Model based on Joint Intentions. In: Jennings, N.R., Wooldridge, M.J., Müller, J.P. (eds.) ECAI-WS 1996 and ATAL 1996. LNCS (LNAI), vol. 1193, pp. 141–155. Springer, Heidelberg (1997)
14. Dunin-Kęplicz, B., Verbrugge, R., Ślizak, M.: Case-Study for TeamLog, a Theory of Teamwork. In: Papadopoulos, G.A., Badica, C. (eds.) IDC 2009. SCI, vol. 237, pp. 87–100. Springer, Heidelberg (2009), doi:10.1007/978-3-642-03214-1:9
15. Lehman, J., Laird, J., Rosenbloom, P.: A Gentle Introduction to Soar: An architecture for Human Cognition. In: Sternberg, S., Scarborough, D. (eds.) Invitation to Cognitive Science. MIT Press (1996)
16. Hilaire, V., Simonin, O., Koukam, A., Ferber, J.: A Formal Approach to Design and Reuse Agent and Multiagent Models. In: Odell, J.J., Giorgini, P., Müller, J.P. (eds.) AOSE 2004. LNCS, vol. 3382, pp. 142–157. Springer, Heidelberg (2005)
17. Schmidt, K.: Analysis of Cooperative Work: A Conceptual Framework. Technical report, Risoe Nat. Lab., Roskilde, Denmark, Risoe Tech. Rep. Risoe-M-2890 (1990)
18. Miao, Y., Khamis, A., Kamel, M.: Applying Anti-Flocking Model in Mobile Surveillance Systems. In: 5th International Conference on Autonomous and Intelligent Systems (AIS 2010), Portugal (2010)
19. Zhu, Z., Khamis, A., Kamel, M.: Applying Emergent Flocking Behaiour to Autoonomous Search adn Rescue. In: Annual Symposium of Systems Design Engineering. University of Waterloo, Waterloo (2009)
20. Khamis, A., Elmogy, A., Karray, F.: Complex Task Allocation in Mobile Surveillance Systems. Journal of Intelligent and Robotic Systems (2011), doi:10.1007/s10846-010-9536-2
21. Rae, A., Khamis, A., Basir, O., Kamel, M.: Particle Filtering for Bearing-Only Audio-Visual Speaker Detection and Tracking. In: International Conference on Signals, Circuits and Systems (SCS 2009), Tunisia (2009)
22. Khaleghi, B., Khamis, A., Karray, F.: Random Finite Set Theoretic Soft/Hard Data Fusion: Application to Target Tracking. In: IEEE 2010 Int. Conf. on Multisensor Fusion and Integration (2010)
23. Elmogy, A., Karray, F., Khamis, A.: Auction-based Consensus Mechanism for Cooperative Tracking in Multisensor Surveillance Systems. Journal of Advanced Computational Intelligence and Intelligent Informatics (JACIIII) 14(1), 13–20 (2010)

Organizing Rescue Agents Using Ad-Hoc Networks

Toru Takahashi, Yasuhiko Kitamura, and Hiroyoshi Miwa

Abstract. When a disaster happens, rescue teams are organized. They firstly search for victims in the disaster area, then share information about the found victims among the members, and finally save them. Disasters often make conventional communication networks unusable, and we employ rescue agents using ad-hoc networks, which enable the agents to directly communicate with other agents in a short distance. A team of rescue agents have to deal with a trade-off issue between wide search activities and information sharing activities among the agents. We propose two organizational strategies for rescue agents using ad-hoc networks. In the Rendezvous Point Strategy, the wide search activities have priority over the information sharing activities. On the other hand, in the Serried Ranks Strategy, the information sharing activities have priority over the wide search activities. We evaluate them through agent-based simulations, comparing to a naïve and unorganized strategy named Random Walk Strategy. We confirm that Random Walk Strategy shows a poor performance because information sharing is difficult. We then reveal the two organizational strategies show better performance than Random Walk Strategy. Furthermore, the Rendezvous Point Strategy saves more victims in the early stages, but gradually the Serried Ranks Strategy outperforms it.

1 Introduction

When a disaster happens, rescue teams are immediately organized to save victims trapped in debris as soon as possible. They firstly search for victims in the disaster area and share information about the found victims. They then gather a necessary number of members required to save the victims. Disasters often damage conventional communication networks such as telephones, cell phones, and even the

Toru Takahashi · Yasuhiko Kitamura · Hiroyoshi Miwa
Kwansei Gakuin University, 2-1 Gakuen, Sanda, Hyogo 669-1337, Japan
e-mail: {toru-takahashi,ykitamura,miwa}@kwansei.ac.jp

J.B. Pérez et al. (Eds.): Highlights on PAAMS, AISC 156, pp. 139–146.
springerlink.com

Internet seriously. Even they are alive, they often become useless because of heavy network congestions. It may be difficult to share information through conventional communication network when a disaster happens. As an alternative way, we employ rescue agents with ad-hoc networks to share information among them. A rescue agent can directly communicate with others in a short distance, and it may be able to communicate with even distant agents by transferring messages through mediator agents like a bucket brigade. If such mediator agents are not located properly among them, it is difficult to make them communicate with each other. Ad-hoc networks, therefore, do not guarantee stable communication because the range of message transmission is limited and the delay and/or disruption of messages occur frequently. We need new organizational schemes to share information among rescue agents using ad-hoc networks.

Rescue operations using ad-hoc networks are more difficult than those using conventional communication networks because it is not easy to share information among rescue agents through ad-hoc networks. This raises a trade-off issue between wide search activities and information sharing activities. If rescue agents disperse widely to search for victims in a disaster area, it is difficult for them to share information because they often are out of communication range. On the other hand, if rescue agents gather narrowly, they can share information easily through ad-hoc networks, but their search area is limited.

We have two strategies to organize rescue agents to deal with this issue. The Rendezvous Point Strategy makes rescue agents repeat to disperse widely to find victims and to get together at the rendezvous point to share information. The Serried Ranks Strategy makes rescue agents disperse only in a short range as far as they can communicate with each other to share information. The rescue agents move to find victims like serried ranks.

In this paper, we evaluate the organizational strategies by using agent based simulations (ABS). ABS is a powerful simulation technique that can represent rescue agents and their interactions [1, 2]. We first simulate a naïve and unorganized strategy named the Random Walk Strategy as the base line of our evaluation, and then evaluate two organizational strategies comparing with the base line.

The rest of this paper is organized as follows. We discuss related works in Section 2. We present our simulation testbed for rescue activities in Section 3. We then evaluate the Random Walk Strategy in Section 4 and the organizational strategies in Section 5. In Section 6, we conclude this paper with our future works.

2 Related Works

We have four important research issues to develop rescue agents with ad-hoc networks: 1) dynamic message routing protocols in ad-hoc networks; 2) saving battery power consumption; 3) strategies for rescue activities using ad-hoc networks; 4) remedies against communication failure.

- **Dynamic Message Routing Protocols:** Since rescue agents move around in the disaster area to save victims, the topology of ad-hoc networks dynamically changes. We must design message routing protocols to cope with it. Johansson et al. [3] discuss a comparison among three protocols; DSDV [4], AODV [5], and DSR [6]. Other dynamic routing protocols for ad-hoc networks have been proposed in [7, 8].

- **Saving Battery Power Consumption:** Since each node in ad-hoc networks is battery driven, power-aware metrics are important to evaluate rescue activities as discussed in [9, 10]. Toh [11] proposed power-aware message routing protocols for a disaster case.
- **Strategies for Rescue Activity:** Rescue strategies to save victims are an important research issue, but there are few works on strategies for rescue activities using ad-hoc networks. Most of models are based on random-based movement [6, 12] of agents or nodes. Aschenbruck et al.[13] presented significant differences of performance between using conventional random-based movement and realistic scenario-based movement in simulations. In this paper, we propose organizational strategies for rescue agents and evaluate them contrasting a random-based strategy.
- **Remedies against Communication Failure:** Since obstacles such as thick concrete shielding often make rescue agents difficult to maintain their ad-hoc communication in real environments, we need remedies to overcome this problem. [14, 15] discuss methods to maintain a communication path between human operators and exploring robots. Ulam[16] discusses a method to recover from communication failures. Since we focus on strategies for information sharing and searching for victims, this paper does not deal with this issue.

3 Simulation Testbed

Our simulation consists of victims that are scattered in a disaster area and rescue agents that search for them. If an agent finds a victim, we assume it needs to gather a necessary number of other rescue agents to save the victim because victims are trapped under debris and more than one agent are required to save a victim. Rescue agents can share information about the victim through ad-hoc networks to gather other agents. The rescue agent can send messages to its neighbors located in the communication range and the neighbors also can pass the messages to their neighbors recursively to share information among rescue agents in the communication range.

We deal with three strategies in this simulation. They are the Random Walk Strategy, the Rendezvous Point Strategy and the Serried Ranks Strategy. The Random Walk Strategy is a non-organizational strategy in which agents move randomly keeping an interval to other agents. The Rendezvous Point Strategy aims to distribute rescue agents widely. It enables them to search for victims in a wide disaster area. We set a rendezvous point where rescue agents return to share their information about victims and to gather other rescue agents. The Serried Ranks Strategy makes rescue agents form serried ranks to keep them share information. They can communicate with each other at any time because all of them are located in the communication range, so it is easy for them to gather other rescue agents to save victims as soon as one of them finds a victim. Since the agents always stay in the communication range, their search activities also are limited in the area. We show the details of simulation below.

The simulator represents the disaster area as a two-dimensional cell model where the distance between two points is defined as the Manhattan distance. Rescue agents and victims are located in the disaster area.

Rescue agent, $A_i(i=1,\cdots,n)$ can stay or move to a neighboring cell in a simulation step following its strategy ($Strategy_i$). All the agents start from the start cell. At every step, the agent can send a message to others in the communication range (r_i) of ad-hoc networks. The message contains the current location of the agent, the location and the rescue cost of victims, and the number necessary of rescue agents if it has found one or more victims.

Victim $R_j(j=1,\cdots,m)$ is located in the disaster area. In order to save a victim, a rescue agent needs to find him/her and to gather a necessary numbers of agents. If a rescue agent and a victim is located in a same cell, the agent can find the victim with probability p. Victim R_j has two parameters; rescue cost c_j and the necessary number of rescue agents t_j where c_j / t_j simulation steps are required to save the victim if the number of rescue agents located at the cell is t_j. Otherwise, they cannot start the saving activity.

3.1 Strategy of Rescue Agents

Random Walk Strategy makes rescue agents move randomly searching for victims. If an agent is located within the distance $interval_i$ from other agents, the agent moves to or stay at a cell where it keeps the longest interval from others.

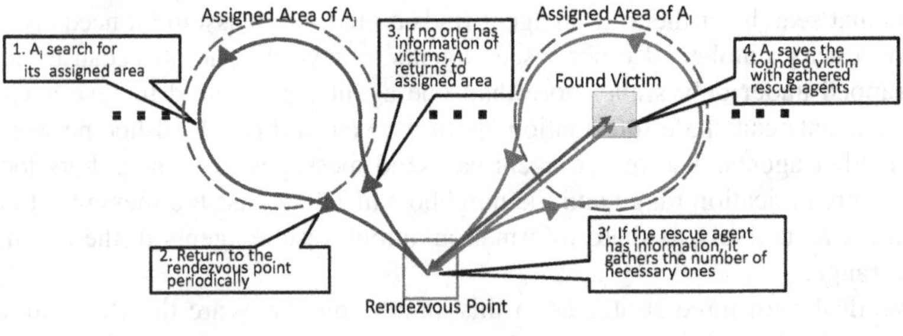

Fig. 1 Rendezvous Point Strategy.

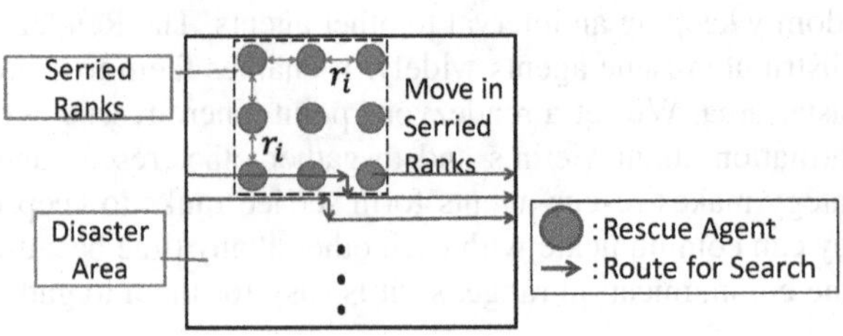

Fig. 2 Serried Ranks Strategy.

As we increase *interval*$_i$ rescue agents are located sparsely. If a rescue agent receives information about victims, it moves to save the nearest victim regardless of the distance to the other agents. When it arrives to a cell with a victim, it waits to save the victim until the necessary number of rescue agents arrive at the cell.

Rendezvous point strategy is an organizational strategy as shown in Figure 1. In this strategy, we assign a search area to each rescue agent without overlapping and set the rendezvous point to share information among agents. Rescue agents repeat to search the assigned areas for victims and to return to the rendezvous point. If an agent returns to the rendezvous point with information about a victim, it waits there to gather the necessary number of other rescue agents. If the agents gather, they move to save the victim.

In the Serried Ranks Strategy, we deploy rescue agents in a rectangle shape, which we call "serried ranks", as shown in Figure 2. They maintain an interval, which is equal to the communication range r_i, among them. They move with keeping the serried ranks to search for victims. The agents move horizontally until the end of the disaster area, then move a step vertically, and move horizontally again.

If a rescue agent finds a victim, it sends the information to rescue agents around it. The necessary number of agent move to save the victim departing from the position in the serried ranks. The rest of agents continue to search for victims keeping their position in the serried ranks. Completing to save the victim, the agents return to the position in the ranks. They can catch up with the serried ranks because the speed of the serried ranks is set to one half of rescue agents.

4 Evaluation of Random Walk Strategy

We evaluate the Random Walk Strategy, especially the performance depending on the distribution of agents defined by *interval*$_i$ and the necessary number of rescue agents. In this simulation, we set the parameters as follows: The number of rescue agents (n) = 20, the number of victims (m) = 20, the probability to find a victim (p) = 0.10, the rescue cost (c_j)=100, and the communication range (r_i)=5. We change *interval*$_i$ from 0 to 5. We set the necessary number of rescue agents to 2 or 5. In the Random Walk Strategy, a rescue agent waits for necessary number of rescue agents at a cell with a victim. When all the rescue agents keep waiting, we define this case as a failure.

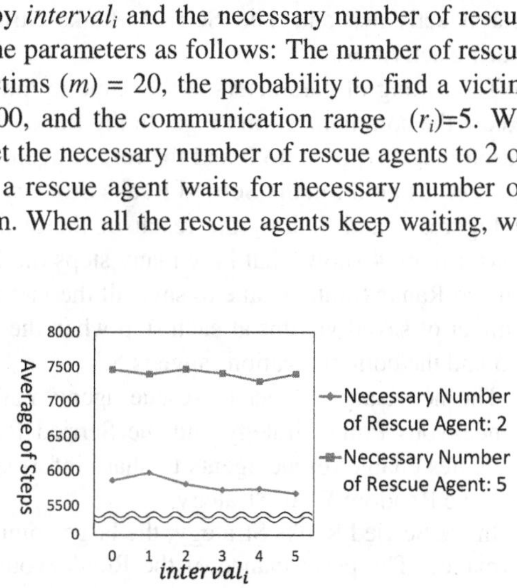

Fig. 3 Steps required to save all the victims without failure using the Random Walk Strategy.

Table 1 Failure rate when the necessary number of rescue agents is 5

$interval_i$	0	1	2	3	4	5
Failure rate(%)	11.1	11.7	11.6	11.5	11.8	15.1

Figure 3 shows the average performance of 1000 simulations, which is simulation steps required to save all the victims without failures. The x-axis represents $interval_i$. Table 1 shows the rate of failures when the necessary number of rescue agents is 5. When the necessary number is 2, we have no failure.

The performance of the Random Walk Strategy is affected by the necessary number of rescue agents. Even if a rescue agent finds a victim, it often cannot gather the necessary number of rescue agents because it is difficult for them to share information about the found victim among the rescue agents through ad-hoc networks.

When the necessary number of rescue agents is 2, the large interval shows a better performance than the small one. When the number is 5, the difference is little but the failure rate increases as the interval does. These results suggest when the necessary number of rescue agents is small, the wide distribution of agents leads to a better performance than narrow distributions. On the other hand when the necessary number of rescue agents is large, the wide distribution of agents leads to a poor performance because it is difficult to gather them through ad-hoc networks.

5 Evaluation of Organizational Strategies

We evaluate two organizational strategies; the Rendezvous Point Strategy and the Serried Ranks Strategy. The Rendezvous Point Strategy emphasizes the wide search activities and The Serried Ranks Strategy emphasizes the information sharing activities.

The setting of parameters is the same to the simulation in the previous section except the communication range. In the Rendezvous Point Strategy, the communication range r_i is set to 0 and the rescue agents can share information only at the rendezvous point. In the Serried Ranks Strategy, we change r_i from 0 to 10.

Figures 4 and 5 show the result of the average performance of 1000 simulations. Figure 4 shows that how many steps the Rendezvous Point Strategy and the Serried Ranks Strategy take to save all the victims. Figure 4 shows the cumulative number of saved victims at each step when the necessary number of rescue agents is 5 and the communication range is 5.

The necessary number of rescue agents makes little difference in both of the Rendezvous Point Strategy and the Serried Ranks Strategy. It means that these strategies enable rescue agents to share information more effectively among them than the Random Walk Strategy.

In the Serried Ranks Strategy, the large communication range improves the performance. The performance of the Rendezvous Point Strategy is close to that of

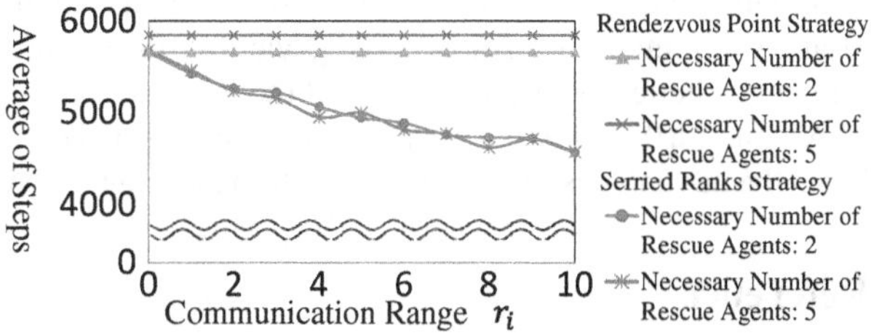

Fig. 4 Steps to save all the victims with the Rendezvous Point and the Serried Ranks Strategy.

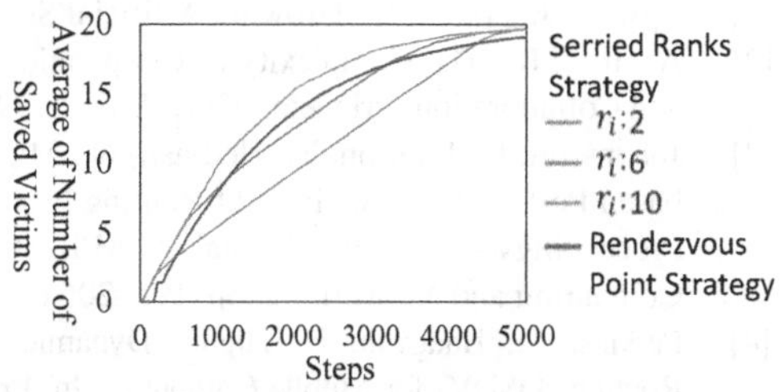

Fig. 5 Cumulative number of saved victims at each step.

the Serried Ranks Strategy when the communication range is 0, but generally speaking, the performance of the Serried Ranks Strategy is better than that of the Rendezvous Point Strategy. In the Serried Ranks Strategy, rescue agents search the narrow area exhaustively and they rarely overlook victims.

Figure 5 shows that the Rendezvous Point Strategy saves more victims than the Serried Ranks Strategy in the middle stage when r_i is less than 6. The Rendezvous Point Strategy can search the disaster area widely in a short time, and the Serried Ranks Strategy searches the area exhaustively but it takes a long time. These results suggest combining two strategies may lead a better performance.

6 Conclusion

We propose organizational strategies for rescue agents using ad-hoc networks in a disaster area. We have a trade-off issue between wide search activities and information sharing activities because the range of ad-hoc communication is limited. If rescue agents disperse widely, they can search a large disaster area, but it is difficult for them to share information, because they often are out of communication range. On the other hand, if rescue agents gather narrowly, they can always share information through ad-hoc networks, but they can search only a narrow area. We propose two organization strategies for rescue and relief operations based on ad-hoc networks. The Rendezvous Point Strategy emphasizes wide search activities and the Serried Ranks Strategy emphasizes information sharing activities.

We first simulate random walking agents and evaluate the effect of the necessary number of rescue agents and the degree of distribution. As a result, we confirm that it is difficult to share information among many agents, especially when they disperse widely.

We then simulate two organizational strategies; the Rendezvous Point Strategy and the Serried Ranks Strategy. As a result, these strategies enable them to share information better than the random walk agents. While the Rendezvous Point Strategy is better in term of saving victims in a short time, the Serried Ranks Strategy is better in term of saving all the victims.

As our further works, we propose a new organizational strategy which combines the advantages of two strategies for rescue agents with ad-hoc networks.

Reference

[1] Epstein, J., Axtell, R.: Growing Artificial Societies. MIT Press (1996)
[2] Axelrod, R.: The Complexity of Cooperation: Agent-Based Models of Competition and Collaboration. Princeton Univ. Press (1997)
[3] Johansson, P., Larsson, T., Hedman, N., Mielczarek, B., Degermark, M.: Scenario-based Performance Analysis of Routing Protocols for Mobile Ad-hoc Networks. In: Proceedings of the 5th Annual ACM/IEEE International Conference on Mobile Computing and Networking, pp. 195–206 (1999)
[4] Perkins, C., Bhagwat, P.: Highly Dynamic Destination-Sequenced Distance-Vector Routing (DSDV) for Mobile Computers. In: Proceedings of the SIGCOM 1994 Conference on Communications Architecture, Protocols and Applications, pp. 234–244 (1994)
[5] Perkins, C.: Ad Hoc On Demand Distance Vector (AODV) Routing. Internet draft, draft-ietf-manetaodv-01.txt (August 1998)
[6] Johnson, D., Maltz, D.: Dynamic source routing in ad hoc wireless networks. In: Imielinski, T., Korth, H. (eds.) Mobile Computing, ch. 5, pp. 153–181. Kluwer Academic Publishers (1996)
[7] Shih, T.-F., Yen, H.-C.: Location-aware Routing Protocol with Dynamic Adaptation of Request Zone for Mobile Ad Hoc Networks. Wireless Networks 14(3) (2008)
[8] Shinoda, K., Noda, I., Ohta, M., Kunifuji, S.: Large-scale Ad-hoc Networking with Social Human Relationship. In: Proc. of IEEE PACRIM 2003, vol. 1, pp. 330–333 (2003)
[9] Singh, S., Woo, M., Raghavendra, C.S.: Power-Aware Routing in Mobile Ad Hoc Networks. In: Proc. ACM/IEEE MOBICOM 1998, pp. 181–190 (1998)
[10] Toh, C.K.: Maximum Battery Life Routing to Support Ubiquitous Mobile Computing in Wireless Ad Hoc Networks. IEEE Communications Magazine 39(6), 138–147 (2001)
[11] Zussman, G., Segall, A.: Energy Efficient Routing in Ad Hoc Disaster Recovery Networks. Ad Hoc Networks 1(4), 405–421 (2003)
[12] Liang, B., Haas, Z.: Predictive Distance-based Mobility Management for PCS Networks. In: Proceedings of INFOCOM 1999 (1999)
[13] Aschenbruck, N., Gerhards-Padilla, E., Martini, P.: Modelling Mobility in Disaster Area Scenarios. Performance Evaluation 66(12), 773–790 (2009)
[14] Sugiyama, H., Tsujioka, T., Murata, M.: Coordination of Rescue Robots for Real-Time Exploration over Disaster Areas. In: Proc. 11th IEEE International Symposium on Object Oriented Real-Time Distributed Computing (ISORC), pp. 170–177 (2008)
[15] Nguyen, H.G., Pezeshkian, N., Gupta, A., Farrington, N.: Maintaining Communication Link for a Robot Operating in a Hazardous Environment. In: 10th Int. Conf. on Robotics and Remote Systems for Hazardous Environments, pp. 28–31 (2004)
[16] Ulam, P., Arkin, R.C.: When good comms go bad: Communications recovery for multi-robot teams. In: Proceedings 2004 IEEE International Conference on Robotics and Automation, pp. 3727–3734 (2004)

Computing Real-Time Dynamic Origin/Destination Matrices from Vehicle-to-Infrastructure Messages Using a Multi-Agent System

Rafael Tornero, Javier Martínez, and Joaquín Castelló

Abstract. Dynamic Origin/Destination matrices are one of the most important parameters for efficient and effective transportation system management. These matrices describe the vehicle flow between different points within a region of interest for a given period of time. Usually, dynamic O/D matrices are estimated from traffic counts provided by induction loop detectors, home interview and/or license plate surveys. Unfortunately, estimation methods take O/D flows as time invariant for a certain number of intervals of time, which cannot be suitable for some traffic applications. However, the advent of information and communication technologies (e.g., vehicle-to-infrastructure dedicated short range communications –V2I) to the transportation system domain has opened new data sources for computing O/D matrices. Taking the advantages of this technology, we propose in this paper a multi-agent system that computes the instantaneous O/D matrix of any road network equipped with V2I technology for every time period and any day in real-time. The implementation was carried out using JADE platform.

1 Introduction

Origin to destination (O/D) matrices are a vital artefact for effective and efficient transportation system safety, operation, design and planning. O/D matrices represent the network user's demands given some network traffic conditions. They contain information about the spatial and temporal distribution of activities between different traffic zones in a determined study area. From a logistic standpoint, long-term average O/D trip demands are needed for transportation design and planning purposes (e.g., future network expansion or urban planning). On the other hand, short-term

Rafael Tornero · Javier Martínez · Joaquín Castelló
Robotics and Information and Communication Technology Institute, Universitat de València, Catedrático José Beltrán 2, 46980 Paterna, Spain
e-mail: {rafael.tornero,javier.martinez-plume}@irtic.uv.es, {joaquin.castello}@irtic.uv.es

J.B. Pérez et al. (Eds.): Highlights on PAAMS, AISC 156, pp. 147–154.
springerlink.com

time-varying O/D demands are important inputs to intelligent transportation systems (ITS) such as advanced traffic information systems (ATIS) and advanced travel management systems (ATMS). For instance, with the information contained in time-varying O/D matrices, it is possible to forecast future traffic conditions and predict congestion so that appropriate control actions (e.g., ramp metering, re-routing) can be determined and effective traffic information can be provided to drivers; thus contribute to improve the safety of the transportation system [11].

O/D trip demands are traditionally obtained from home-interview surveys and/or license plate surveys, which are highly expensive and time consuming. Another economical source of information to infer network O/D demands is automatically recording the traffic counts from induction loop distributed on some links of the road network. Since link traffic counts are measurements of various O/D flows using these links, the information contained in the measured link traffic counts can be used to estimate the unknown O/D demands. The combinination of different sources of information to determine time-dependent O/D demands has been investigated by several authors [16, 11, 15]. Furthermore, the advent of new information and communication technologies (ICT) in the transportation system domain offers new data sources for obtaining short-term time-varying O/D demands [12, 14, 7]. These estimation approaches are classified into two classes: dynamic traffic assignment (DTA)-based approaches [16, 11, 14, 7, 12] and non-DTA-based approaches [15]. In [13], Kattan and Abdulhai formulates the dynamic estimation problem for both families in general terms. The main problem of these methods is the assumption that O/D matrices are constant during some sub-sets of intervals in the period of study, which can be adequate for some traffic applications but not for others, as for instance re-routing. In addition, several important innovations are coming up in the next years; such as the use of new ubiquitous and integral artificial vision based applications or the development of new technologies, allowing direct communication from vehicles to infrastructure (V2I) and among vehicles (V2V) as well [1]. The goal of V2I and V2V integration is to provide a communication link between vehicles on the road (via On-Board Units, OBU) and between vehicles and the roadside infrastructure (via Roadside Units, RSU), in order to increase the safety, efficiency and convenience of the transportation system. These new innovations will generate such a huge amount of data that they will call for distributed processing and storing. For example, let us assume a set of RSUs set up with V2I technology and situated every 500 meters in a 100-kilometres two-way road network with two-lanes in each direction. It results on 200 RSUs. Let us also assume that 1500 vehicles per hour and per lane drive on the network, which results on a total flow of 6000 vehicles per hour. This scenario implies that there are approximately 1.2 millions messages to be processed and/or stored per hour.

Multi-Agent systems have been proposed to solve distributed problems in an extensive number of application domains [10]. In the traffic context, these systems have been applied for solving vehicle routing problems [17, 6] and urban traffic regulation problems [5, 9], among others.

In this paper, we aim at obtaining the dynamic O/D matrix of any road network provided with V2I equipment. To achieve this goal, we propose a multi-agent

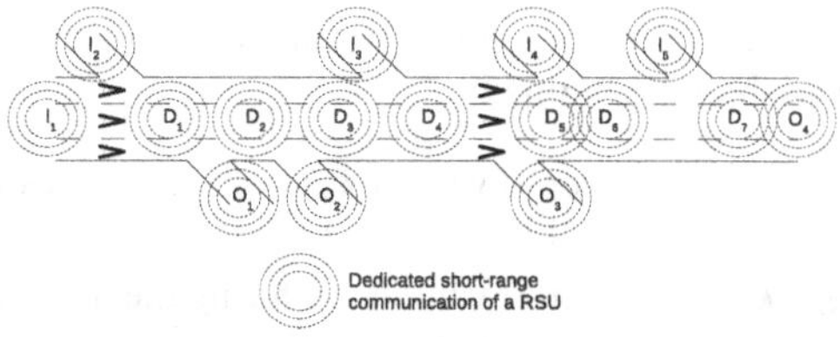

Fig. 1 A theoretical road network

Table 1 Distances (in meters) between road side units

From	I_1	I_2	I_3	I_4	I_5	D_1	D_1	D_2	D_2	D_3	D_4	D_4	D_5	D_6	D_7
To	D_1	D_1	D_4	D_6	D_7	D_2	O_1	D_3	O_2	D_4	D_5	O_3	D_6	D_7	O_4
Distance	500	50	50	50	50	500	500	500	500	500	500	500	500	500	500

system composed of two types of cooperating agents. In the system, agents cooperate hierarchically to compute the instantaneous matrix in a distributed and real-time way for any time period and day. We validate our implementation by means of the SUMO simulator, which is based on a microscopic traffic model simulation and modified for supporting V2I communication technology [8]. The rest of the paper is structured as follows. Section 2 defines the problem of obtaining dynamic O/D matrices. Section 3 describes the multi-agent system implemented to solve the problem. Section 4 shows the results of the evaluation and the discussion. Finally, Section 5 presents the conclusions and put forward the future work.

2 Problem Definition

In this section we define the problem of obtaining the dynamic O/D matrix among a set of different traffic zones on any road network equipped with V2I technology. We focus on the real-time or on-line computation problem, since the on-line or real-time problem involves the O/D computation with regard to real-time traffic management systems.

Network of Study

We compute the dynamic O/D matrix for the road network in Figure 1. This network represents a segment of a theoretical freeway that is composed of 5 entry points and 4 exit points. Table 1 shows the distance between consecutive RSUs. RSU detectors are grouped into three groups: input, exit and intermediate detectors, which are referred to as I, O and D, respectively in Figure 1. The mainline and the on/off ramps consist of three and one lanes, respectively. The vehicles flow from left to right in an one-way unique direction.

Problem Formulation

Time-dependent O/D matrices are obtained by using the data provided by all of the vehicles along a given period divided in a set of intervals of interest. Then an O/D matrix is obtained for each interval. To state the problem formally, the following variables are defined:

$q_i(k)$ number of vehicles entering the network of study using entry I_i during time interval k

$g_{i,j}(k)$ number of vehicles entering the network of study using entry I_i during time interval k that are destined to output O_j

$b_{i,j}(k)$ proportion of vehicles entering on entry I_i during time interval k that are destined to O_j

The cells of an O/D matrix can be specified directly as $g_{i,j}(k)$ or $b_{i,j}(k)$ among others. In this work, we focus on proportions; thus, each cell of the O/D proportion matrix is calculated according to equation 1

$$b_{i,j}(k) = g_{i,j}(k)/q_i(k) \tag{1}$$

3 Multi-agent System

Next, we present a hierarchic cooperative multi-agent system solution for solving the distributed problem of obtaining dynamic O/D trip demands for any road network equipped with V2I technology. Figure 2 shows a diagram of the solution. As observed in Figure 2, our solution is composed of three types of agents mainly, namely *RSUAgent*, *ODMatrixAgent* and *BrokerAgent*.

RSUAgents are always ubicated at RSU devices. The agents of this type communicate with agents of the *ODMatrixAgent* type by means of the request protocol. The cardinality of the relationship is one-to-many, i.e., each *RSUAgent* only communicates with one agent of the *ODMatrixAgent* type. The unique goal of *RSUAgents* is to send the V2I messages received at RSUs to the agent they communicate with. Every V2I message exchanged between two agents is composed of a vehicle identification tag (which is unique for every vehicle) and the timestamp in which the vehicle went through the RSU device.

ODMatrixAgents are usually ubicated at near road equipment distinct of the RSU devices. The main goal of these agents is to compute the partial dynamic O/D matrix. For that purpose, they receive V2I messages from *RSUAgents* or other *ODMatrixAgents*, depending on the level of the hierarchy the agent belongs to. With the data received, these agents fill a data journey structure (a hash table that maps vehicle identification tags with the RSUs the vehicle has gone through) that contains the partial journey carried out by each vehicle so far. This structure is used to update $q_i(k)$, $g_{i,j}(k)$ and $b_{i,j}(k)$ appropriately.

BrokerAgent. There only exists one agent of this type in the system. This agent is just in charge of getting the distributed partial dynamic O/D matrix that each *ODMatrixAgent* owns. It can act as a wrapper for storing the O/D matrix in a database or as a broker for an ITS system that requires this matrix as an input to carry out its normal operation.

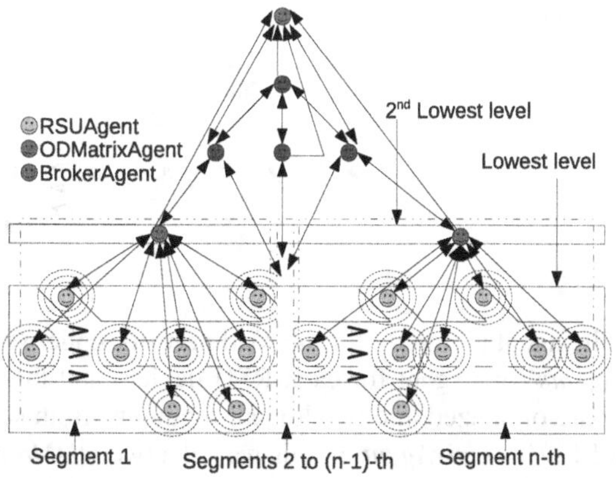

Fig. 2 General agent hierarchy for O/D matrix computation

Description of Agent Behaviour

***RSUAgent*:** In order to convert the raw data received at RSU devices into valuable information (O/D flow or proportion), this type of agent request the corresponding *ODMatrixAgent* agent to compute the O/D matrix every minute (which is the common time in which traffic events are communicated in current ITS systems). For that purpose, the *RSUAgent* agent also sends the new data that the RSU device attached to the agent has received from the vehicles that went through that RSU between two consecutive requests.

***ODMatrixAgent*:** When this type of agent receives the request for computing the O/D matrix, it first try to complete the path for some of the vehicles stored in the agent data journey structure using the V2I messages received with the request. Then, the agent updates $q_i(k)$ when it finds V2I messages received from an *input* RSU. The agent also updates $g_{i,j}(k)$ and $b_{i,j}(k)$ when it finds an ended journey (i.e., journeys initiated by a vehicle at *input* RSU i and ended at *output* RSU j). Once the journey for a vehicle has been processed (i.e., $g_{i,j}(k)$ and $b_{i,j}(k)$ have been updated for that journey), it is removed from the data journey structure of the agent. For the rest of journeys, a negotiation procedure is initiated for determining those journeys that cannot be processed at the level of the hierarchy this agent belongs to and, thus, has to be processed on upper levels, by means of other *ODMAtrixAgent* agents. The data corresponding to those journeys are packed in a message that is sent with the request for computing the O/D matrix to the corresponding *ODMAtrixAgent* agent.

4 Results

In order to validate the multi-agent system proposed, we have developed a prototype using the Java Agent DEvelopment Framework (JADE). JADE provides a middleware which enables the developing and executing a peer-to-peer application based on agent paradigm [2].

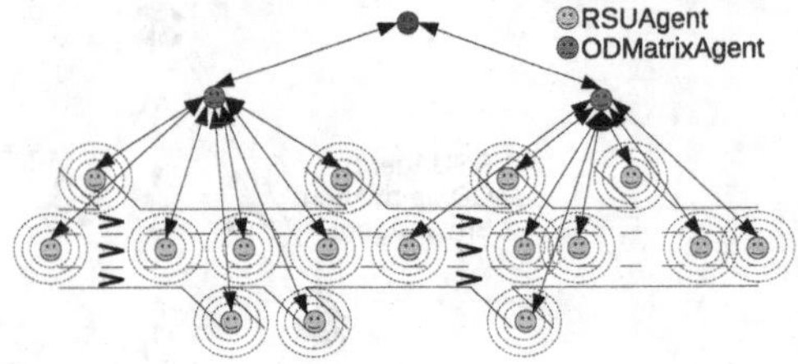

Fig. 3 Agent hierarchy for the theoretical road network

Figure 3 shows the agent hierarchy for the road network under study. As shown, there are 16 *RSUAgents* (one agent for each RSU device) and three *ODMatrixAgents*. These agents are organized into two logical segments. Both segments consist of 8 *RSUAgents* and 1 *ODMatrixAgent*. In this way, the two *ODMatrixAgents* compute the partial dynamic O/D matrix for each segment; meanwhile, the root *ODMatrixAgent* computes the O/D flows for the journeys initiated on one segment and ended on the other one. The agents communicate one time per minute.

SUMO simulator [8] has been used to simulate the road network under study. This simulator has been modified attaching an embedded database to each RSU so as to emulate the real behaviour of V2I technology. Each V2I message only consists of a vehicle identification but, when the message is stored in the database, a timestamp is also added. This allow us to sort these messages and recover the path that each vehicle followed on the road. We have verified and validated the system creating uncongested and congested traffic scenarios. Both scenarios consist of different O/D flows depending on the time of the day, generating more traffic in rush hours.

We have verified and validated the system creating uncongested and congested traffic scenarios with SUMO simulator. Both scenarios consist of different O/D flows depending on the time of the day, generating more traffic in rush hours. However, due to space limitations, we only show here the results for the congested traffic scenario. Nevertheless, the results obtained for uncongested traffic conditions are similar to the ones presented here. Figure 4(a) shows the O/D proportions obtained by SUMO for pairs $I_1 - O_1$, $I_1 - O_2$ and $I_3 - O_3$. Concretely, this figure shows on the X-axis some intervals of 1 hour and on the Y-axis the proportions of vehicles that initiate the journey at I_1 and end the journey at O_1 or O_2 in each interval. Also, the figure shows the proportions of vehicles that initiate a journey at I_3 and end the journey at O_3.

Figures 4(b) and 4(c) show the O/D matrix obtained by the multi-agent system implementation for the intervals 8-9 and 11-12, respectively. These figures show on the X-axis the origin of the journeys, on the Z-axis the destination of the journeys and on the Y-axis the proportions of vehicles that initiate the journey at origin I_i and end the journey at O_j. As it can be seen in these figures, the OD patterns for pairs $I_1 - O_1$, $I_1 - O_2$ and $I_3 - O_3$ in interval 8-9 are 12%, 19% and 60%, respectively. Looking at the interval 11-12, the OD patterns for the same pairs $I_1 - O_1$, $I_1 - O_2$ and $I_3 - O_3$ are 7%, 43% and 75%, respectively. Therefore, these figures show us that the multi-agent system is able to obtain the dynamic O/D matrix exactly and in real-time.

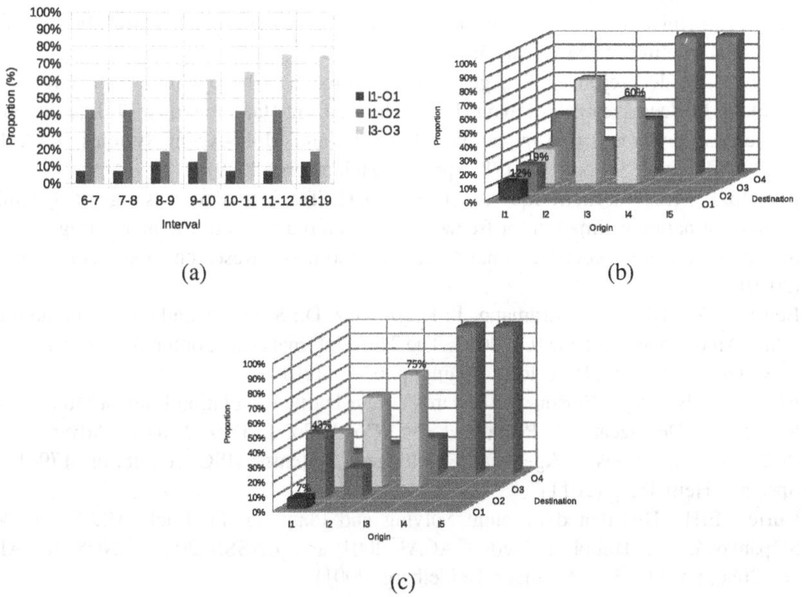

Fig. 4 (a) O/D flows for pairs $I_1 - O_1$, $I_1 - O_2$ and $I_3 - O_3$ (b) instantaneous O/D matrix for interval 8-9 and (c) instantaneous O/D matrix for interval 11-12

5 Conclusions and Future Work

In this paper we have proposed a multi-agent system for obtaining dynamic O/D matrices on any road network equipped with V2I technology. Also, we have shown that when 100% of vehicles are equipped with the adequate technology, the proposed method obtains the exact dynamic O/D matrix for any time period and any day.

As future work, we want to extend the results presented in this paper. Specifically, we are planning to study the penetration rate, transmission errors and RSU/OBU failures with the accuracy of the O/D matrices obtained. We will also present the extension of the multi-agent system so as to obtain the dynamic O/D matrix of a general network of roads, where journeys can be initiated and ended on different roads.

Acknowledgements. This work is part of the INTELVIA project, which is supported by the AVANZA I+D program of the Spanish MITYC, under grant TSI-020302-2009-90.

References

1. Car-to-car communication consortium, http://www.car-to-car.org/
2. Jade: Java agent development framework, http://jade.tilab.com/
3. Ntp: Network time protocol, http://www.ntp.org/

4. Highway capacity manual (2010)
5. Balaji, P., Srinivasan, D.: Multi-agent system in urban traffic signal control. IEEE Computational Intelligence Magazine 5(4), 43–51 (2010)
6. Barbucha, D., Jędrzejowicz, P.: Agent-Based Approach to the Dynamic Vehicle Routing Problem. In: Demazeau, Y., Pavón, J., Corchado, J.M., Bajo, J. (eds.) 7th International Conference on Practical Applications of Agents and Multi-Agent Systems (PAAMS 2009). AISC, vol. 55, pp. 169–178. Springer, Heidelberg (2009)
7. Barceló, J., Montero, L., Marqués, L., Carmona, C.: Travel time forecasting and dynamic origin-destination estimation for freeways based on bluetooth traffic monitoring. Transportation Research Record: Journal of the Transportation Research Board 2175, 19–27 (2010)
8. Behrisch, M., Bieker, L., Erdmann, J., Krajzewicz, D.: Sumo - simulation of urban mobility: An overview. In: SIMUL 2011, The Third International Conference on Advances in System Simulation, Barcelona, Spain, pp. 63–68 (2011)
9. Bhouri, N., Balbo, F., Pinson, S.: Towards Urban Traffic Regulation Using a Multi-agent System. In: Demazeau, Y., Pechoucek, M., Corchado, J., Prez, J. (eds.) Advances on Practical Applications of Agents and Multiagent Systems. AISC, vol. 88, pp. 179–188. Springer, Heidelberg (2011)
10. Durfee, E.H.: Distributed Problem Solving and Planning. In: Luck, M., Mařík, V., Štěpánková, O., Trappl, R. (eds.) ACAI 2001 and EASSS 2001. LNCS (LNAI), vol. 2086, pp. 118–149. Springer, Heidelberg (2001)
11. Hu, S.R., Madanat, S.M., Krogmeier, J.V., Peeta, S.: Estimation of dynamic assignment matrices and od demands using adaptive kalman filtering. ITS Journal - Intelligent Transportation Systems Journal 6(3), 281–300 (2001), doi:10.1080/10248070108903696
12. Jian, S., Yu, F.: A Novel OD Estimation Method Based on Automatic Vehicle Identification Data. In: Chen, R. (ed.) ICICIS 2011 Part II. CCIS, vol. 135, pp. 461–470. Springer, Heidelberg (2011)
13. Kutz, M. (ed.): Handbook of Transportation Engineering, vol. II. McGraw-Hill (2011)
14. Kwon, J., Varaiya, P.: Real-time estimation of origin-destination matrices with partial trajectories from electronic toll collection tag data. Transportation Research Record: Journal of the Transportation Research Board 1923, 119–126 (2005)
15. Lin, P.W., Chang, G.L.: A generalized model and solution algorithm for estimation of the dynamic freeway origin destination matrix. Transportation Research Part B: Methodological 41(5), 554–572 (2007)
16. Sherali, H.D., Park, T.: Estimation of dynamic origin-destination trip tables for a general network. Transportation Research Part B: Methodological 35(3), 217–235 (2001)
17. Zeddini, B., Yassine, A., Temani, M., Ghedira, K.: An agent-oriented approach for the dynamic vehicle routing problem. In: International Workshop on Advanced Information Systems for Enterprises, IWAISE 2008, pp. 70–76 (2008)

Modelling Cryptographic Keys in Dynamic Epistemic Logic with DEMO

Hans van Ditmarsch, Jan van Eijck, Ignacio Hernández-Antón, Floor Sietsma, Sunil Simon, and Fernando Soler-Toscano

Abstract. It is far from obvious to find logical counterparts to crytographic protocol primitives. In logic, a common assumption is that agents are perfectly rational and have no computational limitations. This creates a dilemma. If one *merely* abstracts from computational aspects, protocols become trivial and the difference between tractable and intractable computation, surely an essential feature of protocols, disappears. This way, the protocol gets lost. On the other hand, if one '*merely*' (scare quotes indeed) models agents with computational limitations (or otherwise bounded rationality), very obvious aspects of reasoning become problematic. That way, the logic gets lost. We present a novel way out of this dilemma. We propose an abstract logical architecture wherein public and private, or symmetric keys, and their roles in crytographic protocols, all have formal counterparts. Instead of having encryption and decryption done by a principal, the agent sending or receiving messages, we introduce additional, virtual, agents to model that, so that one-way-function aspects of computation can be modelled as constraints on the communication between principals and these virtual counterparts. In this modelling it does not affect essential protocol features if agents are computationally unlimited. We have implemented the proposal in a dynamic epistemic model checker called DEMO.

1 Introduction

The security of many protocols depends on the computational limitations of the agents involved and on the intractability of inverses for quite tractable computations. *A* standard example is public/private key encryption, e.g., the security of the RSA protocol [6] is (also) based on the complexity of integer factorization. It is easy to multiply primes. It is hard to factorize a number into its prime constituents.

Hans van Ditmarsch · Ignacio Hernández-Antón · Fernando Soler-Toscano
University of Sevilla, Spain
e-mail: {hvd, iha, fsoler}@us.es

Jan van Eijck · Floor Sietsma · Sunil Simon
CWI, Amsterdam, Netherlands
e-mail: {Jan.van.Eijck, S.E.Simon, F.Sietsma}@cwi.nl

J.B. Pérez et al. (Eds.): Highlights on PAAMS, AISC 156, pp. 155–162.
springerlink.com

A goal in information theoretic security is to find abstract notions for cryptographic primitives such as keys and one-way-functions, and there have been several proposals for logical modelling of such primitives [5, 4, 1]. In such approaches it is problematic that logical agents know all logical consequences of their information. If they know two primes, they know their product; and if they know a number, they know if it is the product of primes and what these primes are. There are no one-way-functions in logic.

We present a novel proposal to tackle this problem in the setting of dynamic epistemic logic [2], where the knowledge of the agents involved in protocol execution, including higher-order features (what agents know about each other), is represented in relational structures called multi-agent Kripke models, and change of knowledge is represented by various structural transformations. In this setting agents are also computationally unlimited. However, by means of introducing virtual coding and virtual decoding agents we can simulate computational bounds as communicative restrictions between the authentic agents participating in protocols (the principals) and these virtual agents.

Given a sender A (Alice) and receiver B (Bob), we introduce a coding agent C (Coder) and a decoding agent D (Decoder). The eavesdropper E (Eve) listens in to public communications. For the specific setting of asymmetric public/private key encryption, every agent can communicate in a certain way with the coding agent, who represents the public key, but only one, B, will be able to communicate with the decoding agent, who represents the private key. We can think of the coding and decoding agents as follows. We split each agent into two parts, its *knowledge base*, where we will evaluate the agent's knowledge, and its *computational resources* (cryptographic functions). The first is considered the principal in protocol execution and the latter the virtual agent counterpart. Both can be assumed to be computationally unlimited (this is not a requirement but the assumption to the contrary is what makes *logical* modelling of security so hard and counterintuitive). The coding is performed by C and not by A; the decoding is performed by D and not by B (and both in a way to which no other agent is privy). What the non-(de)coding agents get to know about the outcome of that process is determined by communication, not by computation. On the other hand, the coding and decoding agent only perform that virtual role. They are not privy to the standard communicative aspects of the protocol.

Next, we present the protocol, illustrated by a two-bit secure message passing. Then, an adaptation to symmetric key encryption, and an application, RSA.

2 Public/Private-Key Encryption with Coding and Decoding Agents

- A: Alice (sender)
- B: Bob (receiver)
- C: Coding agent (Bob's public key)
- D: Decoding agent (Bob's private key)
- E: Eve (eavesdropper)

We distinguish five agents. The coding and decoding agents are the *cryptographic agents* and can occur in various protocols. In public-private key encryption the coding agent C stands for Bob's public key and the decoding agent D represents Bob's private key. The agents A, B and E are the *principals* of the protocol (where E may not always be that passive). Principals are proactive agents while cryptographic agents C and D are reactive. We model the knowledge and ignorance of all these agents, represented as uncertainty between the valuations of variables (bits).

The protocol consists of an initialization phase and an execution phase. Information is a bitstring and uncertainty about information is represented as alternative bitstrings. A message can be an *individual message* (to one agent, a.k.a. private – but that would be confusing in our setting with private keys), a *group message* (to more than one and less than five agents), and a *public message* (to all five agents, a.k.a. a public announcement). In the initialization phase background knowledge is incorporated, e.g., that the encoding agent C knows the encryption function. The execution phase consists of the encryption and decryption; its interest to the reader consists of its communicative aspects and properties. As a running example Alice will communicate two bits to Bob.

Table 1 Correspondence between clear and encoded messages

Clear		Encoded	
Number	**Formula**	**Number**	**Formula**
0	$\neg p_1 \wedge \neg p_0$	1	$\neg q_1 \wedge q_0$
1	$\neg p_1 \wedge p_0$	3	$q_1 \wedge q_0$
2	$p_1 \wedge \neg p_0$	0	$\neg q_1 \wedge \neg q_0$
3	$p_1 \wedge p_0$	2	$q_1 \wedge \neg q_0$

With two bits there are four possible secrets. They are shown in Table 1, and also the encoded message that corresponds to each secret. We use p_i to represent *clear messages* and q_i for *encoded messages*. We also use r_i atoms for clear messages. So for example message number 2 is not just $p_1 \wedge \neg p_0$ but $p_1 \wedge \neg p_0 \wedge r_1 \wedge \neg r_0$. We duplicate each p_i with r_i. This is a technical trick to make that the cryptographic agents C and D have different information from the other agents.

In the initialization phase of the protocol (a preprocessing of the relational model to represent the knowledge at the start of protocol execution), A is given a secret value of p_0 and p_1, say 2; and principals A, B and E will be made aware that, for each $i \in \{0,1\}$, p_i and r_i are equivalent (or, in general, that each p_i is either equivalent to r_i or to $\neg r_i$), but not C and D.

What C and D know is shown in Figure 1. Agent C knows the encoding that corresponds to each clear message. Agent D has the information to decode every encoded message. The trick of using both p_i and r_i is that the information of C and D is different, as required for a one-way function. Principals A, B and E will receive a *group message* that $p_i \leftrightarrow r_i$ for $i \in \{0,1\}$. But C and D don't know the correspondence between these two sets of atoms. We can now make a difference between public and private keys.

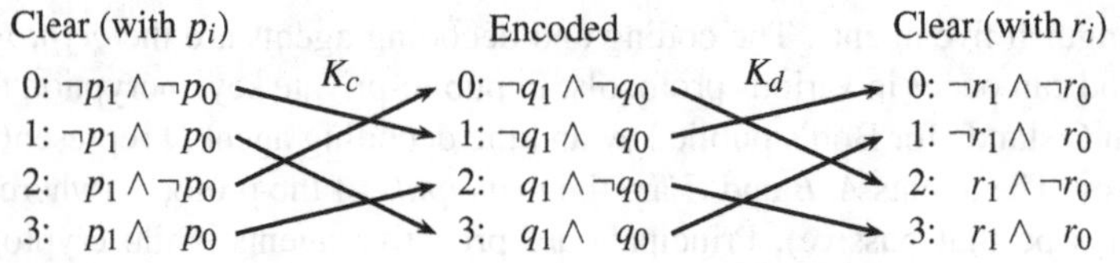

Fig. 1 Encryption and decryption of two bits of information

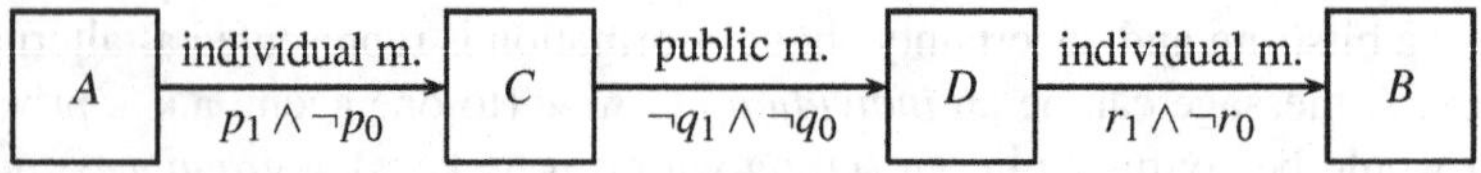

Fig. 2 Alice communicates Bob the number 2

We now get to the execution phase of the protocol. Agent C is Bob's public key. Any of the principals A, B or E can send it a private message consisting of values for p_0 and p_1. Consequently, encoding agent C, who is the only agent knowing the correspondence of these values to values for q_0 and q_1, makes a public announcement of these values (a *public message*). Agent D is Bob's private key. It reacts when a public announcement has been made about q_0 and q_1 by way of informing *only* B (and not just anyone — this is the *private* key part) with a private message about r_0 and r_1. The workflow for A to communicate B the secret 2 is shown in Figure 2.

3 DEMO Implementation

The protocol is implemented in the model checker DEMO [3]. We have employed DEMO Light[1]. First, we declare a module with imported libraries (omitted). As we are going to encrypt two bits, there are four possible secrets. We define a general function `atom` that returns the representation of a given number (in `[0..3]`) with a given propositional letter (`P`, or `Q`). It is the same representation as in Table 1. We also define the logical operator of equivalence `equiv`.

```
atom :: (Num t, Num t1) => (t1 -> Prp) -> t -> Form
atom p 0 = Conj[(Neg (Prp (p 1))), (Neg (Prp (p 0)))]
atom p 1 = Conj[(Neg (Prp (p 1))), (Prp (p 0))]
atom p 2 = Conj[(Prp (p 1)), (Neg (Prp (p 0)))]
atom p 3 = Conj[(Prp (p 1)), (Prp (p 0))]
equiv :: Form -> Form -> Form
equiv a b = Conj[(impl a b), (impl b a)]
```

- **Step 0: Initialization**

We explain the code: `relation_p_r` contains the relation between the `Ps` and `Rs` that is given to the principals A, B and E. The list `secret_pairs` contains all

[1] See `http://homepages.cwi.nl/~jve/software/demolight0/`

pairs with a secret and its encoding, as in Table 1. The `coding_formula` contains the information that is given to the coding agent, and `decoding_formula` to the decoding one, as in Figure 1.

```
relation_p_r :: Form
relation_p_r =
  Conj[(equiv (atom P i) (atom R i)) | i<-[0..3] ]
secret_pairs :: [(Integer, Integer)]
secret_pairs  = [(0,1),(1,3),(2,0),(3,2)]
coding_formula :: Form
coding_formula =
  Conj[(impl (atom P (fst i)) (atom Q (snd i))) |
       i<-secret_pairs ]
decoding_formula :: Form
decoding_formula =
  Conj[(impl (atom Q (snd i)) (atom R (fst i))) |
       i<-secret_pairs ]
```

The model of so-called blissful ignorance (common knowledge of ignorance of the values of all variables) is updated by giving each agent the proper information.

```
e1 :: EpistM Integer
e1 = initM [a,b,c,d,e] [P 0, P 1, Q 0, Q 1, R 0, R 1]
iniM :: EpistM Integer
iniM = upds e1
 [ -- Principals are informed about pi<->ri
   groupM [a,b,e] relation_p_r,
   -- agent C is informed about the coding function
   message c coding_formula,
   -- agent D is informed about the decoding function
   message d decoding_formula]
```

Next, executing the protocol. Alice sends Bob the message number 2, by encrypting it with Bob's public key. As Figure 1 shows, the encoding of 2 is 0. The following formulas represent the chosen secret `sec`, the versions with only `Ps` and `Rs` and the encrypted message `secE`.

```
sec  = Conj[secP, secR]  -- secret to be communicated
secP = atom P 2          -- secret with Ps
secR = atom R 2          -- secret with Rs
secE = atom Q 0          -- encoded secret
```

To demonstrate the execution of the protocol, we show the updates in the epistemic model and the relevant formulas to be checked. Note that we do not announce formulas with the knowlege operators K. Announcing (in a public or individual message) $K_a\alpha$ gives more information than announcing just α, it also authenticates the information α as coming from agent a. Although authentication is crucial in cryptography, here we just send non-authenticated messages and leave authentication for future work.

- **Step 1: Alice chooses the secret message**

We create the `s1` model by sending Alice the secret information. This way we represent the action of choosing number 2. We check that after this action Alice knows the secret information.

```
s1 :: EpistM Integer
s1 = upd iniM (message a sec)
c1 = isTrue s1 (K a sec)
```

- **Step 2: Alice calls the coding function**

Alice sends the secret message to the coding function in order to encrypt it. We check that the coding agent knows the encoding message.

```
s2 :: EpistM Integer
s2 = upd s1 (message c secP)
c2 = isTrue s2 (K c secE)
```

- **Step 3: The encoding of the message is publicly announced**

The coding agent announces the encrypted message. We check that the decoding agent knows the `R`-part of the secret.

```
s3 :: EpistM Integer
s3 = upd s2 (public secE)
c3 = isTrue s3 (K d secR)
```

- **Step 4: Bob calls the decoding function to learn the secret**

We represent this step with a private message to Bob with the `R`-part of the secret. We have three conditions to check:

1. Bob knows the secret.
2. The ignorance of eavesdropper E is common knowledge to the principals.
3. The secret is not common knowledge to A and B.

```
s4 :: EpistM Integer
s4 = upd s3 (message b secR)
c4 = maybe_And[c41,c42,c43]
c41 = isTrue s4 (K b sec)
c42 = isTrue s4 (CK [a,b,e] (Neg (K e sec)))
c43 = isTrue s4 (Neg (CK [a,b] sec))
```

Given uncertainty about communication channels, it is usual in cryptographic protocols that common knowledge is never obtained. For example, agent A doesn't know that B knows the secret, as A is not sure that B has the key and has used it to decode the message.

4 Simplifications and an Application: RSA Encryption

We have modelled public/private key encryption with a coding and decoding agent. That way we simulated one-way functions. In that sort of encryption there is an owner of the key pair, B, who has access to both keys. That means that we can do away with agent D and give B the information of the key, modelled as agent C. This then also removes the need to duplicate `P` atoms as `R` atoms. We can then just use `P` atoms for secrets and `Q` atoms for encoded messages. If we remove both cryptographic agents and give the same key to A and B, we can model symmetric encryption.

As an application we now sketch how to model (for a simple numerical example) the implementation of RSA encryption in dynamic epistemic logic, with DEMO, using the simplified approach with only a coding agent C, as above.

Choose primes 3 and 11, so the modulo of the keys is $n = 3 \cdot 11 = 33$ and $\varphi(n) = (3-1)(11-1) = 20$. Possible key pairs are given by pairs (i, j) of integers in $[2, \varphi(n) - 1]$ such that $ij \mathsf{mod} \varphi(n) = 1$. We choose the pair $(13, 17)$. Then the public key (i, n) is $(13, 33)$, and the private (j, n) is $(17, 33)$. With a public key (i, n), the encoding of a message m, for $0 \leq m < n$, is given by $m^i \mathsf{mod} n$. It is decoded by $(m^i)^j \mathsf{mod} n = m$. For example, with our public key $(13, 33)$ we encode the message 15 as $15^{13} \mathsf{mod} 33 = 9$. It can be decoded with the private key $(17, 33)$: $9^{17} \mathsf{mod} 33 = 15$.

We now show part of the DEMO code. The list `secret_pairs` contains all possible secret messages and their encoded versions.

```
secret_pairs :: [(Int, Int)]
secret_pairs =  [(i, i^13 `mod` 33) | i<-[0..32]]
```

Clear messages are represented by p_m atoms and encoded messages by q_m, for $0 \leq m < n$. We are not using the binary representation as in the previous section.

The ignorance model `e1` contains all possible combinations of the Ps and Qs, now just one `P` and one `Q` in each state. It is updated with the same action as in the previous section, to obtain `iniM`.

```
e1 :: EpistM Integer
e1 =
  let stats =  -- Possible pairs of Ps and Qs
   zip [0..] [ (i, j) | i<-[0..32], j<-[0..32]] in Mo
  [0..((33^2)-1)]   -- States
  [a,b,c,e]         -- Agents
  ([P i | i<-[0..32]]++[Q i | i<-[0..32]]) -- Props.
  [(w,[P (fromIntegral a0), Q (fromIntegral a1)])|
   (w, (a0, a1)) <- stats] -- Content of each state
  [(x,i,j) | x<-[a,b,c,e], -- Accessibility relations
   i<-[0..((33^2)-1)], j<-[0..((33^2)-1)]]
  [0..((33^2)-1)]   -- All states are initially pointed
iniM :: EpistM Integer
iniM = upds e1
 [message c coding_formula,message b decoding_formula]
```

We then execute and check the protocol. Alice sends Bob as message the number 15, encrypting it with Bob's public key. The encryption of 15 is 9. The chosen secret `sec` and the encrypted version `secE` are therefore:

```
sec  = Prop (P 15) -- secret (with P)
secE = Prop (Q  9) -- encoded secret (with Q)
```

The same security conditions as in the previous section are then checked.

5 Conclusions

As counterparts of a sender and a receiver we proposed a virtual coding and decoding agent, to simulate encryption and decryption. Computational restrictions (one-way-functions) can be modelled as constraints on the communication between the principals and these virtual counterparts. This agent-based architecture allows us to verify not only knowledge properties of the principals, but also their ignorance. These are surprising but highly desirable results in a logical setting, where all agents are computationally unlimited.

When using our approach to check protocols with large keys we will have to face the problem of representing epistemic models without the need of creating one state for each possible pair of coding/decoding function, as our DEMO implementation (which uses Kripke models) requires. It will increase the efficiency of our approach.[2]

References

1. Dechesne, F., Wang, Y.: To know or not to know: epistemic approaches to security protocol verification. Synthese 177, 51–76 (2010)
2. van Ditmarsch, H., van der Hoek, W., Kooi, B.: Dynamic Epistemic Logic. Synthese Library, vol. 337. Springer, Heidelberg (2007)
3. van Eijck, J.: DEMO — a demo of epistemic modelling. In: van Benthem, J., Gabbay, D., Löwe, B. (eds.) Interactive Logic — Proceedings of the 7th Augustus de Morgan Workshop. Texts in Logic and Games, vol. 1, pp. 305–363. Amsterdam University Press (2007)
4. Halpern, J.Y., Pucella, R.: Modeling Adversaries in a Logic for Security Protocol Analysis. In: Abdallah, A.E., Ryan, P.Y.A., Schneider, S. (eds.) FASec 2002. LNCS, vol. 2629, pp. 115–132. Springer, Heidelberg (2003)
5. Ramanujam, R., Suresh, S.P.: Information based reasoning about security protocols. Electr. Notes Theor. Comput. Sci. 55(1) (2001)
6. Rivest, R.L., Shamir, A., Adleman, L.: A method for obtaining digital signatures and public-key cryptosystems. Commun. ACM 21(2), 120–126 (1978)

[2] We thank the PAAMS reviewers for their comments.

Assessing Multi-agent Simulations – Inspiration through Application

Catherine Cleophas

Abstract. The application of multi-agent simulations in practical decision support and training gains relevance as technological advances improve computational performance, user interfaces and visualizations. This paper describes the life cycle of one such application, the airline revenue management simulation system REMATE. It highlights the way in which issues of verification, validation and acceptance were treated when implementing and applying REMATE. Feedback loops linking the system, practitioners and researchers are illustrated. Challenges with regard to the required balance of parsimony and realism required from the underlying model are summarized and critically assessed. The paper suggests that through the diverging or additional requirements of practical application, challenges and opportunities for further research in the field of multi-agent simulations arise.

1 Introduction

This paper highlights some opportunities to employ experiences collected through the practical application of multi-agent simulations as a source for research inspiration. For this purpose, it considers the case of a multi-agent simulation implemented for decision support and war gaming in airline revenue management.

Airline revenue management as explained by [14] describes the maximization of revenues through inventory controls. Based on a demand forecast, available products and capacity, revenue optimal inventory controls are computed. The success of revenue management depends on an understanding of dynamic markets influenced by competing suppliers and the resulting demand behavior. As argued in [10], such dynamic markets are a subject well-suited for multi-agent modeling. Multi-agent modeling and simulation implement heterogeneous agent populations that act autonomously ([7]). When implemented as a simulation, such models can be combined with event-based and stochastic concepts (see [9] for further details).

The simulation REMATE is an example of a of a multi-agent simulation implemented and used for training and decision support in revenue management. The

Catherine Cleophas
Freie Universität Berlin, Garystr. 21, Berlin
e-mail: catherine.cleophas@fu-berlin.de

J.B. Pérez et al. (Eds.): Highlights on PAAMS, AISC 156, pp. 163–170.
springerlink.com

motivations for its implementation and use as well as parts of the corresponding process are documented in this contribution. A thin description of REMATE's functionalities and underlying models has been presented in [6]. From the perspective of practical revenue management, REMATE was designed to support the decisions of human analysts interfacing with automated revenue management systems. The need for human experts to analyze and influence the results of these systems has been explained in [12] and [3]. Their task is to help the demand forecast to adapt to short-term changes in the market place and to realize secondary objectives such as competitive productivity.

REMATE incorporates airlines and customers as agents. Markets are described by networks of flights forming itineraries available for booking. On any market several airlines may be competing for customers, who can flexibly decide whether or not to buy the tickets offered and whether and when to cancel. Airlines are differentiated by the products they offer as well as by the revenue management system they implement. Customers' decision rules are based on a rational choice model. The customer agents are individualized through stochastic elements in their preference function and the distributions governing the timing of their requests.

As explained in [15] and recapitulated in [16], simulations may be employed for three types of explanation and prediction: to find out what conclusions may be drawn from complex antecedents ("concept-driven" simulation), to find out what future behavior a target system may display ("qualitative prediction"), and to predict the future state of a target system ("quantitative prediction"). The further text describes the extent to which REMATE attempts to realize each of these purposes and thereby does or does not meet the practitioners' hopes.

The second section of this paper describes the life cycle and current usage of REMATE, focusing on issues of user acceptance. These are framed by and contrasted with the concept of assurance as described in [11]. The third section highlights opportunities for feedback loops observed with regard to REMATE that may be used to improve verification and validation. The final section of this paper offers an outlook with regard to evaluating issues of user acceptance and to improving the assurance of revenue management simulation systems.

2 System Lifecycle: Revenue Management Training for Experts (REMATE)

The simulation REMATE ("Revenue Management Training for Experts") has been presented in [6] for the first time. This section describes the motivation for its development and the process of designing the system, the way in which the system was implemented, and finally its current operation.

2.1 Conception and Design

From 2008 to 2009, REMATE was implemented as a professional software development project by Deutsche Lufthansa AG based on research prototypes described in [5] and [2]. The project was motivated and legitimated based on three objectives: to quantify the effect of new revenue management models and methods (*research function*), to quantify the effect of new revenue management strategies (*decision support function*), and to train new revenue management experts (*training function*).

Three groups of stake holders were driving the development of REMATE: Academic *researchers* cooperating with Deutsche Lufthansa saw an opportunity to develop a simulation including all aspects of several previous prototypes and improving the software quality through professional development. *Experts* and management in charge of the further development of revenue management systems regarded REMATE as a new tool to estimate the (economic) effects of the implementation of new methods. *Analysts* and management of the operational revenue management department hoped to gain a tool both for every-day decision support and training.

To represent all stakeholders, a project team including members from all factions was assembled. The project team collected requirements in several workshops aided by external consultants with a background in software development. The result of this process was a system architecture fit both for stochastic decision support simulations and for deterministic war gaming. The decision support component's primary aim was to answer research questions and help estimating the effects of method changes. The participatory gaming component's focus was to facilitate communication and training.

In participatory games with REMATE, the participants play the role of one or several of the airlines included in the model. Within a single iteration of the simulation, they influence the parameters of their revenue management system in several turns, at several breaking points throughout the booking horizon.

The resulting requirements were detailed in a specification of the simulation model, agent behaviors, and the revenue management methods to be included in the system. While the specification document was written by researchers and experts, it was reviewed and challenged frequently by the analysts.

While some of the specified functionality was original based on new requirements for example with regard to the user interface, other functionalities such as the general simulation framework, agent behaviors and several revenue management algorithms had previously been implemented in the form of research prototypes.

To ensure that the specification of these parts of the system stayed in accordance with previous research, they were specified to replicate the results of the existing prototypes. In this regard, the concept of "docking" as introduced by [1] and quoted with regard to the verification of simulation models by [10] has been realized in the design of REMATE. This way of specifying several components of the model prepared the way for verification testing in the next step, implementation.

2.2 Implementation

REMATE was professionally implemented by a developer company experienced both in the creation of complex software programs and in the design of user interfaces for training simulations. Already during the design phase, developers entered the project team as new members. Due to the originality of the project and the new challenges of implementing a simulation system, the development of REMATE was planned as an iterative rather than as a strictly sequential process.

As part of this process, REMATE was implemented in cycles stages rather than in a single sequence of design, production, and testing. The first cycle included both the construction of the simulation framework and a first version of the agent model. Revenue management algorithms were added in the order of added complexity in the next cycles. The objective was to enable the independent verification of the agent model and the revenue management algorithms.

As part of the implementation, the success of the docking approach was tested. For the simulation framework to be verified, only relational equivalence was required. While the agent behavior could be evaluated for distributional equivalence due to its stochastic elements, numerical equivalence was required for the revenue management algorithms. The system resulting from each iteration of the process was tested (verified) by members of the project team as well as by analysts not part of the team and by professional software testers. As suggested in [11], extreme boundaries and sensitivities of the simulation results were evaluated. In addition to human testers, the developer employed automated test suites for this task.

Especially in later iterations, it became clear that the system had grown too complex to be fully verified in a single testing phase. First indications of this had been encountered in the specification phase, but the issue of realism versus minimalism as described in [10] became urgent at this point: Practitioners refused to consider a simulation system if it could not implement a range of complex customer behaviors. Researcher suggested limiting functionality to what was strictly necessary to properly examine a clearly defined set of questions. However, REMATE was intended as a "universal" revenue management simulation system to be used by both.

Finally, to generate qualitative or even quantitative answers as described by purposes two and three in [15], validity is not limited to the model's structural properties, but also includes its parameters. After the implementation of the system was concluded, a calibration process was planned for REMATE. Calibration was realized by attempting to replicate the indicators observed on real markets through the manipulation of input parameters in the REMATE customer model.

In calibration, a target conflict between the *bias* and the *variance* of simulation results as described in [10] emerged. The numerical deviation of simulation results from indicators observed in the real world determines its bias; the sensitivity of simulation results with regard to changes in the data its variance. An acceptable bias should not lead to a variance that is either too high or too low. If the system is not sensitive to input parameters, it cannot model changes that may occur in the real market place. If it is too sensitive, it may not be robust enough to provide reliable results.

2.3 Operation and Ongoing Development

While the original implementation project was concluded in 2009, operation and implementation of REMATE overlap to this day. The system continues to be iteratively improved both with regard to the correctness of the code, the validity of the model and the extent of its functionality.

REMATE is currently used to train new analysts and to establish a basis for communication with regard to revenue management methods and strategies. Its results on minimalist scenarios are used for ongoing research. Simulation results from calibrated, complex scenarios have already supported the decision process with regard to future revenue management developments and strategic decisions.

In spite of a generally positive reception by all stakeholders, two major sources of conflict have become quite clear throughout the work with REMATE. The antagonism of minimalism and realism also considered in [10] is a constant source of discussions between analysts and researchers. While a system that is minimal can be verified and - to some extent - validated more easily, it may be regarded as useless by field experts desiring specificity. Furthermore, the importance and feasibility of the different purposes of simulations as explicated by [15] is rated quite differently by researchers and analysts. Regardless of cautionary notices attached to any simulations results, practitioners tend to hope for quantifiable predictions of states where researchers are happy to find exploratory models of behavior.

3 Opportunities for Feedback Loops

From the example for the implementation of a multi-agent model for training and decision support introduced in the previous section, several opportunities for feedback loops may be derived. These link the experiences gathered by the designers and the users of agent-based simulations to research questions regarding their verification and validation. In addition, new research questions with regard to the acceptance of simulation results by field experts arise.

3.1 Inspiration for Assurance

As documented by [4], research literature offers diverse approaches to validating simulations. When a multi-agent simulation is applied for decision support, its results have to be reliable for experts to trust them when considering alternative strategies. The importance of the validity of simulation results for systems that extend beyond research purposes is also stressed in [10].

While not necessarily sufficient, verification and validation as described by [11] are certainly necessary conditions for applied simulation systems to be successfully used. Both of these aspects of assurance can be realized in cooperation with practitioners: Analysts can be part of an environment in which vigorous testing of the correctness of the programming as well as the model is feasible. At the same time, they may aid the calibration and thereby the constructive generation of validity.

As analysts are deeply familiar with the behavior of the real system that is to be modeled, they can judge whether the simulation generates results that conform with their expectations. If this is not the case, depending on the state of development, possible explanations for deviation may include faulty implementation (verification failure= or faulty parameterization (validation failure). Simulation experts can then revise the system and thereby start a new iteration of the feedback loop.

In participatory games with a competitive element, any simulation result that does not confirm the players' expectations (and hopes) will likely be challenged. As such a gem can be implemented as an interactive, regularly interrupted simulation run, the participants implicitly test the validity of the intermediate results at every step.

Additionally, analysts may even contribute their knowledge of the way in which the real world functions to improve the model's fit. In the case of REMATE, this feedback was used to generate extensive specifications, support calibration and continuously improve the model's realism. It has to be noted that this way of including practitioners in model generation can endanger the minimalism of the resulting system.

When participating in games based on REMATE, the players get the opportunity to represent different airlines in novel situations. Their decisions can overrule the automated algorithms, prescribing new strategies. Research on revenue management, the modeling of uncertainty and complexity, competition and cooperation may be inspired by observing the ensuing behavior. Research on agent-based modeling may be enhanced by the observation of real-world strategies of communication, learning and decision-making.

3.2 *Inspiration for Acceptance*

Considering the example of REMATE, it becomes clear that even a fully verified and validated simulation system may not be accepted by practitioners if assurance is limited to a macro level. This acceptance is highly relevant for a system to overcome a limitation to pure research.

In order to enable a thorough acceptance of the simulation model by practitioners, it may be beneficial to include them in the specification from the start. As described in the previous section, this inclusion can introduce a bias against the degree of minimalism considered desirable from a validation stand-point. The goal of feedback loops as part of the system specification is therefore a balance between the level of detailed wished for by practitioners and the ceteris paribus conditions and limited quantity of input parameters required for scientific considerations.

As discussed for example in [13], presenting the results of multi-agent simulations in a concise and meaningful way is not a trivial task. At the same time, the accurate presentation not just of the content of the results but also of their applicability to the purposes of simulations as given by [15] may help to avoid dangerous misunderstandings. The result representation must make clear in how far the simulation can provide exploratory models or whether it offers quantifiable predictions of a probabilistic or even deterministic type. For example, a particular challenge of the output presentation in REMATE has been to illustrate the stochastic element included.

Simulation research may therefore benefit from experiences collected in the application of simulations with regard to the interpretation and acceptance of results. When micro-level aspects of the simulations are to be discussed, illustrations of individual agents' choices are required: While every agent's choices are observed and documented, their sum is sometimes difficult to visualize and to critically examine as part of a validation effort.

4 Conclusions and Outlook

This paper has introduced several opportunities for inspiration through the application of agent-based modeling to decision support and participatory games. These opportunities have been categorized by the aspects of agent-based modeling they affect: Assurance and acceptance. While assurance is an issue that is often considered in multi-agent simulation research, the importance of result and model acceptance was stressed here.

By recounting the life cycle of the airline revenue management simulation system REMATE, ways in which concepts such as docking and destructive verification can be used in the application of such systems have been demonstrated. The conflict of minimalism and realism as well as the importance of a common understanding of the purpose of simulations have been emphasised.

This paper intends to show that there do exist opportunities for feedback loops between the application and the theory of agent-based modeling. Such loops can both enhance research and further state-of-the-art practice. They may be used both for the assurance and for the enrichment of models. They depend on flexibility and openness on both sides: Feedback has to be collected through careful observations and existing solutions have to be amenable for improvement.

With regard to the issue of acceptance, further research using the case of REMATE may be conducted. A quantitative analysis into the factors leading to the acceptance of a simulation system and its results may be realized through surveys and laboratory experiments. An attempt at limiting the level of detail included in the revenue management model may consist of implementing nested models as suggested by [11]. This approach could be supported by measuring the degree of complexity included in the various models using a set of metrics as developed in [8].

Acknowledgements. The author would like to thank Deutsche Lufthansa AG for access to REMATE and for supporting the cooperation with revenue management practitioners.

References

1. Axtell, R., Axelrod, R., Epstein, J., Cohen, M.D.: Aligning simulation models: A case study and results. Computational and Mathematical Organization Theory 1(2), 123–141 (1996)
2. Cleophas, C., Frank, M., Kliewer, N.: Simulation-based key performance indicators for evaluating the quality of airline demand forecasting. Journal of Revenue and Pricing Management 8(4), 330–342 (2009)
3. Currie, C.S.M., Rowley, I.T.: Consumer behaviour and sales forecast accuracy: What's going on and how should revenue managers respond? Journal of Revenue and Pricing Management 9(4) (2010)

4. Fagiolo, G., Moneta, A., Windrum, P.: A critical guide to empirical validation of agent-based models in economics: Methodologies, procedures, and open problems. Computational Economics 30(3), 195–226 (2007)
5. Frank, M., Friedemann, M., Mederer, M., Schroeder, A.: Airline revenue management: A simulation of dynamic capacity management. Journal of Revenue and Pricing Management 5(1), 62–71 (2006)
6. Gerlach, M., Frank, M., Cleophas, C., Schroeder, T.: Introducing REMATE: Revenue management simulation in practice. In: Meeting of the AGIFORS Working Group Revenue Management and Cargo, New York City (May 2010)
7. Gilbert, G.: Agent-based models. Sage Publications, Inc. (2008)
8. Klügl, F.: Measuring Complexity of Multi-agent Simulations – An Attempt Using Metrics. In: Dastani, M.M., El Fallah Seghrouchni, A., Leite, J., Torroni, P. (eds.) LADS 2007. LNCS (LNAI), vol. 5118, pp. 123–138. Springer, Heidelberg (2008)
9. Law, A., Kelton, W.: Simulation Modeling and Analysis. McGraw-Hill Higher Education (1997)
10. Marks, R.E.: Validating simulation models: a general framework and four applied examples. Computational Economics 30(3), 265–290 (2007)
11. Midgley, D., Marks, R.E., Kunchamwar, D.: Building and assurance of agent-based models: An example and challenge to the field. Journal of Business Research 60(8), 884–893 (2007)
12. Mukhopadhyay, S., Samaddar, S., Colville, G.: Improving Revenue Management Decision Making for Airlines by Evaluating Analyst-Adjusted Passenger Demand Forecasts. Decision Sciences 38(2), 309–327 (2007)
13. Sanchez, S.M., Lucas, T.W.: Exploring the world of agent-based simulations: simple models, complex analyses. In: Proceedings of the 34th Conference on Winter Simulation: Exploring New Frontiers, pp. 116–126 (2002)
14. Talluri, K.T., van Ryzin, G.J.: Theory and Practice of Revenue Management. Kluwer Academic Publishers, Boston (2004)
15. Troitzsch, K.G.: Social science simulation-origins, prospects, purposes. Lecture Notes in Economics and Mathematical Systems, pp. 41–54. Springer, Heidelberg (1997)
16. Troitzsch, K.G.: Not All Explanations Predict Satisfactorily, and Not All Good Predictions Explain. Journal of Artificial Societies and Social Simulation 12(1), 10 (1997)

Mining Application Development for Research

Johannes Fähndrich, Sebastian Ahrndt, and Sahin Albayrak

Abstract. Nowadays, many research institutes are largely dependent on third party funding. This situation leads to practical work or in other words project work which exceeds the research typical proof of concept implementation. As we were tired of seeing a downward trend in the number of accepted application oriented (full) papers in the major agent conference, we conducted a survey to provide evidence for the thesis that researchers can gain relevant benefits from project work for their research work. Hence, in this paper, we present the results of this survey and discuss different scientific questions researchers should ask themselves during project work.

1 Introduction

The dependence on third party funding [18] of research institutes leads to a lot of practical work that exceeds the research typical proof of concept implementation. As this practical work is mostly some kind of time- and topic-limited collaboration with external partners, we will further refer to it as project work (PW). However, as it is quite common for researchers to publish their achievements in more abstract and theoretical results and adopt these results to the more practical work, the other way around is not [1]. To counter this issue, the IFAAMAS stated in its charter [8] (Point 1, 5, 6) to foster the links between the more theoretical agent community and the more practical community and further to promote applied research. Nevertheless, we were tired of seeing the results of a survey done by *Hirsch et al.* [7] in 2011. The work emphasises a downward trend in the number of accepted application oriented (full) papers in the agent conference *AAMAS (International Conference on Autonomous Agents and Multiagent Systems)*. The survey present results showing that the number of accepted full papers stays almost the same, while the number of accepted application papers drops – from 13 or 10.2% in 2006 to 6.3% in the 2011 edition of the AAMAS.

Johannes Fähndrich · Sebastian Ahrndt · Sahin Albayrak
DAI-Labor, Technische Universität Berlin, Ernst-Reuter-Platz 7, 10587 Berlin, Germany
e-mail: `firstname.lastname@dai-labor.de`

J.B. Pérez et al. (Eds.): Highlights on PAAMS, AISC 156, pp. 171–178.
springerlink.com

However, our thesis is, that application papers are very relevant for the agent community and should address different stakeholders, such as industry, academia and reviewers, to name but a few (see [14, 7]). After discussing the survey results, we conduct a own survey to provide evidence for the thesis that researchers can gain relevant benefits from project work for their research work. Further, this paper shell provide guidance while identifying a good [13] research questions in project work. Commencing with the description of the used survey methodology (See Section 2), we will present the results of that survey (See Section 3). Afterwards we proceed with a discussion through the author guidelines presented by *Hirsch et al.* [7] (See Section 4) and wrap up with a conclusion (See Section 5).

2 Basics and Methodology of the Survey

In order to give a brief introduction to the survey methodology, we will next describe four scientific principles, which are typically taken into account, when researchers find a seminal question [15] and that we used as foundation of the questionnaire:

- *Significance* is quit important for the question of funding. Meaning that if it is foreseeable that the research will not do any good to anyone, then most likely no one will invest into it. The first meta question the researchers should ask themselves about there research could be: "What will the results change?". By finding a good answer to this question, the research can be defended and will not become a pernicketiness. Here lies a grade difference of project and research work. In projects, exceeding pure research, this questions is vital at the beginning to get funding. Arguing why extra time and effort should be invested to create scientific evidence is often hard work. Especially with industry partners, which have to argue their return of investment (ROI) [16].
- *Innovation* means that the results have to extend the *State of the Art* (SotA) or optimise a existing result. Consequently, the principal of innovation is twofold [5]: On the one hand process innovation, on the other hand product innovation. The first optimise the way a solution is created. The latter provides a new solution to a problem. Following *Schumpeter* (e.g. [6]), the so called "creative destruction" is an elementary need of our economic. Taking the differences of product and process innovation into account, the impact of R&D is subject to research [4]. The next meta question to capture the innovation of the possible solution for the research question could be: "What is the SotA regarding this problem?".
- *Traceability* means to uncover and document the steps made towards an solution, referring to all used existing fragments [2] and to follow strict rules on how these references have to be published [18]. Project work has some similar rules. For an example we point to the licensing of software, where the reference has to be established in a research comparable manner.
- *Clearness* is improved by investing additional time to make results better understandable. Due to the lack of resources, this principle is often neglected. Further, one of the reasons why there are less application papers might be, that enterprises

– applying the research results – do not want to publish there knowledge, they rather want to sell it, to increase there ROI [4].

Based on the presented scientific principles, we elaborated a questionnaire that circulated mainly at research institutes which have a scientific focus on distributed artificial intelligence. The questionnaire has been created taking the work of *Pfleeger* and *Kitchenham* [10, 11, 12, 17] into account – among others the questions have to be resistant to bias, appropriate to the respondent and cost-effective. As the objective of the survey lies on the area of conflict between research and project work, we send our questionnaire to institutes which are mainly third party founded. In order to outline the discrepancy between the project and research work, we started the survey with the following explanation to clarify the separation between both:

> To distinguish between project work and research work, we intend research work as the methodical search of new knowledge in order to extend the actual state of the art. In contrast to this, we refer to project work as the appliance of previously available skills to establish approaches for well-formed problems.

In order to provide evidence to our thesis the survey consists of 30 questions separated in nine categories. The answers were given through different provided lickert scales [3]. However, the survey offers an additional tenth category, where the interviewee had the possibility to write down the most important questions they ask themselves during research work – we will further refer to these answers as *scientific questions*.

To provide some structure to the obtained data we applied a research methodology based on *Value-Focused Thinking* [9] (VFT). VFT follows a – especially in AI popular – backward search approach. We had to perform two activities: First deciding which questions are needed to be answered and second figuring out how to get the achieved data with a minimal amount of survey questions. Analysing the survey results we interpreted the likert scale as ordinal scale using the appropriate statistical values like the median (M) and the mean deviation from the median (MD).

3 Survey Results

Overall we received 54 responses during October 2011. Given the breadth of the answers, we feel the results will be usefully for future works, too. However, for this work we will concentrate our interest on the scientific principles we preliminary introduced and the analysis of the collected questions. First of all we want to present the result to the question on experience gained form project work and research (R). Figure 1 shows these results and emphasises that the experience gained during project work exceeds the one gained during research (M = 3, MD = 0.98).

Figure 2 illustrates the distributions of the responses to the questions how often the four preliminary introduced scientific principles. We can see, that almost no researcher disregards the principles (1) in their project work but neither always uses them (5).

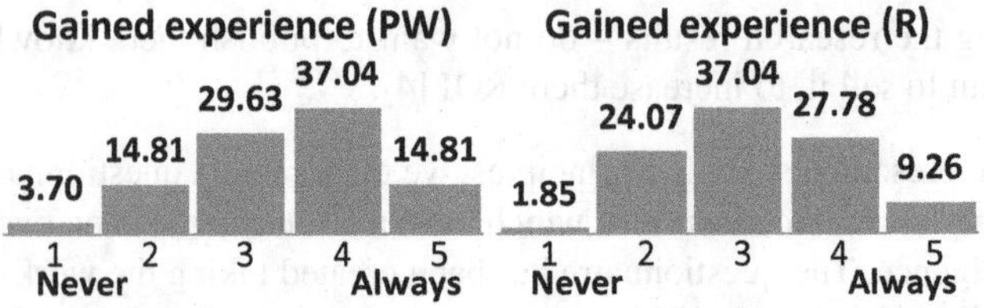

Fig. 1 (1) Gained experience during project work (M = 4, MD = 1.19); (2) Gained experience during research (M = 3, MD = 0.98)

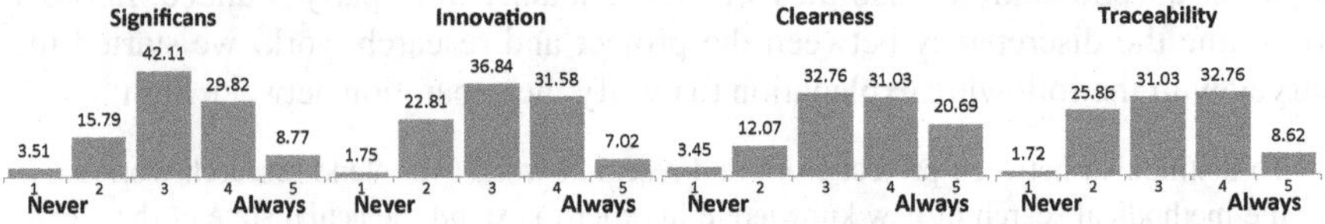

Fig. 2 (1) Significans (M = 3, MD = 0.83); (2) Innovation (M = 3, MD = 0.85) ; (3) Clearness (M = 3, MD = 1); (4) Traceability (M = 3, MD = 0.88)

Analysing Figure 2 and the amount of scientific questions obtained through our survey, we found an interesting discrepancy between the emergence of scientific questions referring to Traceability and Clearness – here we obtained only 4 (2.9%) – and Significance and Innovation, where we obtained 47 or 34.8%. Because of that, Traceability and Clearness seam more important in project work then they are during research. Additionally, we asked about the correlation between project work and research and the researchers motivation.

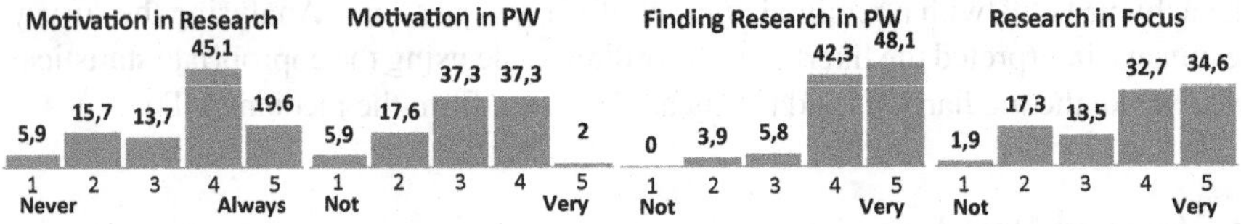

Fig. 3 (1) Motivation during R (M = 4, MD = 1.32); (2) Motivation during PW (M = 3, MD = 1.06); (3) Motivation increase through finding R in PW (M = 4, MD = 1.08); (4) Motivation during PW with R in focus (M = 4, MD = 1.32)

Figure 3 emphasises the impact on the motivation for different topics. One can clearly see, that the motivation during research work is much higher then during project work. Moreover if the researchers find interesting scientific questions in their project work we clearly see an increasing motivation, even more if R and PW are on the same topic. Spending time on thinking about the degree of significance and innovations regarding the results of project work, not only leads to seminal research question which can bear a scientific delta regarding the SotA, but also seams to motivate. Carving out the scientific delta might lead to more acceptance from the scientific community evaluating the research result during a review process for publication.

In the last section of the survey, we asked for scientific questions the researcher ask themselves while working scientifically. We did receive 135 questions, in which we were able to identify six categories. For each of this categories we have verbalised a affiliated question. The initial four principles where covered in the collected questions, but only two of them where addressed frequently.

Significance (25.2%) : 34 questions refer to Significance, which ends in the question: *"Why is my work interesting for others?"*.

Related Work (20.0%) : 27 questions refer to Related Work, we found a wide range of formulation reaching from available conferences to basic knowledge to other approaches and there shortcomings. Our question here is twofold, representing the two types of innovation introduced earlier: *"Are there approaches available for this problem?"* (product innovation) and *"Have the existing solutions shortcomings we can address?"* (process innovation).

Problem Classification (17.0%) : 23 questions refer to the Problem Classification, which ends up in the question: *"What is the problem?"*.

Evaluation (14.1%) : 19 questions refer to the Evaluation and emerged the question: *"How can we validate the results?"*.

Innovation (9.6%) : 13 questions refer to the Innovation adding up to the question: *"What is the scientific contribution?"*.

Resources (5.9%) : 8 questions refer to resources ending in the question: *"How can the work scheduled correctly to the given resources?"*.

In order to give a complete overview over the collected data, we were able to assign 1 (0.7%) question to the scientific principle Clearness and 3 (2.2%) to Traceability. Furthermore, we were not able to assign 7 (5.2%) questions to any of the introduced categories.

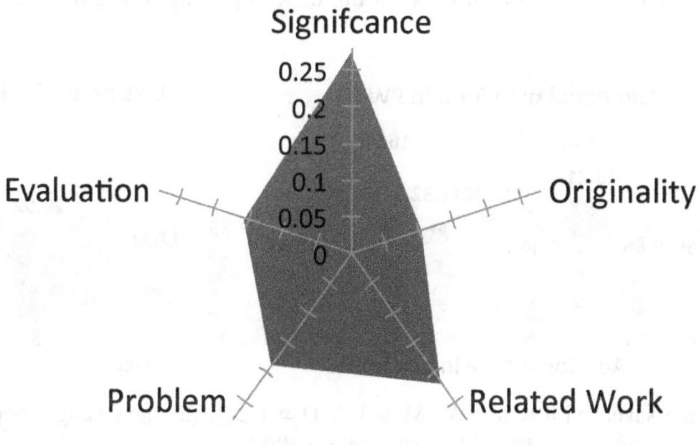

Fig. 4 Relation between the identified research categories and the available resources

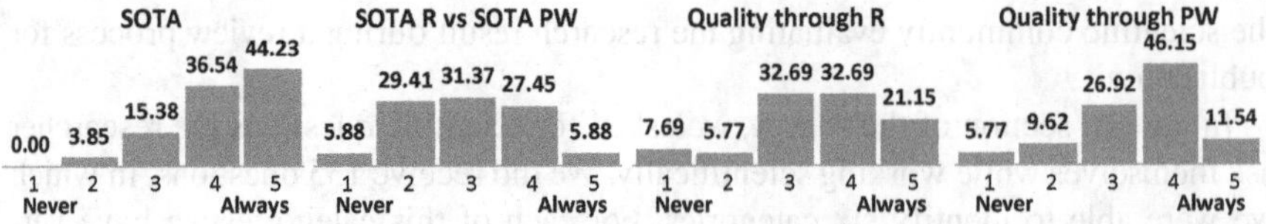

Fig. 5 (1) Amount of SotA in PW (M = 4, MD = 1.14); (2) Difference between the SotA analysis during R and during PW (M = 3, MD = 1.15); (3) Quality gain in R applying PW (M = 4, MD = 1.2) (4) Quality gain in PW applying R (M = 4, MD = 1.32)

Figure 4 illustrates the six categories and emphasise the area of conflict between them (in terms of the available resources) in a multi-dimensional model. Here the number of time a scientific question is asked is generalised to the amount of time spend to answer the question. Because of the practical derivation of application papers and project work, a use case might be evident, which simplifies finding a method of evaluation and some significance to the underlying research question. With only 9.6% of all questions the principle Innovation seams the least important to researchers. Referencing our beginning statement, we conclude that by answering the questions listed above researcher support the identification of an significant and innovative research delta with regard to the SotA. Eventually concluding in to a research question, which could lead to a higher acceptance in a academic community. This on the one hand fosters practical research papers and on the other, motivate researchers.

However, in the survey we asked also about the difference between SotA analysis during project work and during research. Figure 5 illustrates the results and shows that the SotA analysis done in research is about the same as in project work. Taking the data presented in Figure 6 into account, it seams that most researchers experience a quality increase by applying research methods in project work. At the same time, the researchers perceive embedded research methods as time consuming. This leads to the conclusion, that more time is necessary when research is embedded in project work, while seeing research as an additional task opposing to the project work.

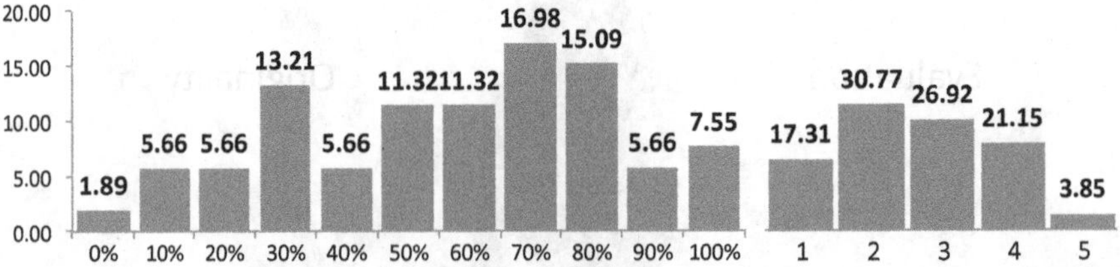

Fig. 6 (1) Time saved with R in PW (M = 3, MD = 1.24); (2) Increasing amount of time required when applying R in PW (M = 70%, MD = 28%)

4 Discussing Guidelines

Through our survey, we were able to identify six categories of questions and to objectify one question for each category. Furthermore we have ranked them through the gained insights from the survey. Consequently, we are able to confirm a major part of the author guidelines presented by *Hirsch et al.* [7] and further extend it with two new questions that an application oriented paper (and probably others too) should answer to emerge a research question from project work:

- Why is my work interesting for others?
- What is the scientific contribution?
- Are there still approaches available for this problem? Have the existing solutions shortcomings we can address?
- What is the problem?
- How can we validate the results?
- How can the work be scheduled correctly to the given resources?

By answering these questions researcher support the identification of an significant and innovative research delta to the SotA and with that on the one hand foster practical research papers and on the other, motivate researchers. As we did not receive a significant amount of questions regarding the Traceability and and Clearness, these two aspects are missing.

5 Conclusion

In this paper, we present the results of a survey regarding the discrepancy between research and project work and the motivation for writing application papers. Overall, 54 researchers answered our questionnaire. The analysis of the questionnaire showed that researcher can gain experience from projects for their research work and emphasises the importance of finding a research question to publish the results of more practical work. Here, we were able to identify six different categories of questions, which we ranked through the gained insights of the survey. Furthermore, we verbalised questions for each category to objectify and extend the author guidelines presented by *Hirsch et al.* [7]. By answering these questions researcher support the identification of an significant and innovative research delta to the SotA and with that, on the one hand foster practical research papers and on the other, motivate researchers. Enabling more insight into the matter, future conferences could classify application papers submitted to evaluate if missing quality might be the reason for a lesser acceptance rate. Furthermore with agents being just one approach, the amount of application papers accepted in other domains should be studied to determine if this is a domain specific problem.

References

1. Balzert, H., Schäfer, C., Schröder, M., Kern, U.: Wissenschaftliches Arbeiten - Wissenschaft, Quellen, Artefakte, Organisation, Präsentation, 1st edn. W3L (2008)
2. Cargill, M.: Writing scientific research articles: Strategy and Steps, 1st edn. Wiley-Blackwell (2009)
3. Carifo, J., Perla, R.J.: Ten common misunderstandings, misconceptions, persistent myths and urban legends about likert scales and likert response formates and their antidotes. Journal of Social Sciences 3(3), 106–116 (2007)
4. Cohen, W.M., Klepper, S.: Firm size and the nature of innovation within industries: The case of process and product r&d. The Review of Economics and Statistics 78(2), 232–243 (1996)
5. Davenport, T.H.: Process Innovation: Reengineering Work through Information Technology. Harvard Business (1992)
6. Elliott, J.E.: Marx and schumpeter on capitalism's creative destruction: A comparative restatement. The Quarterly Journal of Economics 95(1), 45–68 (1980)
7. Hirsch, B., Balke, T., Lützenberger, M.: Assessing agent applications — r&D vs. R&d. In: Weyns, D., Müller, J.P. (eds.) Proceedings of the 12th International Workshop on Agent-Oriented Software Engineering, vol. 1, pp. 93–104 (2011)
8. IFAAMAS: Charter for the international foundation for autonomous agents and multiagent systems
9. Keeney, R.L.: Value-Focused Thinking: A Path to Creative Decisionmaking, 2nd edn. Havard University Press, Cambridge (1996)
10. Kitchenham, B., Pfleeger, S.L.: Principles of survey research part 2: Designing a survey. SIGSOFT Softw. Eng. Notes 27, 18–20 (2002)
11. Kitchenham, B., Pfleeger, S.L.: Principles of survey research part 3: Constructing a survey instrument. SIGSOFT Softw. Eng. Notes 27, 20–24 (2002)
12. Kitchenham, B., Pfleeger, S.L.: Principles of survey research part 4: Questionnaire evaluation. SIGSOFT Softw. Eng. Notes 27, 20–23 (2002)
13. Lipowski, E.E.: Developing great research questions. American Journal of Healthsystem Pharmacy AJHP Official Journal of the American Society of HealthSystem Pharmacists 65, 1667–1670 (2008)
14. Luck, M., McBurney, P., Shehory, O., Willmot, S.: Agent technology: Computing as interaction – a roadmap for agent based computing (2005)
15. Messing, B., Huber, K.P.: Die Doktorarbeit - Vom Start zum Ziel: Lei(d)tfaden für Promotionswillige, 4th edn. Springer, Heidelberg (2007)
16. Pakes, A., Schankerman, M.: The rate of obsolescence of patents, research gestation lags, and the private rate of return to research resources. In: R & D, Patents, and Productivity, NBER Chapters, pp. 73–88. National Bureau of Economic Research, Inc. (1984)
17. Pfleeger, S.L., Kitchenham, B.: Principles of survey research part 1: Turning lemons into lemonade. SIGSOFT Softw. Eng. Notes 26, 16–18 (2001)
18. Stock, S., Schneider, P., Peper, E., Molitor, E.: Erfolgreich promovieren - Ein Ratgeber von Promovierten für Promovierende, 2nd edn. Springer, Heidelberg (2009)

Trust and Transitivity: How Trust-Transfer Works

Rino Falcone and Cristiano Castelfranchi

Abstract. Transitivity in trust is very often considered as a quite simple property, trivially inferable from the classical transitivity defined in mathematics, logic, or grammar. In fact the complexity of the trust notion suggests evaluating the relationships with the transitivity in a more adequate way. In this paper, starting from a socio-cognitive model of trust, we analyze the different aspects and conceptual frameworks involved in this relation and show how different interpretations of these concepts produce different solutions and definitions of trust transitivity.

1 Introduction

Many different approaches and models of trust were developed in the last 15 years [1, 2, 3]: they contributed to clarify many aspects and problems about trust and trustworthiness, although many issues remain to be addressed and some elementary but not so trivial trust properties are left in a contradictory form.

One of them is the problem of trust transitivity. If ***X*** trusts ***Y***, and ***Y*** trusts ***Z***: What about the trust relationship between ***X*** and ***Z***? Different and sometimes diverging answers were given to this problem. The question is not only theoretically relevant; it is very relevant from the practical point of view. In fact one of the main problems of trusting agents in an open and massive multi-agent world is the necessity of exploiting the cumulated trust by other trustees (who we trust) for trusting agents that they know and we do not. But, how does this trust-transfer work? Which are the sophisticated basic cognitive mechanisms for doing this?

In this paper we will present an analysis of the trust transitivity when a socio-cognitive model of trust is taken into consideration. Through this kind of model we are able to evaluate and partially cope with the complexity that the concept of

Rino Falcone · Cristiano Castelfranchi
National Research Council– Institute of Cognitive Sciences and Technologies
Via San Martino della Battaglia, 44 00185 - Roma, Italy
e-mail: {rino.falcone,cristiano.castelfranchi}@istc.cnr.it

J.B. Pérez et al. (Eds.): Highlights on PAAMS, AISC 156, pp. 179–187.
springerlink.com

transitivity introduces when applied to the trust relationship. We cannot in fact just rely on our direct and personal experience; this would restrict too much our interaction possibilities and "market", and also our "trust capital" (how many people and how much trust us, and would be interested in exchanging with us). This precious information not only is preserved in our memory, from previous interactions; but it is generalized (to groups and categories [10, 11]) for making predictions about new people; it is acquired from the others by observation and use of 'signals' (I see that ***X*** exchanged with ***Y***), or by communication (advices, recommendations, reputation, etc); it is acquired from ***Y***'s exhibition of his skills, products, virtues, qualities (signaling and "marketing"). So, trust (and distrust) information is circulating a lot as a precious good, and trust has a very active social dynamics in the sense of *transfer* from one trustor to another, from one trustee to another, from context to context, from task to task. This phenomenon is so important that there is the need of clarifying and explicitly modeling the specific mechanism of this trust "circulation".

2 A Socio-cognitive Model of Trust

In our socio-cognitive model of trust [4, 5] we consider trust as a relational construct between the trustor (***X***) and the trustee (***Y***), about a defined (more or less specialized) task (τ). Introducing a formal operator *Trust*, representing the trust notion, we can write:

$$Trust\,(X\;Y\;C\;\tau\;g_X)$$

where are also explicitly present both the ***X***'s goal (g_X, respect to which trust is tested/activated) and the role of the context (***C***) in which the relationship is going to happen. In fact, the successful performance of the task τ will satisfy the goal g_X. *Task* is the performance ***X*** needs and expects from ***Y***; what ***X*** relies on ***Y*** for achieving g_X.

Using Meyer, van Linder, van der Hoek et al.'s logics [9] we can introduce the ***X***'s mental ingredients of trust. They mainly are the goal g_X, and a set of *main* trustor's beliefs:

$$Bel(X\;Can_{Y,C}(\tau)),\;Bel(X\;Will_{Y,C}(\tau)),\;Bel(X\;ExtFact_Y(\tau))$$

where:

- ***Bel*** is the classical modal operator for representing agents' beliefs;
- $Can_{Y,C}(\tau)$ means that, given the context ***C***, ***Y*** is potentially able to do τ (in the sense that, under the given conditions, is competent, has the internal powers, skills, know-how, etc); (and in the above formula this is what is believed by ***X***);
- $Will_{Y,C}(\tau)$ means that, given the context ***C***, ***Y*** has potentially the attributions for being willing, persistent, available, etc., on the task τ (and in the above formula this is what is believed by ***X***);
- $ExtFact_Y(\tau)$ this factor represents what is determining the set of external conditions (the context ***C***) either favoring or hindering ***Y*** realizing the task τ (and in the above formula is clear that also this is believed by ***X***).

In our model we also consider that trust can be *graded*: ***X*** can have a *strong trust* that ***Y*** will realize the task (maybe 0.95 in the range (0,1)); or even just a *sufficient trust* that ***Y*** will achieve it (maybe 0.6 with a threshold of 0.55; and so on with other possible values). For this we have introduced a trust quantification, calling it the *Degree of Trust*, $\boldsymbol{DoT_{XYC\tau}}$ and, in general, a *threshold* ($\boldsymbol{\sigma}$) to be overcome from this $\boldsymbol{DoT_{XYC\tau}}$. Given the previous analysis of the main components of the trust attitude ($\boldsymbol{g_X}$, $\boldsymbol{Bel\ (X\ Can_{Y,C}\ (\tau)), Bel\ (X\ Will_{Y,C}\ (\tau)), Bel\ (X\ ExtFact_Y\ (\tau))}$), our model is also able to evaluate how a specific degree of trust is, on its turn, resultant from the several *quantifications* of these components:

$$DoT_{XYC\tau} = f\,(DoC_X\ (Opp_Y(\tau)), DoC_X\ (Competence_{Y,C}(\tau)), DoC_X\ (Willingness_{Y,C}(\tau)))$$

where:

- f is in general a function that preserves monotonicity;
- $\boldsymbol{DoC_X\ (Opp_Y(\tau))}$ is the ***X***'s degree of credibility about the external opportunities (positively or negatively) interfering with ***Y***'s action in realizing the task $\boldsymbol{\tau}$;
- $\boldsymbol{DoC_X\ (Competence_{Y,C}(\tau))}$ is the ***X***'s degree of credibility about the ***Y***'s ability/competence to perform $\boldsymbol{\tau}$ in the context ***C***;
- $\boldsymbol{DoC_X\ (Willingness_{Y,C}(\tau))}$ is the ***X***'s degree of credibility about the ***Y***'s willingness to perform $\boldsymbol{\tau}$ in the context ***C***.

We are ignoring the subjective certainty of the pertinent beliefs (how much sure is ***X*** of its evalutative beliefs about that specific quality about either ***Y***'s or the environment, that is a meta-belief; in first approximation we can say that this factor is integrated with the other).

At the same time we are ignoring for now the value of the goal ($\boldsymbol{g_X}$). In fact, this value (its level of relevance) should have a specific and complex influence on the degree of the threshold ($\boldsymbol{\sigma}$): increasing the relevance of the goal both should increase the caution of the trustor of not entrusting a too unreliable trustee (pushing to increase the threshold), and should increase the need of not missing the opportunity to achieve it (pushing to decrease the threshold). Of course, both factors of trustor's personality and the set of potential viable alternatives than that trustee, have also to be considered for defining the complex relationships between threshold and goal.

So, we can say that ***X*** will trust ***Y*** about the task $\boldsymbol{\tau}$ in the context ***C*** if $DoT_{XYC\tau} > \boldsymbol{\sigma}$, that means that a set of analogous conditions must be realized about the other quantitative elements ($\boldsymbol{DoC_X\ (Opp_Y(\tau)), DoC_X\ (Competence_{Y,C}(\tau)), DoC_X\ (Willingness_{Y,C}(\tau))}$).

We do not consider in this paper the detailed analysis of how the degree of trust is resulting by the more elementary components. We also omit of considering the potential correlations among the different components.

In addition, we should also say that all the components shown above (competence, willingness, external factors) are in fact not the more elementary ones; they can be described as resultant by more elementary reasons/components. For example, the competence dimension could be considered as constituted by the *know-how*, the *self-confidence*, and the *ability* sub-components; and so on.

We would like now to introduce the concept of *trustworthiness degree* of an agent: we call $Trustworthiness_{YC}(\tau)$, the Y's trustworthiness about the task τ, in the context C. In general the trustworthiness of an agent is a property of that agent[1]. We have in fact two different meanings of this concept:

- an *objective trustworthiness*, what the agent is actually able and willing to do in standard conditions; his/her actual reliability on that task in the context C.
- a *subjective trustworthiness*, the perceived reliability of the trustee by another agent; it could be different for the different trustors.

The situation is also more complex, because the objective trustworthiness may not be costant with respect to the trustor but the same trustee Y could be differently trustworthy on the same task with different trustors (for example X or Z) (suppose he has different motivations for helping/serving X and Z). For this we introduce also the variable X in the operator: $Trustworthiness_{YXC}(\tau)$.

Supposing the situation in which the context is constant, if we have: $DoT_{XY\tau} > \sigma$ it derives that: $Bel(X\ Trustworthiness_{YX}(\tau) > \Sigma)$

where Σ is the minimum value of trustworthiness for Y (as believed by X) for delegating to him the task τ. In general $\sigma = \Sigma$.

3 Transitivity in Trust

The transitivity property of trust can be presented as: if X trusts Y, and Y trusts Z: What about the trust relationship between X and Z? We are interested to translate this problem in our terms of trust. First of all, we do not consider the unspecified case "X trusts Y" because in our model an agent has to trust another agent with respect to a task (either very well defined or less defined and abstract); this task directly derives from the goal the trustor has to reach with the trust attribution. So we have to transform "X trusts Y" in "X trusts Y about τ" (we are omitting the context C). And given the graded qualification of trust we have that: $DoT_{XY\tau} > \sigma$. This means in particular that X *believes* that Y is potentially *able* and *willing* to do τ and that the *external conditions* in which Y will perform its task are at least not so opposite to the task realization (may be also they are neutral or favorable). So this Y's trustworthiness with respect to X (perceived/believed by X) about τ is based on these specific beliefs. At the same way "Y trusts Z" becomes "Y trusts Z about τ_1" (about the difference between τ and τ_1, see later) with the same kind of particular Y's beliefs about Z and the external conditions.

Also in this case we can say that there is a threshold to be overcome and the condition: $DoT_{YZ\tau_1} > \sigma_1$ is successfully satisfied in case of trust attribution.

In order to transfer/adopt trust as mere evaluation, esteem, potential expectation (disposition) it is necessary to at least examine the task, that is, "about what" X trusts Y, and Y trusts Z; and possibly on which bases (the ascribed qualities). But, in order to transfer trust as 'decision' to rely on, as 'act' of trusting (from the

[1] In fact the trustworthiness (as trust) derives from specific properties of the trustee that could change during the time both for intrinsic reasons and for external conditions.

decision of Y to trust/rely on Z to the decision of X to entrust Z), something more is necessary: the degree of trust and its thresholds.

If we have to consider the trust relationship between "X and Z" as a consequence of the previous trust relationships between "X and Y" and between "Y and Z" we have to specify the task on which this relationship should be based (question of *assimilation* between τ and τ_1) and the degree of trust that must be overcome even from X: $DoT_{XZ\tau} > \sigma_2$ with the consideration of the threshold σ_2.

The role of the trust threshold is quite complex and can have an overlapping with the ingredients of trust. We strongly simplify in this case considering σ as dependent just from the specific intrinsic characteristics of the trustor (those that define an agent intrinsically: prudent, reckless, and so on) independently from the external circumstances (on the contrary, these factors affect the degree of trust, by affecting the more elementary beliefs above showed). In this approximation: $\sigma = \sigma_2$

In the case in which all the agents are defined as having the same intrinsic characteristics (possible for artificial entities), we can also say that: $\sigma = \sigma_1 = \sigma_2$

Moreover, as we just saw, not less important in our approach is that trust is an expectation and a bet *grounded on and justified by* certain beliefs about Y. X trusts Y on the basis of the evaluation of Y's "virtues/qualities", not just on the basis of a statistical sampling, some probability. The *evaluations* about the needed "qualities" of Y for τ are the *mediator* for the decision to trust Y. This mediation role is fundamental also in trust transitivity. Let us now consider the case of the differences between the tasks in the different relationships. In the simplest case of trust transitivity the two tasks should be the same ($\tau = \tau_1$). Is this enough? Suppose for example that there are 3 agents: John, Mary and Peter; and suppose that John trusts Mary about "organizing scientific meetings" (task τ), at the same time Mary trusts Peter about "organizing scientific meetings" (again task τ). Can we deduce that, given the transitivity of trust: John trusts Peter about "organizing scientific meetings"? Is in fact transferable that task evaluation? Given the trust model defined in §2 the situation is more complex and there are possible pitfalls lurking: Mary is the central node for that trust transfer and she plays different roles (and functions) in the first case (when her trustworthiness is about *realizing* the task τ, and in the second case (when her trustworthiness is about *evaluating* the Peter's trustworthiness on the task τ). The situation is even clearer if we split in the example the two kinds of competences: X trusts very much Y as medical doctor; X knows that Y trusts Z as mechanic; will X trust Z as mechanic? Not necessarily at all: if X believes that Y is a good evaluator of mechanics he will trust Z; but, if X believes that Y is a very naive in this domain, and is frequently swindled by people, he will not trust Z. So we have transitivity of trust if it is possible to *assimilate* the *performance of the task* with the *evaluation (eval) of the performance of the task* itself:

$Bel(X\ Trustworthiness_{YX}(\tau) > \sigma)$ ***implies*** $Bel(X\ Trustworthiness_{YX}(eval(\tau)) > \sigma)$.

Here we use the "implication" notion for expressing the following concept: "the reasons for X believing Y as trustworthy on the task τ, are also sufficient for X believing Y as trustworthy on the task to evaluate someone else performing τ".

In words, **X** has to believe that if **Y** is sufficiently trustworthy on the task τ, it is also sufficiently trustworthy on the meta-task of evaluating τ.

More analytically, the qualities (in both the directions: competence and willingness) that **X** is attributing to **Y** about the task τ, support, directly or for analogy, also the qualities necessary for trusting **Y** about the meta-task of evaluating τ (or on a different task τ'). So resuming we have the *more basic case* (case A) of the relationship between *trust* and *transitivity* so defined (we assume $\sigma=\sigma_X=\sigma_Y$):

if
iA) $DoT_{XY\tau} > \sigma$ (**X trusts Y about τ**) and
iiA) $DoT_{YZ\tau} > \sigma$ (**Y trusts Z about τ**) and
iiiA) $Bel(X\ DoT_{YZ\tau} > \sigma)$ (**X believes that Y trusts Z about τ**) and
ivA) $Bel(X\ Trustworthiness_{YX}(\tau) > \sigma)$ ***implies*** $Bel(X\ Trustworthiness_{YX}(eval(\tau)) > \sigma)$
(**Y is sufficiently trustworthy in the realization of the task and in evaluating others performing that task**) ***then***
vA) $DoT_{XZ\tau} > \sigma$ (**X trusts Z about τ**)

We have to underline that (iiA) should not necessarily be true, the important thing is that is true (iiiA).

In the case (B) in which the *tasks are different* ($\tau \neq \tau'$), we have:

if
iB) $DoT_{XY\tau} > \sigma$ (**X trusts Y about τ**) and
iiB) $DoT_{YZ\tau'} > \sigma$ (**Y trusts Z about τ'**) and
iiiB) $Bel(X\ DoT_{YZ\tau'} > \sigma)$ (**X believes that Y trusts Z about τ'**) and
ivB) $Bel(X\ Trustworthiness_{YX}(\tau) > \sigma)$ ***implies*** $Bel(X\ Trustworthiness_{YX}(eval(\tau')) > \sigma)$
then
vB) $DoT_{XZ\tau'} > \sigma$ (**X trusts Z about τ'**)

In the case B, the transitivity essentially depends from the implication reported in the formula (ivB); are there elements in the reasons (believed by **X**) for trusting **Y** on the task τ that (in **X**'s view) are sufficient also for trusting **Y** on the evaluation of a different task τ'?

3.1 Competence and Willingness in Transitivity

The need for a careful qualitative consideration of the nature of the link between the trustor and the trustee, is even more serious. Not only it is fundamental (as we have argued) to make explicit and do not forget the specific "task" (activity, and thus "qualities") **X** is trusting **Y** or **Z** about, but it is even necessary to consider the different dimensions/components of the trust disposition, decision, and relation.

In our model, trust has two basic nucleuses:

(i) *Y*'s *competence*, ability, for correctly performing the delegated task;
(ii) *Y*'s *willingness* to do it, to act as expected.

The two dimensions (and 'virtues' of ***Y***) are quite independent on each other: ***Y*** might be very well disposed and willing to do, but not very competent or unable; ***Y*** might be very expert and skilled, but not very reliable: unstable, unpredictable, not well disposed, insincere, dishonest, etc. Now, this (at least) double dimension affect transitivity. In fact, even assuming that the *competence* is rather stable (and that ***Y*** is a good evaluator of ***Z***'s *competence*) not necessarily ***Z***'s *willingness* is equally stable and transferable from ***Y*** to ***X***. This is a more relation-based dimension. ***Y*** was evaluating ***Z***'s *willingness* to do as expected on the basis of their specific *relation*. Is **Z** a friend of ***Y***? Is there a specific benevolence, or values sharing, or gratitude and reciprocity, or obligation and hierarchical relation, etc.? Not necessarily the reasons that **Z** would have for satisfying ***Y***'s expectation and delegation would be present (or equally important) towards ***X***. ***X***'s relation with **Z** might be very different. Are the reasons/motives motivating **Z** towards ***Y***, and making him reliable, transferable or equally present towards ***X***? Only in this case it would be reasonable for ***X*** to adopt ***Y***'s trustful attitude and decision towards **Z**.

Only certain kinds of relations will be generalized from ***Y*** to ***X***; for example, if ***Y*** trusts **Z** only because it is an economic exchange, only for **Z**'s interest in money, reasonably ***X*** can become a new client of **Z**; or if ***Y*** relies on **Z** because **Z** is a charitable person, generously helping (without any prejudice and discrimination) poor suffering people, and ***X*** is in the same condition of ***Y***, then also ***X*** can trust in **Z**.

3.2 Trust Dynamics Affects Transitivity

Moreover, we have shown ([4, 5]) that ***Z***'s willingness, and even ability, can be affected, increased, by ***Y***'s trust and reliance (this can affect ***Z***'s commitment, pride, effort, attention, study, and so on). ***Z***'s *trustworthiness* is improved by ***Y***'s trust and delegation. And ***Y*** might predict and calculate this in her decision to rely on **Z**.

However, not necessarily the effect of ***Y*** on ***Z***'s trustworthiness will be produced also by another trustor. Thus, also this will affect "transitivity": suppose that ***Y***'s trust and delegation to **Z** makes him more trustworthy, improves ***Z***'s willingness or ability (and ***Y*** trusts and relies on **Z** on the basis of such expectation); not necessarily ***X***'s reliance on **Z** would have the same effect. Thus even if ***X*** knows that ***Y*** reasonably trusts **Z** (for something) and that he is a good evaluator and decision-maker, not necessarily ***X*** can have the same trust in **Z**, since perhaps ***Z***'s trustworthiness would not be equally improved by ***X***'s reliance.

4 Transitivity and Trust: Related Work

The concept of trust transitivity has been considered in other approaches.

A relevant example is given from the Josang's approach; he introduces the subjective logics (an attempt of overcoming the limits of the classical logics) for taking in consideration the uncertainty, the ignorance and the subjective characteristics of the beliefs. Using this approach Bhuiyan et al. addressed the problem of trust transitivity [6], where it is recognized the intrinsic cognitive nature of this phenomenon. However, the main limits of this approach are that trust is in fact the *trust in the information sources*; and the transitivity regards two different tasks (referred to our formalism: $\tau \neq \tau_1$: ***X*** has to trust the evaluation of ***Y*** (task τ) with respect ***Z*** as realizing another task (task τ_1, for example as mechanic). As we showed before, this difference is really relevant for the transitivity phenomenon. In addition, also the first task (***Y*** as evaluator) is just analyzed with respect to the *property of sincerity* (and this is a confirmation of the constrained view of trust phenomenon; they write: "*A*'s disbelief in the recommending agent *B* means that *A* thinks that *B* consistently *recommends the opposite* of his real opinion about the truth value of *x*"; where *A*, *B*, and *x* are, in our terms, respectively ***X***, ***Y*** and τ_1). But in trusting someone as evaluator of another agent (with respect to a specific task), I have also to consider his *competence* as evaluator, not just his *sincerity*. Trust is based on ascribed qualities. ***Y*** could be completely sincere but at the same time completely inappropriate to evaluate that task. Other authors [7, 8, 12, 13], developed algorithms for inferring trust among agents not directly connected. These algorithms differ from each other in the way they compute trust values and propagate those values in the networks.

5 Concluding Remarks

In this paper we have shown: i) how the trust transitivity can be considered a valid property when the analysis of the reasons of the mediator is supported by the belief of the trustor; ii) how the willingness dimension of the trustee's trustworthiness is not easily stable and transferable (and thus transitive).

We have also tried to characterize the basic simplified principles of trust transfer but - of course - this just provides us a frame for future work; these mechanisms should in fact be sophisticated and combined. For example, ***X*** might believe that both ***Y*** and ***W*** trust ***Z*** as for τ, and he has to combine the different deriving "measures" of ***Z***'s trustworthiness due to the different reliability of ***Y*** and ***W*** and their non identical esteems of ***Z***. Similarly: ***X*** might believe that ***Y*** trusts ***Z***, and might also have a personal evaluation of ***Z*** based for example on direct experience or communication with ***Z***; he has to combine this two trusts, also considering (like for ***W***) the reflexive evaluation of himself as a good evaluator!

Moreover, one should also consider the context (which might be non identical in the two trust relations) and in particular the *relation* between ***Y*** and ***Z*** and ***X*** and ***Z***, which might be dissimilar. Is the trustworthiness of ***Y*** in ***Z*** based on ***Z***'s role and sense of duty? Is thus ***Z***'s attitude impersonal and benevolent towards any possible partner or client (like it should be for a public authority or a doctor in hospital)? In fact, ***Y***'s trust might be based (especially as for ***Y***'s reliability) on a special relation between them: personal knowledge, friendship, common values or membership, recommendation by a common friend or "authority", and so on, but this relation will not be there with ***X***, and this will block any transfer; except ***X*** finds some personal link with ***Z*** and thus obtains a "favor". Or, vice versa, is the *relation* between ***Y*** and ***Z*** such that ***Y*** (which in general is a good evaluator in that area) is not objective and credible, since is a friend or a relative of **Z**? As we can see, trust pseudo-transitivity and in general transfer is a rather crucial phenomenon in economic, organizational, political relations; its mechanisms should be clearly disentangled, formalized, and then implemented both for modeling (simulating) and studying social dynamics, and for creating new ICT-based trust devices in HCI and in computer mediated social relations like WEB networks, virtual organizations, and eDemocracy.

References

[1] Marsh, S.P.: Formalising Trust as a computational concept. PhD thesis. University of Stirling (1994), `http://www.nr.no/abie/papers/TR133.pdf`

[2] Yu, B., Singh, M.P.: Searching social networks. In: Proceedings of the Second International Joint Conference on Autonomous Agents and Multi-Agent Systems (AAMAS), pp. 65–72. ACM Press (2003)

[3] Sabater, J.: Trust and Reputation for Agent Societies, PhD thesis. Universitat Autonoma de Barcelona (2003)

[4] Falcone, R., Castelfranchi, C.: The Socio-Cognitive Dynamics Of Trust: Does Trust Create Trust? In: Falcone, R., Singh, M., Tan, Y.-H. (eds.) AA-WS 2000. LNCS (LNAI), vol. 2246, pp. 55–72. Springer, Heidelberg (2001)

[5] Castelfranchi, C., Falcone, R.: Trust Theory: A Socio-Cognitive and Computational Model. John Wiley and Sons, Chichester (2010) ISBN 978-0-470-02875-9

[6] Bhuiyan, T., Josang, A., Xu, Y.: An analysis of trust transitivitå taking base rate into account. In: Proceeding of the Sixth International Conference on Ubiquitous Intelligence and Computing, July 7-9, University of Queensland, Brisbane (2009)

[7] Li, X., Han, Z., Shen, C.: Transitive trust to executables generated during runtime. In: Second International Conference on Innovative Computing, Information and Control (2007)

[8] Golbeck, J., Hendler, J.: Inferring binary trust relationships in web-based social networks. ACM Transactions on Internet Technology 6(4), 497–529 (2006)

[9] Meyer, J.J.C., van der Hoek, W.: A modal logic for nonmonotonic reasoning. In: van der Hoek, W., Meyer, J.J.C., Tan, Y.H., Witteveen, C. (eds.) Non-Monotonic Reasoning and Partial Semantics, pp. 37–77. Ellis Horwood, Chichester (1992)

[10] Falcone, R., Piunti, M., Venanzi, M., Castelfranchi, C.: From Manifesta to Krypta: The Relevance of Categories for Trusting Others. ACM Transaction on IST (in press, 2012)

[11] Burnett, C., Norman, T., Sycara, K.: Bootstrapping trust evaluations through stereotypes. In: van der Hoek, Kaminka, Lesperance, Luck, Sen (eds.) 9th International Conference on Autonomous Agents and Multi-Agent Systems (AAMAS 2010), Toronto, Canada, pp. 241–248 (2010)

[12] Tavakolifard, M., Herrmann, P., Öztürk, P.: Analogical Trust Reasoning. In: Ferrari, E., Li, N., Bertino, E., Karabulut, Y. (eds.) IFIPTM 2009. IFIP AICT, vol. 300, pp. 149–163. Springer, Heidelberg (2009)

[13] Christianson, B., Harbison, W.: Why Isn't Trust Transitive? In: Lomas, M. (ed.) Security Protocols 1996. LNCS, vol. 1189, pp. 171–176. Springer, Heidelberg (1997)

A Reputation Framework Based on Stock Markets for Selecting Accurate Information Sources*

Ramón Hermoso

Abstract. Reputation has been presented as a cornerstone in recent years to foster trust relationships among individuals in multiagent systems. Moreover reputation, as an aggregation of opinions, has become a major topic in research. We are particularly interested in e-commerce scenarios, where capital drives decision-making processes of the agents. More concretely, our work focuses on how to facilitate the search of good reputation sources. In this paper we put forward a theoretical framework to manage reputation of a company by using a dynamic reputation market mechanism based on econometric stock markets analogy.

Keywords: Trust, reputation, economy, stock markets.

1 Introduction

Trust is a key concept when dealing with interpersonal interactions in any kind of social system. Estimating trust becomes a serious issue when trying to decide whether delegate certain task to other individual or requesting a service to certain entity [3]. Multi-Agent Systems (MAS) have shown to be a useful paradigm to come up with solution for complex problems. One of the most important characteristics of MAS is the sociability that the participants may present. For that reason, during the last 15 years a lot of different

Ramón Hermoso
Centre for Intelligent Information Technologies (CETINIA), University Rey Juan Carlos, Tulipán s/n, 28933, Madrid, Spain
e-mail: `ramon.hermoso@urjc.es`

* The present work has been partially funded by the Spanish Ministry of Education and Science under project OVAMAH-TIN2009-13839-C03-02 (co-funded by Plan E) and Agreement Technologies (CONSOLIDER CSD2007-0022, INGENIO 2010).

J.B. Pérez et al. (Eds.): Highlights on PAAMS, AISC 156, pp. 189–196.
springerlink.com

approaches have been presented in the literature, most of them focusing on trust models and reputation mechanisms. Even when a lot of effort has been done so far, still in the research community there is not a universally accepted definition neither for the concept of trust nor reputation [6]. Studying related literature one can identify two main schools of research. On the one hand, *cognitive trust* claims that trust forms part of the agent's mental state, based on generated beliefs [1]. On the other hand, *probabilistic trust*, that attempts to give a probabilistic meaning for trust as an estimate to measure how successful an individual is performing certain action [5, 7]. In this work we adhere to the latter trend, where reputation is generally combined with local past experience to come up with a trust value for a specific agent or situation. Reputation is then defined as a metric that may be used by the agents to calculate trust. When dealing with individuals, an agent's opinions are formed by using its trust values concerning the agents or situations requested. However when working on business environments, the task of calculating reputation of a company is not straightforward. More concretely, playing in the same court with big companies, one could suspect that his own opinion is actually tiny or even worthless, due to the small set of interactions carried out with the company taking into account the universe of them that the company provides and the possibly enormous set of relevant variables. For that reason the use of distributed reputation mechanisms could be presented as a poorly solution to gather opinions, due to the huge number of different requests that an agent (or a company) would have to do in order to come up with valuable information [1]. In this paper we intend to cope with the problem of how select accurate reputation sources (opinion sources) when is needed.

From the 16th century until these days, stock markets have been used as meeting places where sellers and buyers have exchanged their stocks. From a social point of view a stock measures the value of the ownership of (a very small) part of a company. Since stocks are available for any kind of purchaser (individuals or companies can invest in it), current values actually represent the amount of money that investors (or buyers) are willing to pay for them. Many authors claim that besides money, stock markets deal with investors' trust on different companies, and thus high values in the long-term denote high confidence of investors about the well functioning of the company (and vice versa). The aim of this paper is to take this stock market metaphor to build a framework in which the value of stocks represents the reputation of a company as a reputation source.

The paper is organised as follows: Section 2 gives more details of the framework. Then in Section 3 the reputation framework is presented. A brief discussion, stressing the problems that underlie the model is explained in Section 4. Section 5 summarises the paper and points out some future work.

---

[1] Not to mention the added complexity to select so many "accurate" opinion sources.

2 Preliminaries

In order to present a formal reputation framework based on the use of stock markets, let us firstly introduce some fundamental issues. As previously mentioned, this work deals with the concept of reputation in which e-commerce companies are involved. We are not interested in the type of service that companies offer or the type of goods they sell, but how these companies and, specially the opinions they can exchange, may support others on better choosing reputation sources. A reputation source is typically an agent that is requested to provide its opinion about a third party. Consumers of this information, in a real world scenario might be manufacturers, retailers or final customers.

Firstly it is important to say that the only service that is provided in our market is reputation provision. We do not focus on the activities the system was built for, but only in the reputation exchange management. Let's denote with $\mathcal{C}$ the set of potential companies (or agents) participating in the system. The following assumptions are required for the correct understanding of the framework:

1. The number of reputation requests that a company c_i could reply to at a given time t is finite.
2. A company c_i will have an objective utility function, based on financial gain/loss of monetary capital (sum of assets and liabilities). So more money means higher utility (and vice versa).
3. A company c_i will have an initial capital ic_i that represents the assets when the company enters the system.

With these assumptions we address the problem of building a framework that supports to choose the best possible counterpart for obtaining accurate opinions from third parties.

3 A Formal Framework for Reputation Markets

3.1 Basic Definitions

In order to better put forward the functioning of the model we firstly need to define some key concepts:

Definition 1. *A stock (k_i) is the set of shares of certain company c_i. The size of the stock will represent the number of possible reputation requests (opinions) that a company may process at a given time. Thus, $|S_i| = |k_i|$ means that a share is equivalent to a service instance that can be request to the company at time t. In other words the stock of a company will be equivalent to the amount of reputation queries that is willing to reply. As in real markets, the stock is issued by the company itself, being its size also decided by the company.*

Definition 2. *A company's stock will have a price* $p : C \times t \rightarrow \mathbb{R}$*, consisting of last selling price at time* t.

This framework is based on the purchase/sale of reputation requests. That is, when A purchases a stock k' (with $|k'| \leq |k_B|$) of company B, A ensures the right to send $|k'|$ reputation requests to B and be replied. It could be seen as hiring the service of a trading agency to give you advices (or tips). Thus a company could buy shares for short-term purposes (e.g. next time-step needs) or, anticipating the behaviour of the market, could buy more stock than needed just for gambling. A value of a stock then represents the reputation of a company in that precise moment[2].

Definition 3. *We define* ***reputation capital*** *of company* c_i *at time* t *as:*

$$\Theta^{Rep} = |k_i| \cdot p(c_i, t)$$

that is, the reputation capital measures how wanted is the company at time t. It is considered as capital since, although it does not represent actual assets, it denotes prices that others are willing to pay to ask for reputation to it in the current situation (potential incomes).

Definition 4. *A* ***portfolio*** $\Pi_A = \{\pi_1, ..., \pi_n\}$ *is defined as the collection of stocks that a company* A *owns at a given time* t*, in which* $\pi_i = (c_j, q_j)$*. That is, the portfolio is formed by different stocks from distinct companies, where* $q_j \leq |k_j|$ *states for the quantity of shares owned of a company.*

Analogously to the definition of reputation capital the following defines the pay-off function:

Definition 5. *A company* A*'s* ***pay-off*** *is defined by the function* $\Lambda : \mathbb{R} \times \mathbb{R} \times t \rightarrow \mathbb{R}$*, where* t *is time. Pay-off function for company* c_A *will consist of the difference between the cost (purchases) and the revenues (sales) paid and obtained, respectively, by company* c_A *at time* t*:*

$$\Lambda_{c_A} = [p(c_A, t) \cdot \sum_{(c_A, q_k) \in \Pi_i}^{\Pi} q_k] - \sum_{(c_j, q_j) \in \Pi_A} q_j \cdot [p(c_j, t)]$$

Definition 5 inherently entails that a company's pay-off at a given time t is independent of the actual price the company paid for the stocks in the past, but relies on the current prices for those stocks in the market. It is reasonable to think that if the company sells all its assets at time t the described function would perfectly represent its revenue.

3.2 Buying and Selling Policies

A company A joins a reputation market in order to share and request its opinions over time. How a company decides whether to use reputation to estimate

[2] The value of a stock is updated with the price paid in the last stock sale corresponding to that company.

trust on other companies is something already discussed in the literature. Researchers agree that there must exist a trade-off between the use of past experiences and the use of reputation [2]. It is out of the scope of this paper to go into this topic. However, we are interested in the reasoning process of how a company decides whom to ask about reputation. Thus assuming that A needs opinions from others to form its own trust value at time t:

Definition 6. *The purchase of stocks is defined by the function* $buy_{c_i} : \mathcal{C} \times \Pi \times \mathbb{R} \rightarrow [0, 1]$*, or commonly* $buy_{c_i}(c_j, \pi_r, p')$ *that means that company* c_i *executes a purchase of* $\pi_r = (c_j, q_k)$ *of company* c_j *at price* p'*. The result of the function will be* 1 *for successful transactions and* 0 *to indicate failures.*

A company can buy stock of another company *iff* it has enough liquid assets to execute the purchase. Different policies (most of them modified from the econometric science) can be derived for buying a stock:

Growth-based policies: the use of this policy takes into account the size of different companies for buying reputation. The size of a company and its growth may be measured as the number of different requests it can attend at a given time t. This number might change over time by means of recapitalization processes (more shares might be issued).

Risk aversion-based policies: this type of policy deals with the attempt of minimising the risk when purchasing reputation. One of the most used functions to estimate risk in investing is VAR (*Value-at-Risk*), defined by:

$$Prob[\Delta\tilde{P}(\Delta t, \Delta\tilde{x}) > -VAR] = 1 - x$$

where $\Delta\tilde{P}(\Delta t, \Delta\tilde{x})$ is the change in the market value of the portfolio, expressed as a function of the *forecast horizon* Δt and the vector of changes in the random state variables $\Delta\tilde{x}$. The parameter α is the confidence level. The interpretation is that, over a large number of training days, the value of the portfolio will decline by no more than VAR α % of the time.

Statistical-based policies: these typically use statistical methods such as Monte-Carlo simulation techniques to calculate the optimum (lower cost) purchasing policy.

Hybrid policies: uses a mixture of different policies.

When to sell and at which price is also an interesting issue when facing the construction of a reputation stock market. An agent may decide to sell stock due to diverse reasons:

- Has enough historical data recorded (past experiences) and considers that the system has stabilized (behavioural changes on agents rarely occur). Thus it takes the decision of reducing its investment in reputation. This means that needs to sell some of the stock (or the whole) it has.
- Decides to gamble and considers that selling some of the stock at time t may result in a revenue. For instance, the company may consider that the stock has reached its maximum price of sale, so it becomes worth to be sold.

Table 1 Example of a RSI

Co. name	Value
B	9.24
A	1.27
C	3.99

Different selling protocols coming from economics could be used to govern a purchase. As a preliminary study we claim that an auction-based mechanism could be used[3]. For instance, a Vickrey auction protocol [8] might be used when a company decides to sell its stock. It is proved that truthful bidding dominates the other possible strategies (underbidding and overbidding), so it is an optimal strategy. This property ensures that companies will bid with honest purchase prices.

3.3 Reputation Stock Index

In every stock exchange market there exists a public record of data, consisting of a list of prices for different companies' stocks, historical data, and so forth. In the framework presented in this paper we present a Reputation Stock Index (RSI), that is in charge of displaying the effects of different transactions in the system. RSI will contain a list with different companies stocks, associated to their current prices; i.e. a pair $(A, 10.21)$ informs that last sale of stock of company A was made at the price of 10.21 per unit. Table 1 shows an example of RSI.

Analogously as it is done in real stock markets, some interesting indexes can be generated from the RSI information. For instance, it would be worth to calculate the average value of the RSI as it happens with NASDAQ or S&P 500 in USA or with EuroStock in EU. This information would permit a company to check in a glimpse which stocks are more promising to return good revenues. Another important index is Stock Turnover (ST), that is, how much money is exchanged per time unit. This last feature is crucial if we want to observe if the system stabilizes.

Updates in the RSI are made after stock purchase, replacing the stock value in the table by the last sale price.

4 Discussion

The framework presented in this paper tries to tackle some of the problems derived from the nature of the reputation itself. Reputation is an aggregation of subjective opinions and often a distributed value. How to choose proper reputation sources to ask about third parties is a non-trivial task [9]. This framework supports that task by gathering reputation of different companies

[3] Actually, some stock markets such as NYSE market follow this model.

by using a metaphor of stock markets. In this framework we propose a reputation stock market transforming reputation requests into a tangible asset companies can negotiate with. Even if opinions are still subjective, prices for buying reputation requests are not, so estimating purchasing or selling values becomes easier. This framework allows to test with different types of company profiles (different business strategies), such as: conservatives, companies that only seek regular accurate opinions for low prices; gamblers, that are only interested in making money playing on the reputation market (not only use it as a way to select good business counterparts); non-rational companies, e.g. more human-like entities that do not base their decisions on rational reasoning. We claim that experiments with different profiles may help to better explore the advantages and drawbacks of the model.

Bounded resources, such as initial capital and limitation on the number of requests that a company can attend at a given time (number of shares that the company issues) is something interesting from MAS designer point of view. This feature keeps the system from being a monopoly of opinion generation. The idea is that this framework could be useful for any kind of open MAS that needs trust for its well functioning, even if the system is resource-bounded. We are specially interested in apply this framework to high dynamic supply chains, more concretely in the formation phase.

Stock price estimation is one of the most complex problems in econometry. Low prices do not always mean low level of outcomes (in our case inaccurate opinions about third parties) but also a high offer to provide opinions. It is reasonable that a company A with $|k_A| = 10$ that provides regular accurate opinions will be (in long term) as expensive as company B with $|k_B| = 10000$ providing the same level of accuracy. In our approach, stock pricing for reputation requests could be directly supported by applicable econometric theories. We plan to use different pricing techniques in future experiments.

Stock markets to manage the use of reputation sources is an incentive-based mechanism to keep agents from cheating when giving their opinions. A similar approach in the literature was proposed by Jurca and Faltings [4]. In their work they put forward an incentive mechanism based on buying and selling reputation with a credit-based system. Although the idea of fostering honest opinions is shared with our work, we do use market metaphor to maintain the reputation and to self-adapt changes in the system.

5 Conclusions

The paper addresses the problem of control and maintenance of reputation exchange between agents in a e-commerce environment. We put forward a framework based on stock exchange markets to monitor and manage reputation information over time. This framework is based on the construction of a stock market in which stocks represent reputation requests for an agent. This preliminary work presents the basis for a mid-term research on reputation market mechanisms. We strongly believe that this type of mechanisms is suitable for open distributed e-commerce societies, where agents represent companies and are able to exchange information automatically. As future work we plan to theoretically explore in detail some of the open issues that come up with this type of framework, such as gambling, collusion of companies or pricing techniques. We also have in mind to implement the framework and experiment with different scenarios, with diverse company strategies or open and closed populations.

References

1. Castelfranchi, C., Falcone, R.: Principles of trust for mas: Cognitive anatomy, social importance, and quantification. In: ICMAS 1998: Proceedings of the 3rd International Conference on Multi Agent Systems, pp. 72–79. IEEE Computer Society, Washington, DC (1998)
2. Fullam, K.K., Suzanne Barber, K.: Dynamically learning sources of trust information: experience vs. reputation. In: AAMAS 2007: Proceedings of the 6th International Joint Conference on Autonomous Agents and Multiagent Systems, pp. 1055–1062. ACM, New York (2007)
3. Hermoso, R., Billhardt, H., Ossowski, S.: Role evolution in open multi-agent systems as an information source for trust. In: 9th International Conference on Autonomous Agents and Multi-Agent Systems (AAMAS 2010), pp. 217–224. IFAAMAS (2010)
4. Jurca, R., Faltings, B.: Towards Incentive-Compatible Reputation Management. In: Falcone, R., Barber, S., Korba, L., Singh, M. (eds.) AAMAS 2002. LNCS (LNAI), vol. 2631, pp. 138–147. Springer, Heidelberg (2003)
5. Marsh, S.P.: Formalising Trust as a Computational Concept. PhD thesis, University of Stirling (1994)
6. Harrison Mcknight, D., Chervany, N.L.: The meanings of trust. Technical report, University of Minnesota (1996)
7. Luke Teacy, W.T., Patel, J., Jennings, N.R., Luck, M.: Travos: Trust and reputation in the context of inaccurate information sources. Journal of Autonomous Agents and Multi-Agent Systems 12 (2006)
8. Vickrey, W.: Counterspeculation, auctions and competitive sealed tenders. Journal of Finance 16, 8–37 (1961)
9. Yu, B., Singh, M.P.: Detecting deception in reputation management. In: Proceedings of the 2nd International Joint Conference on Autonomous Agents and Multiagent Systems, AAMAS 2003, pp. 73–80. ACM, New York (2003)

Contextual Norm-Based Plan Evaluation via Answer Set Programming

Sofia Panagiotidi, Javier Vázquez-Salceda, and Wamberto Vasconcelos

Abstract. There is a recent trend on agent-oriented methods and abstractions in the field of Service Engineering to tackle governance of distributed (Semantic) Web systems. Some approaches are based on the creation of a social level where actors' behaviour is regulated by means of computational norms. We present a framework where norm-enabled agents can, at runtime, 1) enter an organisational context, 2) get the organisational specification, including norms, and translate it into the agents' internal representation and 3) determine the quality of a to-be-adopted plan taking into account the incentives derived from the norm-regulated context. A normative model formalisation is provided using Semantic Web elements. Then a translation of the formalism into Answer Set Programming and a full implementation of a normative plan evaluator are presented.

1 Introduction

As distributed electronic systems grow to include thousands of components, from grid to peer-to-peer nodes, from (Semantic) Web services to web-apps to computation in the Cloud, governance of such systems is becoming a real challenge. Although some standard approaches to governance have tried to solve it by means of predefined workflows and operational modes, it is foreseen that achieving and maintaining governance by static methods will be difficult (if not impossible), as the participating entities, their modes of interaction or the intended purpose of the distributed system may change over time. Therefore there is a need to create more flexible approaches. One approach for governance of distribute systems, which

Sofia Panagiotidi · Javier Vázquez-Salceda
KEMLG Group, Univ. Politecnica de Catalunya
e-mail: {panagiotidi,jvazquez}@lsi.upc.edu

Wamberto Vasconcelos
Department of Computing Science, Univ. of Aberdeen, UK
e-mail: w.w.vasconcelos@abdn.ac.uk

J.B. Pérez et al. (Eds.): Highlights on PAAMS, AISC 156, pp. 197–206.
springerlink.com

comes from multi-agent systems research, is to add social order mechanisms to the system, where the individual computational entities' behaviour is guided or even restricted in order to ensure certain behaviour.

In this paper we focus on the use of organisational norms as the basis of social order and we focus on how to implement agents that can use the norms' incentives to explore potential future actions and strategically decide which path to follow. There is much theoretical and applied research on languages for norms and normative agents (see Section 7). Our approach differs from existing work in two basic aspects: 1) the norm representation used (which comes from [9]) is independent from the ASP (Answer Set Programming)-based representation used in normative reasoning, thanks to the use of a Model-Driven engineering approach; 2) our normative reasoner is able to explore, at execution time, not only the effect of the norms on the next step of execution, but along several paths in the future, each path being a plan composed by several actions.

2 System Architecture

In our architecture all agents can enter and operate into organisational contexts which include i)an explicit representation of the system's organizational structure (including a model of the stakeholders, their relationships, their goals, responsibilities and organisational norms); ii) operational descriptions of the actions in the domain; and iii)a domain ontology, describing the concepts in the domain. For the agent to be organisational-aware, it should understand and reason about the structure, work processes, and norms of the organisation. Following a Model-Driven Engineering (MDE) approach, we have implemented mechanisms to automatically transform the organisational model (including its norms), the ontological elements and action descriptions into the representations that can be used by our agents in their plan generation and evaluation.

The agent core is programmed in Agentscape, a BDI implementation which, as many others (JADEx, Jason, JACK or 2APL) does not support automatic plan generation within the BDI cycle, typically using pre-computed plans. To solve this our agents perform their means-ends-reasoning by 1) requesting plans from a JSHOP2 [7] planner, and 2) perform a "norm-driven plan evaluation", i.e. evaluating the "goodness" of the generated plan, taking into account the incentives in the normative environment. The plan evaluator component receives a plan generated by the planner, an initial state description, a set of instantiated norms and a set of instantiated domain actions and, taking into consideration the norms and the current state of affairs, evaluates the given plan. The evaluation of a plan mainly concerns the violation of the restrictions imposed by the norms. The implementation of the normative plan evaluator presented in this paper is based on the operational semantics of normative elements' lifecycle defined in [10] and on a normative environment implementation [11] in ASP.

3 Example

We will use an example throughout the paper in order to illustrate the utility of our framework. The example is a simple emergency scenario where a building evacuation has been ordered due to some threat. To ensure the occupants safety, they need to be transferred to another building with the use of an available vehicle (or more, but in our simplified case we assume only one). The evacuation takes place as follows: the vehicle takes a number of occupants from the building to be evacuated, drives to the other building, leaves the passengers there and returns to continue with the procedure. The goal is to have no remaining occupants in the initial building.

We model actions that can be executed depending on their preconditions: load people on the vehicle, move vehicle from one building to another, unload people at a building. In addition to this, a restrictive norm is imposed: every vehicle has a capacity of passengers associated to it, which, in case of emergency evacuation, cannot be exceeded by one.

4 Preliminary Definitions

This section lays out some basic definitions to ground our discussion. We represent first-order variables generically as x, y, z, possibly with subscripts to differentiate them; constants are represented as a, b, possibly with subscripts. Actual sample variables follow the Prolog [2] convention, that is, they are any string starting with an upper case letter, and sample constants are any string starting with a small case letter.

Initially, we define ***atoms***:

Definition 1. *An atom α is one of the following:*

- $c(x)$*, where c is an OWL*[1] *class*
- $p(x,y), \neg p(x,y)$*, where p is an OWL DL individual-valued property (with d_x and r_y being the domain and range of the property)*
- $q(x,a), \neg q(x,a)$*, where q is an OWL DL data-valued property (with d_x and r_a being the domain and range of the property and r_a of a built-in type such as integer or string)*
- $\geq(x,y), =(x,y), +(x,y,z), -(x,y,z), *(x,y,z), \div(x,y,z)$ *special built-in atoms representing mathematical functions*

We denote $c(x)$ as α^c. Informally, $c(x)$ asserts that x an instance of class c and $p(x,y)$ represents the fact that x is related to y by property p. Atoms may refer to individuals, data literals, individual variables or data variables. We also refer to properties as *fluents*[2].

Example 1. *We represent the set of fluents of the example of Section 3 as*

$$F = \{\, hasOccupants(B,Num), capacity(B,C), at(V,B), peopleOnBoard(V,Num), evacuation(B,V)\,\} \qquad \square$$

---

[1] Web Ontology Language, available at http://www.w3.org/TR/owl-ref

[2] As in Event Calculus, since we are interested in describing these properties throughout time.

We make use of sets of atoms to represent an expression:

Definition 2. *An expression* Ξ *is a possibly empty and finite set of atoms* $\{\alpha_0,\ldots,\alpha_n\}$.

An expression stands for the existential quantification[3] of the conjunction of its atoms, that is, $\{\alpha_0,\ldots,\alpha_n\} \equiv \exists \mathbf{x} \bigwedge_{i=0}^{n} \alpha_i$, where $\mathbf{x}$ are all variables in $\alpha_i, 0 \leq i \leq n$.

We employ explicit substitutions [2, 6] to store the results of our computations. A substitution, denoted as σ, is a possibly empty and finite set of pairs x/τ, where x is a variable and τ is a term (that is, a variable, a constant or a functional symbol with any number of nested terms). We denote as $\alpha \cdot \sigma$ the application of the substitution σ to α, whereby we replace all occurrences of x with $\tau \cdot \sigma$, if $x/\tau \in \sigma$; $a \cdot \sigma = a$, for a constant a. We also define $\Xi \cdot \sigma = \{\alpha_0,\ldots,\alpha_n\} \cdot \sigma = \{\alpha_0 \cdot \sigma,\ldots,\alpha_n \cdot \sigma\}$.

We represent ***states*** of the world (or states of affairs) as a (partial) instantiation of an expression via a substitution σ, that is, $\Xi \cdot \sigma$, and we shall denote these as s. If the state is such that all variables are ground, that is, they all get assigned a constant via σ, then we denote this as $\bar{s}$. We represent as $\bar{s}_0$ a special *initial* state. The initial state is not allowed to contain any of the special built-in atoms.

Example 2. *The initial state of our example is:*

$$\bar{s}_0 = \left\{ \begin{array}{c} building(X_1), building(X_2), vehicle(X_3), peopleOnBoard(X_3,X_6), evacuation(X_1,X_3), at(X_3,X_1), \\ hasOccupants(X_1,X_4), hasOccupants(X_2,X_5), capacity(X_3,X_7) \end{array} \right\}$$

with substitution $\sigma = \{X_1/maternity, X_2/hclinic, X_3/jeep, X_4/10, X_5/3, X_6/0, X_7/5\}$ □

We introduce a representation for ***actions***:

Definition 3. *An action* κ *is the tuple* $\langle a, \Xi_c^I, \Xi_P, \Xi_A, \Xi_D^{prop} \rangle$ *where:*

- *a is the unique label/name of the action;*
- $\Xi_c^I = \{c_0^I(x_0^I),\ldots,c_n^I(x_n^I)\}$ *is a set class atoms, representing the inputs of the action;*
- Ξ_P *is a set of atoms (class or otherwise) representing the preconditions of the action;*
- Ξ_A *is a set of atoms (class or otherwise) to be added to the current state;*
- $\Xi_D^{prop} = \{p_0^D(x_0^D,y_0^D),\ldots,p_m^D(x_m^D,y_m^D)\}$ *is a set of property atoms to be deleted from the current state.*

Example 3. *A sample action is:*

$$\kappa_1 = \left\langle \begin{array}{c} loadPeople, \{vehicle(V), building(B), int(Num)\}, \\ \{hasOccupants(B,O), \geq (O,Num), peopleOnBoard(V,P), -(O,Num,O2), +(P,Num,P2)\}, \\ \{hasOccupants(B,O2), peopleOnBoard(V,P2)\}, \{hasOccupants(B,O), peopleOnBoard(V,P)\} \end{array} \right\rangle$$ □

We formally define the meaning of an action with respect to a given state as a function $\mathbf{f}(\kappa,s) = s'$, s of the form $\Xi \cdot \sigma$, with two cases:

1. if $((\Xi_P \cdot \sigma') \cup \Xi_c^I) \sqsubseteq (\Xi \cdot \sigma)$, that is, the precondition expression is semantically entailed in the current state, for some σ', then $s' = ((\Xi \cdot \sigma) \cup (\Xi_A \cdot \sigma')) \backslash (\Xi_D^{prop} \cdot \sigma')$;
2. otherwise the precondition does not hold and $s' = \perp$ is a "failed state" and the action is labeled as "impossible".

The first case caters for actions which "trigger" given a state, that is, their preconditions hold (via semantic subsumption), hence the effects (additions and deletions)

[3] A universal quantification may be used as default but the notation for expressions would become more complex.

are carried out in the current state, yielding the next state resulting from executing the action. The second case is the exception of the first one, and we make use of a "failed" state as its output, so as to highlight that the action was not performed (the empty set, the obvious candidate, does not work, as it is possible to obtain an empty set as a result of case 1 above).

We define ***plans*** as a sequence of actions.

Definition 4. *A plan* π *is a sequence of actions* $\{\kappa_1, \ldots \kappa_n\}$. *An empty plan is* $\{\}$.

We define the application of a substitution on a plan, $\{\kappa_1, \ldots, \kappa_n\} \cdot \sigma$, as $\{\kappa_1 \cdot \sigma, \ldots \kappa_n \cdot \sigma\}$, that is, the sequence of actions each of which with the application of the substitution. We define the meaning of a plan with respect to a given initial state s_0, as the sequence of states arising from its individual actions, that is, $\mathbf{f}^*(\langle \kappa_1, \ldots, \kappa_n \rangle, s_0) = \langle s_1, \ldots, s_n \rangle$, where $s_i = \mathbf{f}(\kappa_i, s_{i-1})$, that is, the previous state of affairs is input for the next action, and the outcome is obtained via $\mathbf{f}$, and $\mathbf{f}(\kappa, \perp) = \perp$, that is, an attempt to execute an action on a failed state is a failed state.

Example 4. *Two sample plans and the actions substitutions are below.* □

π_1	$\left\{\begin{array}{c} \mathit{loadPeople}(V,B1,N), \mathit{move}(\mathit{jeep},B1,B2), \mathit{unloadPeople}(V,B2,N), \mathit{move}(\mathit{jeep},B2,B1), \\ \mathit{loadPeople}(V,B1,N), \mathit{move}(V,B1,B2), \mathit{unloadPeople}(V,B2,N) \end{array}\right\}$	$\left\{\begin{array}{c} V/\mathit{jeep}, B1/\mathit{maternity}, \\ B2/\mathit{hclinic}, N/5 \end{array}\right\}$
π_2	$\left\{\begin{array}{c} \mathit{loadPeople}(V,B1,N1), \mathit{move}(\mathit{jeep},B1,B2), \mathit{unloadPeople}(V,B2,N1), \mathit{move}(\mathit{jeep},B2,B1), \\ \mathit{loadPeople}(V,B1,N2), \mathit{move}(V,B1,B2), \mathit{unloadPeople}(V,B2,N2) \end{array}\right\}$	$\left\{\begin{array}{c} V/\mathit{jeep}, B1/\mathit{maternity}, \\ B2/\mathit{hclinic}, N1/7, N2/3 \end{array}\right\}$

Since norms may have normative force only in certain situations, we associate norms with an activating condition. ***Norms*** are thus typically abstract, and are instantiated when the norms activating condition holds. Once a norm has been instantiated, it remains active, irrespective of its activating condition, until a specific expiration condition holds. When the expiration condition occurs, the norm is assumed no longer to have normative force. Finally, independent of these two conditions is the norms normative condition, which is used to identify when the norm is violated. It is important to note that, although our norm representation does not explicitly include deontic operators (O, P, F), the combination of the activating, deactivating and maintenance conditions is as expressive as conditional deontic representations since they can be reduced to equivalent deontic expressions (for more see [1]). Formally a norm consists of:

Definition 5. *A norm* ν *is a tuple* $\langle b, t, \Xi_A, \Xi_D, \Xi_M \rangle$ *where* b *is the unique name,* t *is the type (obligation or prohibition),* Ξ_A *is the activating expression,* Ξ_D *is the deactivating (or expiration) expression and* Ξ_M *is the maintenance expression of the norm.*

Example 5. *A sample prohibiting norm, as in the working example is:*

$\langle \mathit{prohib1}, \{\mathit{evacuation}(B)\}, \{\neg \mathit{evacuation}(B)\}, \{\mathit{capacity}(V,C), \mathit{peopleOnBoard}(V,\mathit{Num}), +(C,1,C1), \geq(C1,\mathit{Num})\}\rangle$

That is, that whenever the evacuation of a building with a vehicle has been ordered, the number of people on board of the vehicle cannot exceed the vehicle's capacity by more than one. □

Having defined the elements of a norm, deontic expressions are reduced to the operational representation based on [10], where detailed semantics over the norm's lifecycle can be found. Here we explain briefly the norm lifecycle semantics.

In order to know whether a norm is active or violated, it is not enough to be aware of the current state of affairs. An additional knowledge concerning the norm status at previous states is required. For example, in order to derive that a norm gets deactivated one needs to know not only that the deactivating condition holds at a specific state but also that the norm was active previously (that the activating condition has occurred at some point in the previous states). Therefore, such properties need to be defined with respect to an entire sequence of states (trajectory path) visited during the execution of a plan, starting from an initial point. Thus, the entailment for norm lifecycle properties is defined with respect to a trajectory path produced by a plan.

1. wrt a trajectory of states $\langle s_1,\ldots,s_n\rangle$, a norm is *active* at state s_n, denoted as $active(\nu,s_n)$ if *i)* $active(\nu,s_{n-1})$, or if *ii)* $(\Xi_A\cdot\sigma)\sqsubseteq s_n$ and $(\Xi_D\cdot\sigma)\not\sqsubseteq s_n$ for some σ.
2. wrt a trajectory of states $\langle s_1,\ldots,s_n\rangle$, a norm has been *violated* at a state s_n, denoted as $viol(\nu,s_n)$, iff *i)* $active(\nu,s_n)$, and *ii)* $(\Xi_M\cdot\sigma)\not\sqsubseteq s_n$, for some σ.
3. wrt a trajectory of states $\langle s_1,\ldots,s_n\rangle$, a norm has been *deactivated* at a state s_n, denoted as $deactivated(\nu,s_n)$, iff *i)* $active(\nu,s_{n-1})$ and *ii)* $(\Xi_D\cdot\sigma)\sqsubseteq s_n$, for some σ.
4. wrt a trajectory of states $\langle s_1,\ldots s_i,\ldots,s_n\rangle$, a norm has been *complied* with, denoted as $complied(\nu,s_n)$, iff *i)* $active(\nu,s_j)$ for all $i\leq j<n$, *ii)* $deactivated(\nu,s_n)$, and *iii)* it is not the case that $viol(\nu,s_j)$ for any $i\leq j<n$

We explore the *utility* of a plan, providing a simple metric to compare plans with respect to their norm compliance. We associate a norm ν with two non-negative integers, $\langle\nu,u,w\rangle$, representing a a punishment (if the plan violates the norm) and a reward (if the plan complies with the norm).

Definition 6. *Given a state* s_0*, a set of norms* $N=\{\nu_1,\ldots,\nu_i\}$*, an association of norms to punishments and rewards* $\langle\nu_i,u_i,w_i\rangle$*, the utility of plan* π*, denoted as* $util(\pi,s_0,N)$ *is computed as:*

$util(\pi,s_0,N)=\sum w_i-\sum u_j$*, for all* $\nu_i\in N$ *such that* $complied(\nu_i,s_n)$ *and for all* $\nu_j\in N$ *such that* $viol(\nu_j,s_n)$ *wrt* $\mathbf{f}^*(\pi,s_0)=\langle s_1,\ldots,s_n\rangle$

5 Transformation Procedure to ASP

As there is no reasoner that supports the language in Section 4, we translate it into ASP rules, extending our previous work in [11].

The domain knowledge (ontology) contains the generic knowledge about the elements described as above. Figure 1 (left) depicts the ASP rules that occur from the ontology.

We define a generic transformation process for atoms (to be used in the action and norm representation), and how each one is represented at a state in ASP code. Figure 2 details this process.

Atom	ASP Rule
p(x,y)	fluent(p(X,Y)) :- d_x(X), r_y(Y).
q(x,y)	fluent(q(X,Y)) :- d_x(X), r_y(Y).

Atom in Ξ	ASP Rule
$c(x_i)$	c(X_i) :- X = τ_i.
$p(x_i,x_j)$	holds(p(X_i,X_j),1) :- $X_i = \tau_i, X_j = \tau_j$.
$q(x_i,x_j)$	holds(q(X_i,X_j),1) :- $X_i = \tau_i, X_j = \tau_j$.

Fig. 1 Domain Knowledge and Initial State Transformation

Assume that we have an initial state $\bar{s} = \Xi \cdot \sigma$ where Ξ is a set of atoms and $\sigma = \{x_1/\tau_1, \ldots, x_n/\tau_n\}$. Figure 1 (right) depicts the ASP rules that occur from the initial state transformation.

Atom in Ξ	ASP Rule
$c(x_i)$	c(X_i)
$p(x_i,x_j)$	holds(p(X_i,X_j),S), fluent(p(X_i,X_j)), $d_{x_i}(X_i), r_{x_j}(X_j)$
$\neg p(x_i,x_j)$	$\neg$ holds(p(X_i,X_j),S), fluent(p(X_i,X_j)), $d_{x_i}(X_i), r_{x_j}(X_j)$
$q(x_i,x_j)$	holds(p(X_i,X_j),S), fluent(p(X_i,X_j)), $d_{x_i}(X_i), r_{x_j}(X_j)$
$\neg q(x_i,x_j)$	$\neg$ holds(p(X_i,X_j),S), fluent(p(X_i,X_j)), $d_{x_i}(X_i), r_{x_j}(X_j)$
built-in	$X >= Y$ or $X = Y$ or $Z = X + Y$ or $Z = X * Y$

Fig. 2 Atoms Transformation Function (*atom_trans*)

Actions transformation is done in two steps. The actions preconditions and the actions effects. We assume an action κ is the tuple $\langle a, \Xi_c^I, \Xi_P, \Xi_A, \Xi_D^P rop\rangle$ as in Definition 3.

Preconditions. For every action, a rule as such is created:

precondition($\alpha(X_0^I,\ldots,X_n^I)$,S) :- $c_0^I(X_0^I)$), ..., $c_n^I(X_n^I)$, *state*(*S*), *atom_trans*$_{a_1}$, ..., *atom_trans*$_{a_k}$.

where $c_0^I(X_0^I)), \ldots, c_n^I(X_n^I)$ are the inputs of the action and all $a_1, \ldots, a_k$ are the atoms in the preconditions Ξ_P of the action and *atom_trans* is the atoms transformation function for every atom to ASP code according to Figure 2.

Effects. All variables used in the atoms in the effects need exist be at least once inside an atom in the precondition, or, be have the same name as one of the input parameters of the action. This is to make sure that no unbound variables exist in the add or delete effects. For every non negated property atom $p(x,y)$ in the add effects of an action, such rule is created:

holds(p(X, Y),S+1) :- $c_0^I(X_0^I)$), ..., $c_n^I(X_n^I)$, state(S), *atom_trans*$_{a_1}$, ..., *atom_trans*$_{a_k}$, fluent(p(X,Y)), action_performed ($\alpha(X_0^I,\ldots,X_n^I)$,S).

where $c_0^I(X_0^I)), \ldots, c_n^I(X_n^I)$ are the inputs of the action and all $a_1, \ldots, a_k$ are the atoms in the preconditions Ξ_P of the action and *atom_trans* is the atoms transformation function for every atom to ASP code according to Figure 2. For every negated property atom in the delete effects of an action, a similar (with negated head) rule is created respectively.

For every action (described as above) in a plan, a rule as such is created:

action_performed($\alpha(X_0^I,\ldots,X_n^I)$, step) :- $X_0^I = \tau_0^I$, ..., $X_n^I = \tau_n^I$.

That is a straightforward translation of the execution of an action at a discrete point of time (state), providing groundings to all input variables of the action. Thus step is an integer representing the state at which the action is executed and increases every such rule that occurs.

The norms transformation is done in a similar fashion, in order to produce ASP rules for the activating-condition, deactivating-condition and maintenance-condition. Thus for a norm of type "prohibition" defined as in Definition 5 firstly, the following rule is generated[4]: `prohibition_id(`d`)`. Then, for the activating condition the transformation produces:

`activating_condition(`d`,`S`) :- state(S),` $atom\_trans_{a_1}$, ..., $atom\_trans_{a_k}$.

where all $a_1, \ldots, a_k$ are the atoms in Ξ_A. Similarly, a rule `deactivating_condition` and `maintenance_condition` are produced for the sets Ξ_D and Ξ_M.

The complete code can be found at "http://www.lsi.upc.edu/ panagiotidi/asp-code".

6 Plan Evaluation

The evaluation is done in two layers (Figure 3 depicts the rules used to reason about norms). The one is to verify that the plan is executable. That is, for each of the actions in the plan, to check whether its preconditions are satisfied at the state that the action gets executed. This is implemented by defining the `possible` and `impossible` predicates. The second is to detect whether the violations of norms that might occur throughout the execution of the plan. Thus, the reasoner "simulates" the execution of the actions and follows the states that the execution leads to, checking at the same time the norms' lifecycle, i.e. whether norms are activated, deactivated and violated. This is expressed by the predicates `prohibition_activated`, `prohibition_deactivated`, `prohibition_violated`, `active` and `violation`. Currently we restrict the evaluation to the detection and counting of the number of violations that occur.

```
int(0..100).  state(0..50).
%---------------------Reasoning Rules-------------------
impossible(Action, S) :- action_performed(Action, S), not precondition(Action, S), state(S).
possible(Action, S) :- action_performed(Action, S), precondition(Action, S), state(S).
-holds(F, S+1) :- -holds(F, S), not holds(F, S+1), fluent(F), state(S).
 holds(F, S+1) :- holds(F, S), not -holds(F, S+1), fluent(F), state(S).
%------------------Norm lifecycle--------------
norm_activated(Id, 0) :- activating_condition(Id, 0), prohibition_id(Id).
norm_activated(Id, S) :- not active(Id, S-1), not deactivating_condition(Id, S),
  activating_condition(Id, S), prohibition_id(Id), state(S).
norm_deactivated(Id, S) :-active(Id, S-1),deactivating_condition(Id, S), prohibition_id(Id), state(S).
active(Id, S) :- norm_activated(Id, S), prohibition_id(Id), state(S).
active(Id, S) :- not norm_deactivated(Id, S), active(Id, S-1), state(S), prohibition_id(Id).
violation(Id, S) :- not maintenance_condition(Id, S), active(Id, S), state(S), prohibition_id(Id).
-violation(Id, S) :- active(Id, S), not violation(Id, S), state(S), prohibition_id(Id).
```

Fig. 3 General reasoning rules in ASP

The result when validating π_2 is partially depicted in Figure 4. We can see that the norm is activated at the beginning of the execution of the plan (state 1) as the evacuation has been ordered since the initial state (state 1). The plan is fully executable

---

[4] Similarly for "obligation" type norms.

```
norm_activated(prohibl,1)
possible(loadPeople(jeep,maternity,7),1) possible(move(jeep,maternity,hclinic),2)
possible(unloadPeople(jeep,hclinic,7),3) possible(move(jeep,hclinic,maternity),4)
possible(loadPeople(jeep,maternity,3),5) possible(move(jeep,maternity,hclinic),6)
possible(unloadPeople(jeep,hclinic,3),7)
-violation(prohibl,1)  violation(prohibl,2)  violation(prohibl,3) -violation(prohibl,4)
-violation(prohibl,5) -violation(prohibl,6) -violation(prohibl,7) -violation(prohibl,8)
```

Fig. 4 Evaluation result for π_2

(as mentioned above this acts as a verification of the executability of a plan), since no occurrence of `-possible` exists. Concerning the respectfulness of the norms, during the first round of loading people to the vehicle, the plan breaks the norm by exceeding the capacity by two (loads 7 people), thus leading to a violation.

ASP allows a wider variety of queries than is typically provided in Event Calculus implementations but space constraints do not allow such an illustration here. Having the ASP reasoner embedded in the framework, the implementation of reasoning over the impact of the compliance or non-compliance to the norms and the calculation of the utility function can happen over the returned ASP results (counting each norm's violations and compliances that appear in the results and calculating the utility function as in Definition 6). It is easy to show that such a the translation of the domain to ASP guarantees the correctness of the results and the termination of the execution in all cases.

We have currently experimented with domains of up to 100 actions (and subsequently approximately 100 states) and although the execution becomes somehow slow as the size of the domain gets larger, the total calculation time still remains within reasonable (real time) boundaries.

7 Related Work and Conclusions

While there exists a rich background in the study of how legal, or normative, systems impact on the activities of social individuals, most of the work concentrates on the theoretical aspects of the normative concepts and little of this work focuses on developing norm-based societies where agents are able to take normative positions into account during practical reasoning.

In [4] the authors map institutional models to ASP, still, they do not provide a way to reason about complete plan paths during the execution of which norms follow a lifecycle of possible activation, deactivation and compliance or violation. In [5] the authors create and extend a programming language that implements normative agents. They consider norms as being represented by counts-as rules and sanctions as rules of the opposite direction. In [3] the authors extend the previous by defining the properties enforcement and regimentation and a model checking component which verifies these properties. Although the approach seems promising, they still does not cover reasoning and decision making over the norms. Close to our work is also [8] and [1].

One step leading our work further would be to extend the action and state representation semantics and implementation (taking advantage of the non-monotonic nature of ASP) under open world assumption in order to treat incomplete knowledge. Another interesting feature to be explored is the treatment of concurrency - actions can be executed in parallel, which can lead to reasoning over plans that contain concurrent interacting processes. Further to this, in modern technology, where actions are mostly represented by services, often it happens that there is no single service capable of performing a task, but there are combinations of existing services that could do so. For this reason a representation of an action synthesis (composite action) is needed. Thus, an extension of our framework could be towards evaluating plans and domains which additionally consist of complex actions. Finally, more advanced, context-dependent plan evaluation methods are being considered. These should include varying penalties and rewards, according to the state of the world.

References

1. Alvarez-Napagao, S., Aldewereld, H., Vázquez-Salceda, J., Dignum, F.: Normative Monitoring: Semantics and Implementation. In: De Vos, M., Fornara, N., Pitt, J.V., Vouros, G. (eds.) COIN 2010. LNCS, vol. 6541, pp. 321–336. Springer, Heidelberg (2011)
2. Apt, K.R.: From Logic Programming to Prolog. Prentice-Hall, U.K (1997)
3. Astefanoaei, L., Dastani, M., Meyer, J.J., de Boer, F.S.: On the Semantics and Verification of Normative Multi-Agent Systems. International Journal of Universal Computer Science 15(13), 2629–2652 (2009)
4. Cliffe, O., Vos, M.D., Padget, J.: pecifying and analysing agent-based social institutions using answer set programming, pp. 99–113 (2006) ISBN: 3-540-35173-6
5. Dastani, M., Tinnemeier, N.A., Meyer, J.J.: A Programming Language for Normative Multi-Agent Systems. Information Science Reference, Hershey (2009)
6. Fitting, M.: First-Order Logic and Automated Theorem Proving. Springer, New York (1990)
7. Ilghami, O., Nau, D.S.: A general approach to synthesize problem-specific planners (2003)
8. Kollingbaum, M.J.: Norm-governed practical reasoning agents. Univ. of Aberdeen (2005)
9. Okouya, D., Dignum, V.: Operetta: a prototype tool for the design, analysis and development of multi-agent organizations. In: Proceedings of the 7th International Joint Conference on Autonomous Agents and Multiagent Systems: Demo Papers, pp. 1677–1678 (2008)
10. Oren, N., Panagiotidi, S., Vázquez-Salceda, J., Modgil, S., Luck, M., Miles, S.: Towards a Formalisation of Electronic Contracting Environments. In: Hübner, J.F., Matson, E., Boissier, O., Dignum, V. (eds.) COIN@AAMAS 2008. LNCS, vol. 5428, pp. 156–171. Springer, Heidelberg (2009)
11. Panagiotidi, S., Nieves, J.C., Vazquez-Salceda, J.: A Framework to Model Norm Dynamics in Answer Set Programming. In: Proceedings of FAMAS 2009, vol. 494 (2009)

Trust and Normative Control in Multi-agent Systems: An Empirical Study

Joana Urbano, Henrique Lopes Cardoso, Ana Paula Rocha, and Eugénio Oliveira

Abstract. Despite relevant insights from socio-economics, little research in multi-agent systems has addressed the interconnections between trust and normative notions such as contracts and sanctions. Focusing our attention on scenarios of betrayal, in this paper we combine the use of trust and sanctions in a negotiation process. We describe a scenario of dyadic relationships between truster agents, which make use of trust and/or sanctions, and trustees characterized by their ability and integrity, which may influence their attitude toward betrayal. Both agent behavior models are inspired in socio-economics literature. Through simulation, we show the virtues and shortcomings of exploiting trust, sanctions and a combination of both.

1 Introduction

Computational trust models are important when addressing social relationships, both in the real world and in artificial agent societies. Trust is used to tackle the problem of social control. Additionally, norms and sanctions also play an important part by exerting pressure on individuals to conform. A number of studies from the social sciences look into the interrelationships between these issues.

Social relations are often associated with uncertainty and vulnerability of interacting partners, especially while no trust relationship is formed yet. This eventually leads to opportunism, which can be defined as "some form of cheating or undersupply relative to an implicit or explicit contract" [11]. Governance mechanisms used to reduce opportunism include control and monitoring. In fact, one approach to break out low trust dynamics in bilateral relationships is to use legalistic remedies, including the use of formal contracts [7]. Luhmann [5] suggests that the existence of legal norms is one of the most effective remedies to confine the risk associated with lack

Joana Urbano · Henrique Lopes Cardoso · Ana Paula Rocha · Eugénio Oliveira
LIACC / DEI, Faculdade de Engenharia, Universidade do Porto, Rua Dr. Roberto Frias, 4200-465 Porto, Portugal
e-mail: {joana.urbano,hlc,arocha,eco}@fe.up.pt

J.B. Pérez et al. (Eds.): Highlights on PAAMS, AISC 156, pp. 207–214.
springerlink.com

of trust, making it more comfortable for a potential truster to decide to invest trust in a relationship: legal regulations and sanctions reduce the risk of being betrayed. Furthermore, legal norms can foster the constitution of trust [1]. However, designing detailed contracts involves substantial drafting costs [11, 12], as does monitoring. According to Mayer *et al.*, legalistic remedies may bring organizational legitimacy but are often ineffective, being described as "impersonal substitutes for trust" [6].

At this point, it seems reasonable to think that trust can be used to reduce transaction costs associated with negotiation, monitoring and enforcement [4]. On the other hand, monitoring and enforcement have an important role when there is not appropriate evidence on which trust can be built, making the act of trust a risky engagement [5]. These concepts are therefore interconnected.

Das and Teng [2] state that trust denotes expectancies about other agents' motivations; it is not meant to influence or affect their behavior. Control mechanisms, on the other hand, may be used with the intent of deterring opportunism. Das and Teng distinguish trust in a partner from confidence in a transaction, that is, the certainty about cooperative behavior regardless of the possible motivations of the trustee. Accordingly, confidence in a transaction may be obtained as a combination of trust and control: for the same level of confidence, if we trust less, we use more control mechanisms. Furthermore, trust and control are seen as parallel and supplementary notions: they contribute independently to the level of confidence. Any one of these mechanisms may be used if an increase in transaction confidence is needed.

Despite these relevant insights from socio-economics, little research in multi-agent systems has addressed the interconnections between trust and normative notions such as contracts and sanctions. In this paper we try to cover this gap by combining, in a negotiation process, the use of trust and contractual sanctions in scenarios of betrayal, where violations of commitments are voluntary and harmful for the betrayed entity. More specifically, we formalize a model of agents (inspired in socio-economics literature) for dyadic relationships between truster agents that make use of trust and/or sanctions as a control mechanism, and trusted agents (or trustees) that are characterized by their ability and integrity, which dictate their bias toward betrayal. We demonstrate in the paper that several forces must be weighted in the interrelation between trust and norms, such as the ability and integrity of the trustees, the sanctioning costs and the motivation to betray, and that the quantification of these forces is not a trivial task.

This paper is structured as follows. Section 2 describes the scenario underlying our study and presents the behavior models for both trusters and trustees. Section 3 describes a set of experiments we have performed and interprets the results obtained. Finally, Section 4 concludes the paper.

2 Scenario and Agents Behavior Model

We here describe our interaction scenario and present the behavior model of the agents considered in this study. We address dyadic relationships with clients and providers of services. Clients are trusting agents (i.e. the *trusters*) who need to select the best providers for interaction, and providers are trusted agents (i.e. the *trustees*).

A truster starts by sending a call-for-proposals for a particular service, for which each trustee will provide its own proposal. When assessing proposals, trusters take into account their utility and (optionally) the perceived trustworthiness of each proponent. The truster will try to establish a contract with the proponent of the better assessed proposal, for which it may decide to include control mechanisms in the contract. If, for some reason, the trustee is not able to accommodate this contract, the truster will try with the proponent of the second best proposal, and so on. At the contract enactment phase, each hired trustee will have the opportunity to fulfill the contract or to violate it, according to the behavior model described later.

2.1 Trusters

The truster behavior model is based on the interplay between trust and control, as discussed in [2] and [8]. When considering the establishment of a contract with a trustee, the truster computes a confidence threshold Ct that indicates the minimum confidence he needs for entering into that particular transaction. This value is calculated by weighting the perceived risk R by the agent's risk aversion Ra. Risk, in turn, is modeled as a function of the weight of the transaction volume Tv on the agent's overall production volume Pv and the perceived trustworthiness T of the trustee, computed dynamically using a computational trust model. We thus have that $Ct = R * Ra$, where risk $R = Tv/Pv * (1 - T)$.[1] Risk aversion ranges from 0 (a risk lover agent) to 1 (totally risk averse).

Having a minimum confidence threshold, the truster will propose, to a selected trustee, a contract that includes a level of control (represented as a sanction to apply in case of violation) computed according to the general notion from [2] that *Confidence = Trust + Control*. By suggesting an appropriate sanction, the truster tries to raise his confidence on the contract that is to be established with a particular trustee, of which it has some trustworthiness assessment.

2.2 Trustees

The model of behavior of the trustees is inspired in the model of betrayal in organizations of Elangovan and Shapiro [3]. In our model, trustees of low integrity tend to enter in new contracts even when they do not have enough resources to satisfy them, i.e. they are aware that they may have to (voluntarily) violate one or more of their active contracts. On the contrary, trustees with high integrity may refuse the contract if they do not have enough resources to satisfy the deal without violating previous agreements. We also assume that all trustees have a predefined level of *competence*,

[1] In the experiments, we use T/ζ instead of T, due to the fact that computational trust models typically overrate the trustworthiness estimations, as they tend to aggregate the outcomes of past evidence using statistical methods, without taking into consideration the *relationship* that was active between interacting partners at the evidence time [10].

i.e. an innate ability to provide products of good quality. Violations that are due to (lack of) ability are not voluntary, and thus are not considered *betrayals* [3].

In the equations that follow x denotes a trustee, y denotes a truster, c denotes a contract, and p denotes a contract proposal. When time-stamping terms using a superscript, we assume a discrete time line. Unless otherwise noted, variables are assumed to be universally quantified.

The decision to betray vs. keep the status quo is made when the trustee is awarded a new contract for which it does not possess enough resources (e.g. a stock that is periodically replenished). The new contract is considered a *business opportunity* if it presents significant higher utility than at least one of the trustee's ongoing contracts; this assessment also depends on the integrity of the trustee, as shown in Equation 1, where $\delta \in [0,1]$ is the integrity parameter.

$$\exists c\ Contract(c,x,y)^t \wedge ContractProposal(p,y,x)^t \wedge Utility(c,x) < Utility(p,x) * (1-\delta) \wedge Resources(p,x) > FreeResources(x)^t \Rightarrow NewOpportunity(p,x) \quad (1)$$

After identifying a new opportunity, the trustee is going to assess the current situation, namely: i) the benefits of betraying; and ii) its relationship with the potential victim of betrayal.

Assessing the Value of Betraying. The trustee assesses the benefits of betraying by taking into account both the utility associated with the new opportunity and the existence of a relevant sanction associated with the potential contract to betray. This sanction is considered *irrelevant* to the trustee if its value is smaller than a given (adjustable) percentage γ of the utility associated with the new opportunity. In this case, the value of betrayal is high. In order to reduce the complexity of the model, we chose three qualitative values for the value of betraying, as illustrated in Equation 2.

$$\begin{aligned} Utility(p,x) - Sanction(c,x) < \gamma_1 * Utility(c,x) &\Rightarrow VBetrayal(c,x,low) \\ \gamma_1 * Utility(c,x) \leq Utility(p,x) - Sanction(c,x) < \gamma_2 * Utility(c,x) &\Rightarrow VBetrayal(c,x,medium) \\ \gamma_2 * Utility(c,x) \leq Utility(p,x) - Sanction(c,x) &\Rightarrow VBetrayal(c,x,high) \end{aligned} \quad (2)$$

Assessing the Value of the Relationship. The trustee assesses the relationship with the potential victim by considering the number of past contracts between both partners in the last σ units of time (perspective of continuing the relationship, cf. Equation 3 for $\sigma = 3$), and the existence of other contracts (if the trustee is currently engaged in at least ξ contracts with other trusters, cf. Equation 4 for $\xi = 2$). The perceived value of the relationship is given in Equation 5.

$$\exists c_1, c_2, c_3\ Contract(c_1,x,y)^{t-1} \wedge Contract(c_2,x,y)^{t-2} \wedge Contract(c_3,x,y)^{t-3} \Rightarrow PerspContinuity(x,y,high)^t \quad (3)$$

$$\exists c_1, c_2\ Contract(c,x,y)^t \wedge Contract(c_1,x,y_1)^t \wedge Contract(c_2,x,y_2)^t \wedge c_1 \neq c_2 \wedge y_1 \neq y \wedge y_2 \neq y \Rightarrow HasOtherContracts(x)^t \quad (4)$$

$$\neg PerspContinuity(x,y,high)^t \wedge HasOtherContracts(x)^t \Rightarrow VRelationship(x,y,low)^t$$
$$\neg PerspContinuity(x,y,high)^t \wedge \neg HasOtherContracts(x)^t \vee$$
$$PerspContinuity(x,y,high)^t \wedge HasOtherContracts(x)^t \Rightarrow VRelationship(x,y,medium)^t \quad (5)$$
$$PerspContinuity(x,y,high)^t \wedge \neg HasOtherContracts(x)^t \Rightarrow VRelationship(x,y,high)^t$$

The decision to betray a partner (accepting the new contract) or instead to keep its trust takes into consideration the assessment made by the trustee concerning the values of betrayal and relationship. In case there is more than one contract which is deemed to be betrayed, the trustee will only betray the one with less utility, provided that its allocated resources are enough to take into account the new contract. In no contract is deemed to be betrayed, the trustee declines to accept the new contract.

It is important to note that even new contracts may be betrayed later on if another opportunity arises. Contracts are violated at enactment time, which means that the decision to betray is made much earlier than the *act* of betray.

3 Experiments

In this section, we present the experiments we have run as support to our study. We used the trading scenario described in detail in the previous section.

3.1 Experimental Setup

We ran all the experiments using the computational trust model described in [9]. Each experiment was composed of 80 rounds, and at every round each buyer started a new negotiation cycle by issuing a new call for proposals. At the first round of each experiment, the repository of trust evidence for every supplier was cleaned. We used 80 buyers and 120 suppliers. Every experiment was run 30 times. Table 1 shows the values we assigned to the model's variables in the experiments.

Table 1 Configuration of parameters

	δ_{low}	δ_{medium}	δ_{high}	γ_1	γ_2	σ	ξ	ζ
value	0.3	0.6	0.9	0.3	0.9	3	1	4

The effective betrayal of contracts was configured probabilistically taking into consideration the assessed values of the benefits of betraying and of the relationship. This probability was 1.0 for high benefits and low and medium relationship values; for high benefits and high relationship values, or medium benefits and low relationship values, the probability of betrayal was 0.5; finally, a betrayal happened with probability 0.2 when both the values of the benefits of betraying and of the relationship were medium. In all the remaining cases, the trustees did not betray the trusters.

Configuration of Trusters. The sanction value was calculated as $S = Ct - T/\zeta$. This formula provides the relationship between the trustworthiness of a trustee and the level of sanctions S that a truster will propose to be included in the contract. We start from the formulation of $Ct = T + S$, where for the reason explained before we reduce the weight of the trust parcel. Every truster has a value $Ra \in [0,1]$ picked randomly at setup, and a value Pv also picked up randomly from a range of fixed minimum and maximum values. Tv is a dynamic value proposed by a trustee resulting from a specific contract negotiation.

After a betrayal, the truster enters a resent period toward the offender trustee, meaning that the trustworthiness of the latter drops to zero immediately after the betrayal and is gradually softened with time. This effect on the trustee's trustworthiness T is given by $T = T * \frac{\Delta_t}{\rho}$, where Δ_t represents the elapsed time (in these experiments, $\rho = 3$ rounds) since the time of the offense.

The population of trusters follows an uniform distribution over the possible types:

None	The truster does not use sanctions nor trust.
Sanctions	The truster use sanctions but does not select partners based on the trust.
Trust	The truster uses trust to select partners but does not use sanctions.
S&T	The truster uses trust both to select partners and to compute sanctions.

Configuration of Trustees. In order to emulate the existence of a potential new opportunity (cf. Equation 1), all suppliers had a limited stock within a simulation round. The utility of a contract for a trustee is calculated by multiplying the dimension of the proposal (i.e. the quantity of material agreed over the available stock of the trustee) by the relevance of the price in the proposal.

The innate ability of trustees was configured as a random value ascribed at setup in range $[0.5, 1]$, and their integrity were also randomly picked up considering the possible δ values (Table 1).

Evaluation Metrics. In these experiments, we used eight different performance metrics: Δ_{sup} (no. of different suppliers selected by all buyers in one round); τ, α and ι (average trustworthiness, ability and integrity of the suppliers selected by the buyers in one round, respectively); o^+ (number of contracts with positive outcome in a round); Σ (average sanction applied by all buyers to the contracts they establish with the suppliers, in one round); O and β (opportunities to betray faced by the suppliers, and the effective betrayals occurring in one round, respectively). All metrics took values in $[0, 1]$, all averaged over all rounds and all runs of the experiments.

3.2 Results

Experimental results are shown in Table 2. From the table, we verify that there is a clear difference in results when we consider the use of trust in negotiation and when we do not. In fact, approaches that do not use trust (None and Sanctions) select the partners based only on the utility of their proposals. Consequently, there is a wider choice of suppliers (Δ_{sup}) and the average values of trustworthiness (τ) and ability (α) of the selected suppliers is rather low. Also, as more suppliers are selected per round, the less the chance that they overpass their stock, and the opportunities of betrayal (O) are, thus, smaller than in the approaches that use trust (Trust and S&T). In contrast, the latter are more exposed to effective betrayals (β), due to the just mentioned effect.

Table 2 The results of the experiments

	Δ_{sup}	τ	α	ι	o^+	Σ	O	β
None	0.92	0.27	0.73	0.42	0.66	0.00	0.17	0.09
Sanctions	0.93	0.27	0.73	0.42	0.69	0.38	0.18	0.05
Trust	0.77	0.82	0.88	0.43	0.74	0.00	0.34	0.16
S&T	0.80	0.81	0.88	0.43	0.76	0.18	0.40	0.13

The most effective approach to prevent betrayals is to use sanctions standalone, although this has the disadvantage of not caring about the ability dimension of trustees, thus decreasing o^+. Comparing approaches S&T and Trust, we verified that although the former raised the number of betrayal opportunities, the addition of sanctions allowed to reduce the number of effective betrayals in almost 20%.

3.3 Discussion

The experiments we have run allowed to shed some light on the complex interrelation between trust and sanctions. We observed that sanctions are more effective in helping on preventing betrayals, and that trust is more effective in distinguishing between entities with different abilities. However, our study shows that in a realistic world, where trustees have different ability and integrity values, the interrelation between both governance mechanisms is less evident. The supplementary use of trust and sanctions seems to be the more balanced approach in terms of contracts with positive outcome ($o+$), at the expense of a moderate usage of sanctions ($\Sigma = 0.18$) and a (not irrelevant) rate of betrayals ($\beta = 0.13$). Finally, the results were not conclusive about the best governance mechanisms to prevent selection of trustees with low values of integrity.

4 Conclusions

In this paper, we presented an empirical study about the interrelations between trust and norms. This study was grounded on solid theory from diverse research areas concerning trust and norms. The novelty of this study concerns the experimental exploration of the complex relationships between factors such the ability and integrity dimensions of trustworthiness, risk, sanctions and betrayals. From the gained experience in this work, we conclude that the conjunctive use of trust and norms – described in theoretic works as a promising interrelated governance mechanism – is not a trivial task.

For future work, we intend to refine our approach to get a better insight on the problem and reach more significant results.

Acknowledgements. This research is supported by Fundação para a Ciência e a Tecnologia (FCT), under project PTDC/EIA-EIA/104420/2008. The first author is supported by FCT under grant SFRH/BD/39070/2007.

References

1. Bachmann, R.: Trust, power and control in trans-organizational relations. Organization Studies 22(2), 341–369 (2001)
2. Das, T.K., Teng, B.: Between trust and control: Developing confidence in partner cooperation in alliances. Academy of Management Review 23(3), 491–512 (1998)
3. Elangovan, A.R., Shapiro, D.L.: Betrayal of trust in organizations. The Academy of Management Review 23(3), 547–566 (1998)
4. Ireland, R.D., Webb, J.W.: A multi-theoretic perspective on trust and power in strategic supply chains. Journal of Operations Management 25(2), 482 (2007)
5. Luhmann, N.: Trust and Power. John Wiley & Sons, New York (1979)
6. Mayer, R.C., Davis, J.H., Schoorman, F.D.: An integrative model of organizational trust. The Academy of Management Review 20(3), 709–734 (1995)
7. Sako, M.: Does trust improve business performance? In: Lane, C., Bachmann, R. (eds.) Trust within and between Organizations: Conceptual Issues and Empirical Applications, Oxford University Press (1998)
8. Tan, Y.-H., Thoen, W.: An Outline of a Trust Model for Electronic Commerce. Applied Artificial Intelligence 14(8), 849–862 (2000)
9. Urbano, J., Rocha, A.P., Oliveira, E.: Computing Confidence Values: Does Trust Dynamics Matter? In: Lopes, L.S., Lau, N., Mariano, P., Rocha, L.M. (eds.) EPIA 2009. LNCS, vol. 5816, pp. 520–531. Springer, Heidelberg (2009)
10. Urbano, M.J., Rocha, A.P., Oliveira, E.: A Dynamic Agents' Behavior Model for Computational Trust. In: Antunes, L., Pinto, H.S. (eds.) EPIA 2011. LNCS, vol. 7026, pp. 536–550. Springer, Heidelberg (2011)
11. Wathne, K.H., Heide, J.B.: Opportunism in interfirm relationships: Forms, outcomes, and solutions. The Journal of Marketing 64(4), 36–51 (2000)
12. Williamson, O.E.: Transaction-Cost Economics: The Governance of Contractual Relations. Journal of Law and Economics 22(2), 233–261 (1979)

Transparent Multi-Robot Communication Exchange for Executing Robot Behaviors

Carlos Agüero and Manuela Veloso

Abstract. Service robots are quickly integrating into our society to help people, but how could robots help other robots? The main contribution of this work is a software module that allows a robot to transparently include behaviors that are performed by other robots into its own set of behaviors. The proposed solution addresses issues related to communication and opacity of behavior distribution among team members. This location transparency allows the execution of a behavior without knowing where is located. To apply our approach, a multi-robot distributed receptionist application was developed using robots that were not originally designed to cooperate among themselves.

1 Introduction

Service robots help humans in many tasks, including cleaning tasks, medical applications, surveillance and guiding. Service robots solve a specific problem and consequently their sensors and actuators are designed accordingly. A natural consequence of this design is that these robots are not reusible in other tasks they were not originally conceived for.

We find that these robots are very good at performing the task for which they have been developed; we could say that they are experts in one or more specific tasks. This specialization can be understood as an ability to perform an action.

As new applications appear, humans demand more functionality from service robots. One way to meet these challenges is to expand the capabilities of the robot, whether in hardware or software. Instead, we explore another approach for the robot

Carlos Agüero
Universidad Rey Juan Carlos, Camino Molino, 28943, Madrid, Spain
e-mail: `caguero@gsyc.urjc.es`

Manuela Veloso
Carnegie Mellon University, 5000 Forbes, Pittsburgh, PA 15213, USA
e-mail: `veloso@cs.cmu.edu`

J.B. Pérez et al. (Eds.): Highlights on PAAMS, AISC 156, pp. 215–222.
springerlink.com

to cooperate with other robots that are able to solve the new task demands. The latter approach requires a shared effort among all the robots, since we have gone from working with a single robot to a robot team. Quoting Doug McIlroy, the inventor of Unix pipes, "This is the Unix philosophy: Write Programs that do one thing and do it well. Write Programs to work together"[11].

In this paper we present an approach that helps to make available to the whole team all the different individual behaviors of each robot for solving tasks. This aggregation is done transparently, so that all the robot share the same set of behaviors. These behaviors could be performed by the robot itself or they could require cooperation with other robots that would run them in their place. The net effect is that the behavior is done opaquely, regardless of which of the robots have completed it.

The modular solution described in this work addresses issues of communication between robots with different operating systems, programming languages and integration with the host infrastructure on which it needs to operate.

The rest of this paper is organized as follows. First, we discuss the related state of the art. Then, the design of the software module and its main components are presented. Next, we detail the distributed receptionist application developed using our approach. Finally, some discussion and future lines are presented.

2 Related Work

As we witness the explosion of cloud computing, the idea of using other resources on the Internet has spread to robotics with the creation of Cloud Robotics[7]. This approach provides virtually unlimited resources alleviating the limited features of robots. For example, the Google Goggles[6] application allows the user to send a photograph of an object and, if it has been previously processed by someone else, the object is recognized. The cloud can also store knowledge and models. The RoboEarth project[12] describes how an articulated arm equipped with sensing capabilities can create a model to open a drawer. Then, another articulated arm with rudimentary sensors can request the information previously stored in the cloud and use it to open the drawer by adjusting the model to its actuator skills.

The key idea of reusing knowledge is fundamental to our approach and a prevailing concept in Cloud Robotics. However, the module we present here also aims to reuse the behaviors of other robots, reaching a higher level of cooperation based solely on knowledge reuse from the cloud.

The term *Robot as a Service*[2] was created using the concept of *Service-Oriented Architecture*, which provides a communication mechanism through standard interfaces and standard protocols. The idea that each robot maintains a common layer for offering services is shared with the work presented here. However, our approach seeks to make transparent the fact that the services may require communication with another robot to get started, achieving an even higher level of abstraction.

Swarmnoids project[3] is an example of how a team of heterogeneous robots can solve tasks that are beyond their individual capabilities. UAVs (Unmanned Aerial Vehicles) cooperate with ground vehicles in highly-coupled tasks, generating self-organized cooperative behaviors. The work presented in [9] for the treasure hunt domain uses the very popular market-based mechanism of coordination to determine which robot performs each behavior in an environment that requires the cooperation of heterogeneous robots. In [4] other work is described based on this principle. For a more detailed analysis and a Multi-Robot Task Allocation taxonomy consult [5]. In all of these works, communication between team members is made explicitly, losing the level of transparency that our novel alternative offers.

3 Multi-Robot Task Module

Multi-Robot Task Module (MRTM) follows a distributed approach and, accordingly, each robot should run an instance of MRTM. The main objective of the MRTM is to encapsulate all access to the behaviors the team can offer. The behaviors do not necessarily have to be implemented within the module itself, but the interface to access them does. Some interesting features offered by this design are the behavior location transparency and its cross-platform availability.

The behavior location transparency concept is defined as the ability to hide which robot is providing a behavior for solving a given task. In practice, this solution allows us to maintain a robot team, where all robots share a single interface. In addition, this interface brings together all the behaviors that each robot can perform. Essentially, the work of the MRTM module is to invoke the behavior that has been requested, either by running an algorithm or a behavior on the robot itself, or by requesting the behavior of another robot's MRTM component. Figure 1 shows the internal structure of the MRTM, whose main components will be explained below.

When there is a robot team where all members have a different behavior repertoire, it is necessary to know which robot can perform which behavior. A possible solution is that each application queries all the robots, and obtains a list of available behaviors, along with the robot that provides that behavior. Location transparency takes care of this tedious operation and delegates all the job to MRTM module, so applications just use the behaviors without knowing where they actually reside.

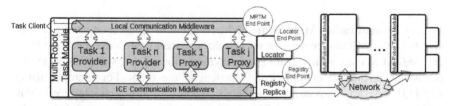

Fig. 1 Overview of the Multi-Robot Task Module

Ice Communication Middleware. All communication among robots is performed by Internet Communications Engine[8] (Ice). Ice is an object-oriented middleware used in distributed systems. Ice's main virtues include its multi-platform and multi-language support, and efficiency.

An Ice object is an entity in a local or remote robot that can reply to client requests. An interface declares a set of operations (behaviors) that an Ice object can perform and clients can invoke. Operations performed between client and server are declared using an independent of a specific programming language called Slice.

A proxy is an object used to contact with a remote Ice object. The proxy emulates a specific Ice object's interface and its operations can be invoked. In MRTM we use indirect proxies, which are composed by an Ice object identifier and object adapter identifier. An object adapter is a container of Ice objects in a given robot. An indirect proxy does not contain any addressing information. To find the correct server, the client-side Ice run time passes the proxy information to a location service. In our system, to eliminate this single point of failure and ensure the highest possible availability, we have included a *Registry* for each MRTM, thanks to the master-slave replication supported on Ice.

Ice Communication Middleware (ICM) combines all operations related to Ice. These operations are the creation of the object adapter and the registration of Task Providers within the adapter. In addition, this sub-module is responsible for resolving indirect proxies to perform remote invocations and receiving calls from other robots, as well as initializing the registry process.

In MRTM, an Ice object is realized by creating a class that implements the interface to that Ice object. In our approach these classes are called Task Providers and consist of related behaviors for a given task.

To receive requests on a given MRTM, they must be called by a different MRTM. When a particular behavior is requested, if it cannot be executed by any local Task Provider, the corresponding Task Proxy is used. The proxy is just a binding in the form of an object, which hides the communication with another MRTM. To invoke the method of the object that triggers the behavior, the proxy has to be resolved. This is to find the End Point for the object adapter that includes the Task Provider of interest. *Locator* is responsible for implementing this work. *Locator* communicates with one of the *Registries* and resolves the indirect proxy. Once the proxy is set up, the remote invocation can be performed and will be received by the ICM's object adapter of another robot. In turn, this object adapter will distribute the request to one of its Task Providers, which will launch the behavior.

Local Communication Middleware. When a client or application invokes a specific behavior, MRTM receives the request through the Local Communication Middleware (LCM) sub-component. The implementation of this sub-component depends on the internal communication system of the robot. For example, using ROS[13], this sub-component would be a node on the robot. The interface of this LCM node would form a set of ROS services. In short, the function of this sub-module is to integrate into the host and serve as an infrastructure gateway for clients of behaviors. There are two steps to implement this component: To instantiate an MRTM object and to create an interface to use all the behaviors that the robot team is able to perform. Depending on the local middleware this could be a set of methods declared in a header file, a series of services, or a series of published topics.

Task Providers. The Task Provider is another sub-component of the MRTM. Its mission is to trigger the execution of a particular behavior on the robot. Inside MRTM, there will be as many Task Providers for the local behaviors that the robot can perform. MRTM does not intend to modify the internal architecture of the robot and make it monolithic. It is the opposite, as these local behaviors implementations would remain independent modules distributed within the robot (e.g., other ROS nodes). MRTM simply groups all Task Providers to manage calls to these local behaviors more easily.

Task Proxies. Like the Task Provider sub-component, which start the execution of local behaviors, the Task Proxy sub-component run behaviors remotely requiring execution by another robot. To implement each Task Proxy you must get an indirect proxy to MRTM that is capable of performing the behavior, and then, to perform the remote invocation using the indirect proxy as an object and the behavior as a method.

4 Multi-Robot Receptionist

To illustrate the MRTM design made in section 3, we applied the proposed MRTM to a team of robots performing a distributed people reception task. The task required a receptionist to welcome visitors and ask for the name of the person to visit. The receptionist retrieves and offers the office location associated with that person and suggests the option of being escorted to the office.

To conduct the experiment, two different robots and a directory program that ran on a conventional computer were used. The first is the Nao robot[1]. It is a medium-sized humanoid robot running GNU/Linux. Nao features two cameras, Wi-Fi, bumpers and hi-fi speakers, among other sensors and actuators. Its mission in the experiment was to meet the people who came and to offer directory and escort tasks.

The second member of the team was the CoBot robot[10]. It is equiped with a LIDAR, two cameras, one kinect and an omnidirectional base. CoBot is a visitor companion robot able to navigate through the environment, transport objects, deliver messages, and escort people. Its goal in this experiment was to escort a visitor from a known area (the exit of the elevator) to the office of the person he wanted to visit.

The third and final team member was a directory program that ran on a laptop. It contained a directory of people and their associated offices.

Figure 2 shows the overall architecture of the experiment (the elements *Registry* and *Locator* have been deliberately omitted to simplify the diagram). All agents have the same interface, which allows the deployment of directory and escort behaviors. While in this experiment the application of user interaction was only deployed on the Nao, any other robot could technically have offered the same functionality.

The LCM sub-component of the CoBot robot was integrated into a specific ROS node responsible for managing the behavior of the robot. In turn, the LCM modules of the Nao receptionist robot and directory program were included in objects that communicated with each host infrastructure through regular method invocations.

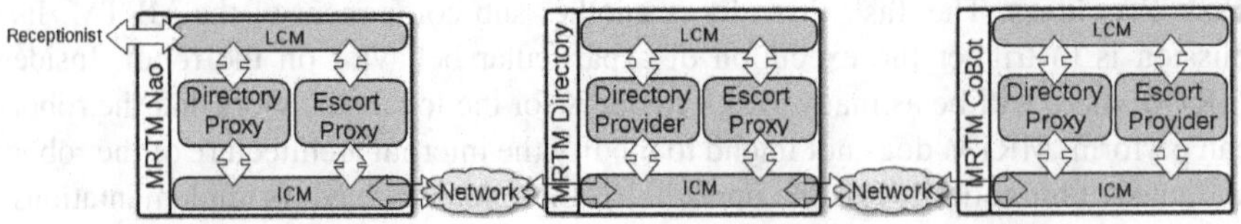

Fig. 2 Overview of the MRTMs on the Receptionist experiment

The Slice interfaces shared among the agents of this experiment are shown below. The *Directory* consists of a single function that returns the office associated with a person. In turn, the *Escort* interface includes the method *escort*, which selects a specific office and a time window and returns a task ID. This value can be used to cancel the booking in a future calling to the *cancelTask* method.

```
module MultiRobotReceptionist {
 interface Directory  {
   string people2office( string personName ); };
 interface Escort {
   string escort( string room, string startDate, string startH, string startM,
       string startP, string endDate, string endH, string endM, string endP );
   int cancelTask( string taskId ); }; };
```

After greeting and interacting with the visitor, Nao uses its Directory Proxy to consult the directory program and ask for the destination. Note that the Nao does not even know which robot or agent actually implements this service. The receptionist application invokes the method *person2office* as if it is implemented in itself. Once the visitor is informed of the office number, Nao offers the visitor the option of being escorted to the office. If the proposal is accepted, Nao uses the *escort* method offered by its MRTM to start escorting the visitor.

As we described before, the MRTM module takes over and triggers the escort behavior using the Escort Proxy of the CoBot robot. CoBot has a scheduler for requesting different behaviors and when they should be executed by the robot. The Escort Provider of the CoBot robot makes a reservation for the escort task before offering visitors the option of being escorted. If the reservation is successful, the receptionist Nao offers the possibility of escorting the visitor. Then, if the visitor rejects the offer, Nao uses the *cancelTask* method to remove that task from the scheduler. Figure 3 shows a sequence of frames illustrating the key moments of the experiment.

The resources consumed by MRTM were measured during the experiment. CPU overhead was completely negligible, 40MB of memory was consumed and there was no continuous bandwidth used. The spikes on the bandwidth consumption were at a maximum of 1Kb/sec. and occurred during the Directory and CoBot requests. A video of the experiment can be downloaded at this link[1].

[1] http://dl.dropbox.com/u/2831576/PAAMS12_caguero.mov

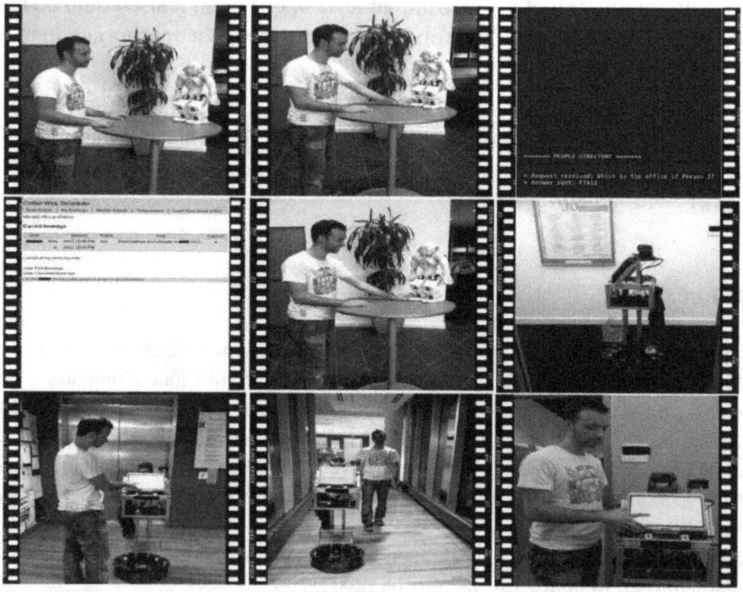

Fig. 3 Sequence of shots extracted during an experiment escorting a person. Upper left: Visitor is greeted by the Nao receptionist. Upper central: Visitor and Nao interact using voice and bumpers for select a person. Upper right: Directory program receives the request and answers with the office number. Middle left: The escort task is reserved on the CoBot scheduler. Middle central: Visitor accepts to be escorted. Middle right: CoBot starts navigating towards the meeting spot. Lower left: CoBot waits at a known location until the visitor arrives. Lower central: The visitor is escorted by CoBot. Lower right: The visitor arrives at the destination's office.

5 Conclusions and Future Work

We have presented the MRTM as a solution to the problem of the remote execution of behaviors with location transparency. All the work for localizing behaviors and interoperability with different platforms has been solved using the tools provided by the Ice middleware. This is one of the advantages over other middleware for robots. Our approach allows for the integration of a MRTM module into more devices such as robots with different operating systems or even mobile phones, with a very low impact on CPU overhead, memory consumption and network bandwidth.

An experiment with robots and humans was conducted successfully for a multi-robot receptionist task. The receptionist program running on the Nao robot was ignoring where was the people directory located or which robot was escorting the visitors. All the low level details of exchanging information were hidden by MRTM.

Thanks to the common interface provided by MRTM, programs that use behaviors are easier to design because there is no explicit negotiation among the robots.

Despite the results obtained, there are open questions for future research. One of the main future lines is the inclusion of a Multi-Robot Task Allocation sub-module used by multiple robots that provide the same Task Provider. It also could be responsible for negotiating with multiple robots to choose the most appropriate in terms of some utility function.

Acknowledgments. The authors also would like to thank all the members of the CORAL group, Robotics Institute and URJC Robotics group who are collaborated in this work.

References

1. Aldebaran Robotics (2011), http://www.aldebaran-robotics.com
2. Chen, Y., Du, Z., García-Acosta, M.: Robot as a Service in Cloud Computing. In: Proceedings of the 2010 Fifth IEEE International Symposium on Service Oriented System Engineering, SOSE 2010, pp. 151–158. IEEE Computer Society, Washington, DC (2010)
3. Ducatelle, L.G.F., Di Caro, G.: Cooperative Self-Organization in a Heterogeneous Swarm Robotic System. In: Proceedings of the Genetic and Evolutionary Computation Conf. (2010)
4. Gerkey, B.P., Mataric, M.J.: Sold!: auction methods for multirobot coordination. IEEE Transactions on Robotics 18(5), 758–768 (2002)
5. Gerkey, B.P., Mataric, M.J.: A formal analysis and taxonomy of task allocation in Multi-Robot systems. The International Journal of Robotics Research 23(9), 939–954 (2004)
6. Google. Google Goggles (2011), http://www.google.com/mobile/goggles/
7. Guizzo, E.: Cloud Robotics: Connected to the Cloud, Robots get Smarter (2011), http://spectrum.ieee.org/automaton/robotics/robotics-software/cloud-robotics
8. Henning, M.: A New Approach to Object-Oriented Middleware. IEEE Internet Computing 8, 66–75 (2004)
9. Jones, E., et al.: Dynamically Formed Heterogeneous Robot Teams Performing Tightly-Coordinated Tasks. In: Int. Conf. on Robotics and Automation, pp. 570–575 (May 2006)
10. Rosenthal, S., Biswas, J., Veloso, M.: An effective personal mobile robot agent through symbiotic human-robot interaction. In: AAMAS 2010, vol. 1, pp. 915–922 (May 2010)
11. Salus, P.H.: A quarter century of UNIX. ACM Press, New York (1994)
12. Waibel, M., et al.: RoboEarth. IEEE Robotics Automation Magazine 18(2), 69–82 (2011)
13. Willow Garage. ROS (Robot Operating System) (2011), http://www.ros.org/

Using Multi-Agent Systems to Facilitate the Acquisition of Workgroup Competencies

Diego Blanco, Alberto Fernández-Isabel,
Rubén Fuentes-Fernández, and María Guijarro

Abstract. The acquisition of competencies for teamwork is gaining relevance in higher education curricula. However, lecturers face increasing difficulties to monitor and guide the student evolution regarding these issues. As opposed to traditional face-to-face interactions in classrooms, the current educational context promotes students' autonomy to organize their learning activities, with an intensive use of Learning Management Systems (LMS) and other online facilities. This paper presents an architecture for Adaptive Multi-Agent Systems (AMAS) that supports workgroup learning in this setting. It embeds expert knowledge about participant profiles in teams and their evolution, and resources to work the related competencies. An AMAS uses this knowledge to carry out assessments of students and continually monitor their activities looking for patterns in behaviours and interactions. This information allows notifying to lecturers events that need their attention and suggesting them resources to improve students' competencies. The system is currently undergoing the validation of its knowledge basis, which this paper also reports.

Keywords: Workgroup, Acquisition of competencies, Higher education, Multi-agent system, Application support, Adaptation to user, User evaluation.

1 Introduction

Higher education is undergoing important changes triggered by emergent educational needs. Current students are supposed to acquire the knowledge and technical skills of their disciplines of choice, but also competencies required for a successful

Diego Blanco · Alberto Fernández-Isabel · Rubén Fuentes-Fernández · María Guijarro
Facultad de Informática,
Universidad Complutense de Madrid, Madrid, Spain
e-mail: {diego.blanco,afernandezisabel,
ruben,mguijarro}@fdi.ucm.es

J.B. Pérez et al. (Eds.): Highlights on PAAMS, AISC 156, pp. 223–230.
springerlink.com

performance in the modern information societies. Among others, students also need to get capabilities for an autonomous and long-life learning, and to use their skills in multidisciplinary groups that address problems in innovative ways [11].

In order to facilitate this learning, universities are adopting student-centred approaches focused on tailoring the learning experience to the specific needs of each student [12]. They are also increasing the relevance of group activities, as means to create communities of practice between students that encourage their constructive and critical peer collaboration and where they can practice teamwork [6].

The intensive use of software tools, such as Learning Management Systems (LMS) or web 2.0 applications, facilitates the adoption of these new learning strategies, though their support is still insufficient. These tools are intended to facilitate interactions but not to reflect on their results. They provide raw information (e.g. use statistics or authored contents) that lecturers need to analyse to establish the student and group states. From that analysis, lecturers have to find out the best way to guide the evolution of the class. These tasks demand from lecturers increased efforts and analysis skills on teamwork that are not so common in their background.

Our work addresses these issues with an architecture for Adaptive Multi-Agent System (AMAS) intended to boost class teamwork. It supports monitoring, analysis and guiding of the class applying expert knowledge extracted from Social Sciences. These AMAS are integrated with LMS to get information and set up activities using their infrastructure. Lecturers act as the final supervisors that accept and customize the AMAS suggestions on learning resources.

The applied expert knowledge includes participant profiles [1], interaction patterns [5], and resources to expose students to certain experiences [9]. It crystallizes in questionnaires and *social properties*. *Social properties* [4] formalize the description of group settings and workflows following the Activity Theory (AT) [8, 14]. The properties are used to document learning resources and situations, and to detect relevant observations from LMS using pattern-matching techniques. In both cases, the theoretical background of AT supports lecturers interpreting the information.

The architecture includes four main types of agent. *Student* and *group assistant* agents monitor and evaluate individual students and their groups respectively. They try to determine if their targets are accomplishing the proposed activities and fulfilling the related goals. *Lecturer assistants* use the reports provided by other assistants to produce digests for lecturers. They also communicate with *resource advisor* agents to get information on the resources available to articulate and guide teamwork. For instance, advisors can suggest a setting useful to initiate a student in the role of group leader. All the assistants learn from the feedback of their users the kind of information that they should push forward in order to reduce their workload.

This work is illustrated with the discussion of the current validation of the expert knowledge in the architecture. Experiments are undergoing in a Software Engineering course of a Computer Science School in a Spanish university using Moodle[1].

The rest of the paper is organized as follows. Section 2 introduces the basis of our work in AT and social properties. Sections 3 and 4 describe the proposed AMAS,

[1] `http://moodle.org`

focusing on the expert knowledge used and the architecture respectively. Then, Sect. 5 reports the preliminary results with the AMAS, that Sect. 6 compares with related work. Finally, Sect. 7 discusses some conclusions on the approach and future work.

2 Social Properties

The Activity Theory (AT) [8, 14] is a paradigm from Social Sciences. Its key concept is the *activity*, which is a contextualised human act.

An *activity* [8] is a social process carried out by a *subject*. The subject uses *tools* to transform *objects* into *outcomes*. These *outcomes* can satisfy some subject's *objectives*. There is also a social dimension whose central concept is the *community*. The *community* is a set of subjects that share social meanings and artefacts. Communities are related to the previous elements through *divisions of labour* and *rules*. Both of them represent the knowledge, artefacts and procedures of the community, but have different scopes. The *division of labour* focuses on the current activity. It covers aspects such as roles and skills required in tasks or team organization. *Rules* affect the current activity but are not specifically targeted to it. Examples of them are laws, economic systems or religions. The instances of all these concepts related to an activity constitute its *activity system* [3]. The activity system includes the physical and social context of the activity and its historical development. Social systems are represented as networks of activity systems interconnected by shared artefacts. At this analysis level, there is no difference between physical and mental activities.

The original AT studies are not well suited for engineering settings: they make an intensive use of textual discussions and require a high-level of expertise in social studies to understand and apply them. *Social properties* [4] address this issue. They replace free text with a template that combines diagrams and structured text.

The template describes individual settings and their *related properties*. These links explain the relation between the related properties and the correspondences of their concepts. This allows, for instance, connecting a conflict with a potential solution and explaining how to rework the original setting to get the fixed one.

Diagrams are specified with UML-AT. It is a UML profile with stereotypes for the concepts and relationships of AT. It also adds extra primitives such as the generic *artefact* concept or the *change of role* relationship to indicate that an artefact adopts a new role in a system (e.g. the outcome of an activity becomes the tool of another).

The analysis of information with social properties is made by identifying them in specifications described with UML-AT using pattern-matching algorithms. This process is semi-automated, as an expert makes the final validation of its results.

3 Expert Knowledge for Teamwork

The proposed architecture pursues supporting lecturers and students in the learning activities for teamwork skills, trying to alleviate their workload and improve the results. It performs assessments of the situation of students and groups, and proposes

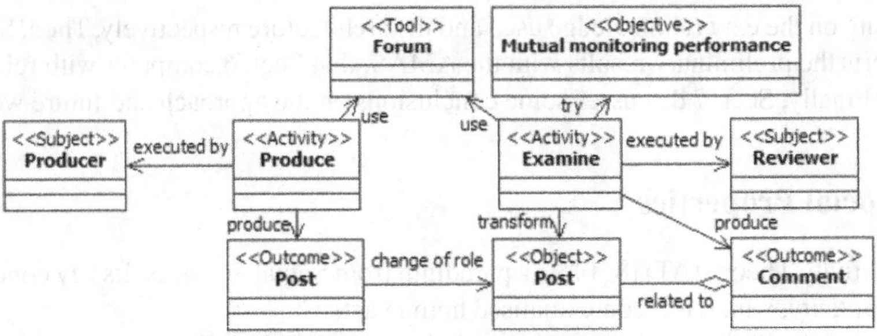

Fig. 1 UML-AT representation of mutual members monitoring in forums.

resources for the learning activities according to the pursued goals. These activities use different types of knowledge.

Our approach mainly considers two works on student's profiles to assess their capabilities. The Belbin's model [1] identifies nine roles that should appear in a well-balanced team that works smoothly and effectively. Our teams are composed following this model and considering that some students are able to play several roles. The *Knowledge, Skill, and Ability requirements for Teamwork* (KSAT) [13] is a catalogue of desirable features of people working in teams. The work in our teams is aimed at giving students a suitable environment to improve their KSAT. For each team, lecturers and students collaborate to provide 360-degree evaluations that include superiors, peers and subordinates. When students gain new competencies, lecturers reconfigure teams to work out other KSAT.

Lecturers facilitate the previous evolution in order to get a successful learning environment. For this purpose, they gather feedback from the interactions happening in LMS, interpret it, and use different resources to promote new dynamics. The knowledge for this process is obtained from literature on teamwork, adapted to LMS functionality, and represented with social properties for a semi-automated use.

Fig. 1 shows an example of property extracted from [7] and adapted to forums about the group behaviour *identifying mistakes and lapses in other team members' actions*. The UML-AT diagram shows two activities, *produce* and *examine*. Both of them make use of the *forum* tool to share information in a team: *produce* to post new *content*, and *examine* to get *content* and include new information as *comments*. The diagram makes explicit the objective pursued by the *reviewer* subject when executing the *examine* activity. The subject tries to achieve the *mutual monitoring performance*, which according to [7] is one of the aspects that make up teamwork.

The architecture uses these properties to define what can be observed and interpret it. For instance, Fig. 1 shows the use of a forum: when a student posts a new element and opens a thread, s/he is executing the *produce* activity with certain artefacts (e.g. information, tools and collaborators). The student assistant agents must provide an implementation able to recognize this activity. When there is a match between actual observations and a property, that match is stored as a *fact* about students and their teams. Additional information is inferred from observations, e.g. the

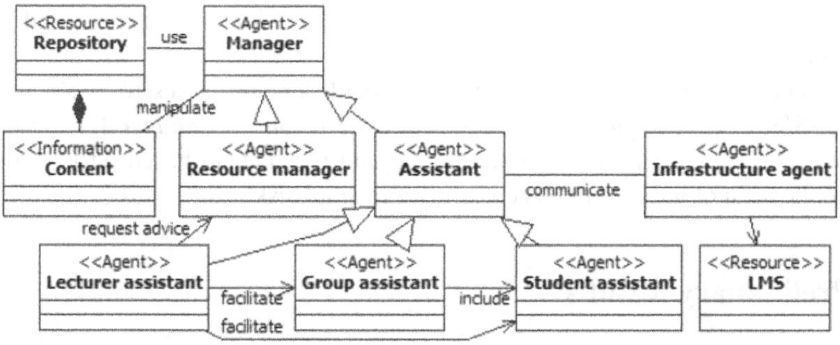

Fig. 2 AMAS architecture.

related objectives. In this example, when a student performs the *examine* activity, it is assumed that s/he is pursuing the *mutual monitoring performance* objective.

Social properties are also used to document the workflows used to achieve certain learning objectives. For this purpose, a property specifies the activities involved in the workflow, their requirements as artefacts, and the expected learning results.

4 The AMAS Architecture

The architecture for AMAS includes two layers of agents. The *infrastructure layer* is the interface with the LMS. The *user layer* includes the four types of agents mentioned in the introduction. Fig. 2 summarizes this architecture.

Infrastructure agents receive events, get information and act on LMS. They use these capabilities to process information and requests on behalf of *manager* agents. For the LMS events, they support a simple subscription and notification mechanism. They also translate information between LMS interfaces and the AMAS internal representation with UML-AT. For instance, in the case of the property in Fig. 1, the infrastructure agent accesses a forum and represents each comment there as an execution of either *produce* or *examine* activities with their related activity systems. Then, the agent notifies the new information to the interested *assistant* agents.

Assistant agents support participants in the class. They store and manage the information required for this purpose in *repositories*. Current types of information include: student *profiles* and their *evolution* [1, 13]; *questionnaires* to determine these profiles and evaluate student and group performance and work [9]; and *social properties* [4] and *facts* observed in LMS described with UML-AT. The *evolution* information includes data about the history of users' interactions with the system, e.g. preferences or activities already performed. The management of this information includes its preliminary interpretation through matching with social properties and deciding the relevant events to notify to users. The *student assistants* forward metrics on their users' activities, e.g. frequency and size of their contributions. The *group assistants* also consider member interactions to detect potential isolation issues. The *lecturer assistant* receives the previous notifications and general digests about course activity.

The lecturer studies this information to decide the resources to apply in the course. S/he uses a *lecturer assistant* to ask the *resource manager* for these resources. The retrieval relies on a simple tagging mechanism based on keywords. The application of the resource is currently a manual task.

5 Preliminary Results

The proposed architecture is undergoing preliminary evaluation in a Computer Science School of a Spanish university. The target course belongs to a 4-year degree and the subject is Software Engineering, which is delivered in the second year. The course has 46 students enrolled, but only 26 assisting to class, 23 men and 3 women, with an average age of 20 years. The LMS used is a tailored version of Moodle.

The initial questionnaires had heterogeneous results, with students that did not consider themselves good for any role and others that were good for several roles. The distribution of Belbin's profiles [1] among the later was: 9 *completers*, 10 *implementers*, 9 *team workers*, 2 *specialists*, 6 *evaluators*, 8 *coordinators*, 3 *resource investigators*, 2 *plants* and 12 *shapers*. This uneven distribution made impossible to set up groups with students that played all the 9 profiles. Lecturers decided to make up 4 teams of 6-7 members with the maximum possible of different roles. These groups have 9, 7, 8 and 7 out of the 9 profiles. The number of members is close to that of required profiles in an effective team according to [1], but still manageable by students with little experience in teamwork.

First observations on the dynamics of these groups showed some malfunctions. After the initial outburst of activity, communication in forums sharply declined as it was expected. More interesting is that the application of the social property in Fig. 1 about monitoring inside teams warned lecturers about 43% of students not posting at all and 21% with no-comment on any post of their team fellows. Among students commenting, it showed that they interacted with other 2 members of their teams in average. This pointed out low values of the *KSA to communicate openly and supportively* [13].

Lecturers considered these rates did not help to start building the required social bonds in teams. The resource suggested for this case was a general post in the forum advising to improve communication in teams. No specific figures on interactions were provided in order to avoid fake messages that were useless for teams.

6 Related Work

This work is related to two main lines of research: analysis of teamwork and semi-automated tools to support it.

Our research tries to provide a general architecture able to incorporate different analysis techniques for teamwork. The current prototype of the AMAS incorporates: works on profiles [1] and skills [13] required for teamwork with the purpose of assessing student evolution; and resources based on good practices for teamwork in learning settings [9]. Nevertheless, other works regarding, for instance, team evolution [3] or cooperative learning [6], have not been included. A key problem here is building a cohesive theoretical framework for the support provided by the AMAS. The research area is highly fragmented, with works focused on very specific aspects of teamwork and relying on different and partial theories. Moreover, most of works make only public their results, but not the complete description of the tools they used to elicit information, e.g. questionnaires or interviews. This makes difficult applying their analysis in other works.

Regarding tools, literature mainly focuses on the study of standard applications such as LMS. The common conclusion is that lecturers need to make an important effort with these tools in order to get the proper teamwork settings [10]. There are some aids, but their utility and scope are limited. The review in [2] highlights that acquiring teamwork competencies is a multi-faceted problem, and the related support tools need to incorporate and integrate its requirements. Our work addresses several of the open problems pointed out there. It tries to provide an open and flexible architecture that supports new collaboration and analysis methods. Other interesting work from the perspective of monitoring and analysis is [7]. It reports different visualizations for the use information of tools such as wikis or subversion systems. These visualizations provide insights on who is interacting with whom, and the frequency and size of people contributions.

7 Conclusions

This paper has presented an architecture for AMAS intended to support learning of teamwork competencies. It pursues several goals discussed below.

The architecture is designed to facilitate the integration of different theoretical perspectives and support infrastructures. Agents encapsulate well-defined tasks and the open AMAS easily support replacing them by others with similar skills. Knowledge is declaratively specified with UML-AT and social properties. This allows an iterative and incremental definition of the events and workflows to consider, grounding them in a widespread social theory. It also supports the description of resources applicable in a course and their goals and learning outcomes.

Providing agents the ability to manipulate the previous knowledge makes possible the adaption of the overall system to the evolution of the course. This is the case, for instance, of analysing the events to notify or the resources to apply to certain students according to the current state of their teamwork competencies. This adaptation and the flow of information in the system reduce the workload required to set up and carry out this kind of activities. Lecturers need to devote less time to monitor students, analyse their work and propose activities, and students get feedback from the system on potential fails or tips for their teamwork.

The proposed architecture is undergoing validation. Preliminary results have shown the usefulness of students' profiles and monitoring of their interactions to prevent some common malfunctions in their teams.

The approach still presents multiple open issues that need further work. First, the integration with LMS is quite limited. It is focused on the public exchange of information in forums and uploaded/downloaded resources. Additional interactions happen through mail or in the contents of resources, and they are relevant for teamwork. Second, the applied knowledge is basic and fragmentary. Research is needed to include a richer model on how teams interact and students learn from it. This would allow designing specific activities for teams and individuals to acquire certain competencies. Finally, additional courses are going to be enrolled in the experiments to validate the architecture and knowledge in a wider setting.

Acknowledgements. This work has been done in the context of the project "SOCIAL AMBIENT ASSISTING LIVING - METHODS (SociAAL)", supported by Spanish Ministry for Economy and Competitiveness, with grant TIN2011-28335-C02-01. Also, we acknowledge support from the Programa de Creación y Consolidación de Grupos de Investigación UCM-BSCH GR58/08.

References

1. Belbin, M.: Team roles at work. Butterworth-Heinemann (1993)
2. Dimitracopoulou, A.: Designing collaborative learning systems: Current trends & future research agenda. In: Proceedings of the 2005 Conference on Computer Support for Collaborative Learning (CSCL 2005), pp. 115–124 (2005)
3. Engeström, Y.: Learning by expanding: An activity-theoretical approach to developmental research. Orienta-Konsultit (1987)
4. Fuentes-Fernández, R., Gómez-Sanz, J., Pavón, J.: Understanding the human context in requirements elicitation. Requirements Engineering 15(3), 267–283 (2010)
5. Gersick, C.: Time and transition in work teams - toward a new model of group development. The Academy of Management Journal 31(1), 9–41 (1988)
6. Johnson, D., Johnson, R., Smith, K.: Cooperative learning: Increasing college faculty instructional productivity. ASHE-ERIC Higher Education Report 4 (1991)
7. Kay, J., Maisonneuve, N., Yacef, K., Reimann, P.: The Big Five and Visualisations of Team Work Activity. In: Ikeda, M., Ashley, K.D., Chan, T.-W. (eds.) ITS 2006. LNCS, vol. 4053, pp. 197–206. Springer, Heidelberg (2006)
8. Leontiev, A.: Activity, Consciousness, and Personality. Prentice Hall (1978)
9. Michaelsen, L., Knight, A., Fink, L.: Team-based learning - A transformative use of small groups. Praeger Publishers (2002)
10. Nijhuis, G., Collis, B.: Using a web-based course-management system: an evaluation of management tasks and time implications for the instructor. Evaluation and Program Planning 26(2), 193–201 (2003)
11. Reichert, S., Tauch, C.: Bologna four years after: Steps towards sustainable reform of higher education in Europe. European University Association (2003)
12. Scott, P.: The Meanings of Mass Higher Education. Open University Press (1995)
13. Stevens, M., Campion, M.: The knowledge, skill, and ability requirements for teamwork: Implications for human resource management. Journal of Management 20(2), 503–530 (1994)
14. Vygotsky, L.: Mind in society (1978)

Adaptive Vehicle Mode Monitoring Using Embedded Devices with Accelerometers

Artis Mednis, Georgijs Kanonirs, and Leo Selavo

Abstract. Monitoring of specific attributes such as vehicle speed and fuel consumption as well as cargo safety is an important problem for transport domain. This task is performed using specific multiagent monitoring systems. To ensure secure operation of such systems they should have autonomous and adaptive behaviour.

The paper is describing an adaptive agent for vehicle mode monitoring using embedded devices with accelerometers. Data processing algorithm and adaptive functionality are discussed and their evaluation is presented with vehicle standing mode detection as high as true positive rate of 97% using real world data. Optimization of parameters for data processing algorithm is performed as well as suggestions for their application described.

1 Introduction

Transportation of passengers and cargo is performed using different vehicle types. To ensure economical transportation hardware/software systems are used for monitoring of specific attributes such as selected route, vehicle speed, fuel consumption, drivers' shift time, cargo safety [3] etc.

These systems are designed and developed in manner that minimizes possible manipulations by vehicle driver. One paradigm to achieve this goal is system with maximum autonomy and independence from the monitored vehicle. For example, data about the vehicle speed is acquired not from a vehicle electronic onboard

Artis Mednis · Georgijs Kanonirs · Leo Selavo
Digital Signal Processing Laboratory,
Institute of Electronics and Computer Science,
14 Dzerbenes Str., Riga, LV 1006, Latvia
e-mail: `firstname.lastname@edi.lv`

Leo Selavo
Faculty of Computing, University of Latvia,
19 Raina Blvd., Riga, LV 1586, Latvia

J.B. Pérez et al. (Eds.): Highlights on PAAMS, AISC 156, pp. 231–238.
springerlink.com

system but from a separate GNSS ([4], p.2) receiver included in monitoring system. Unfortunately there exist possibilities to bypass or sabotage monitoring system, for example, using GNSS jammers. Therefore data acquisition solutions or a "black box" that are hardened against bypassing or sabotage are necessary.

Our developed vehicle mode monitoring solution is meant for use as a single adaptive agent in a multiagent system. It is based on adaptive algorithm and uses as input data only measurements from 3-axis accelerometer.

Related work is discussed in Section 2. Technical requirements are listed in Section 3. Main algorithm and adaptive functionality are described and analyzed in Section 4. The final section presents the conclusion that the proposed solution detects vehicle standing mode with 97% reliability and adaptive functionality makes it suitable for different vehicle types.

2 Related Work

There are several accelerometer based adaptive multiagent systems for monitoring of different types of objects such as elderly people in their own homes [5, 7, 8] as well as power transformers in transmission substations [1]. All these systems share a common paradigm - a large and complex task is splitted into many small and simple subtasks and distributed between separate agents. Some of these subtasks may require from corresponding agent an adaptive behaviour.

There are also several accelerometer based systems which purpose is monitoring of vehicle mode. In the simplest case the main goal of the system is distinguishing between two main vehicle modes - standing and driving [2, 9]. Deciding about actual vehicle mode is performed using predefined signal patterns and thresholds as well as input from additional data sources such as GNSS receivers. More advanced systems not only detect actual vehicle mode but also assist or even temporarily replace common GNSS based vehicle position determination systems [10]. In both cases there is a need for calibration of the system for specific vehicle. Our approach includes not only detecting of actual vehicle mode but also an adaptive functionality what allows the usage of the system in different types of vehicles such as passenger cars and busses.

3 Technical Requirements

The research described in this paper was carried out as a feasibility study of a wider in scope industrial research project. This project assumes design of multiagent system with the aim to restrict fuel misappropriation. Therefore, following list of technical requirements was chosen as a basis for vehicle mode monitoring subsystem:

1. The system should be able to detect two vehicle operation modes - driving mode and standing mode, including standing mode with working engine.

2. The system should be able to detect vehicle operation mode while driving on roads covered with asphalt such as city streets as well as intercity highways.
3. The system should be able to detect vehicle operation mode in real time with granularity at least 10 seconds (in optimal case - not exceeding 1 second).
4. The system should be able to run on embedded device with a microcontroller characterized by the following parameters and using not more than 30% of its resources: CPU speed 16 MIPS, program memory 128 KB, RAM 3862 bytes and data EEPROM 1024 bytes.
5. The system should use acceleration measurements as input data from 3-axis accelerometer with range configurable from ± 2 g to ± 8 g. Connection to vehicle electronic onboard system or use of GNSS receiver are not intended.

4 Our Approach

To examine the suitability of 3-axis accelerometer as the only sensor for vehicle mode monitoring a real world experiment was performed. During this experiment passenger car BMW 323 Touring carried out 13.5 km long distance (3 rounds x 4.5 km, Fig. 1 - on the left) in 34 minutes.

Accelerometer data acquisition was performed 37x per second using a slightly modified LynxNet collar device developed during our past research activities related to wild animal monitoring using sensor networks [11]. Among other hardware items this embedded device includes Analog Devices 3-axis accelerometer ADXL335. Parallel to accelerometer data acquisition vehicles position and speed data (Fig. 2 - on the left) were collected 1x per second using SBAS capable GNSS receiver Magellan eXplorist XL.

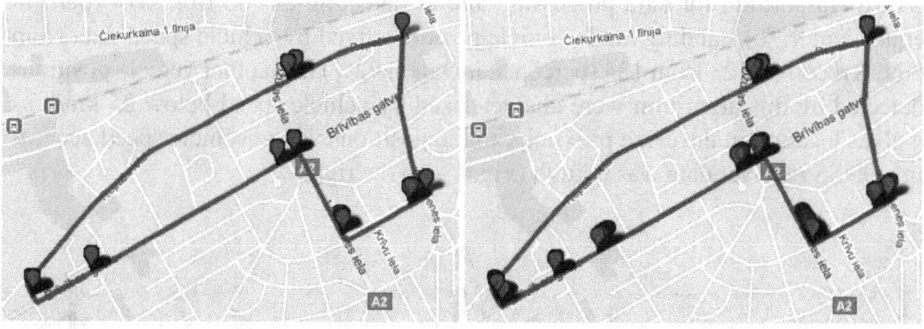

Fig. 1 4.5 km long road network fragment used for real world experiment. On the left - marked positions correspond to coordinates where GNSS receiver fixed vehicle speed 0 km/h, on the right - marked positions correspond to coordinates where STDEV algorithm fixed activity level less than 0.04 g.

4.1 Main Algorithm

Processing of accelerometer data was performed using slightly modified STDEV algorithm from our past research activities related to real time pothole detection using Android smartphones with accelerometers [6]. Original version of the STDEV algorithm includes calculation of accelerometer vertical Z-axis data standard deviation as well as thresholding of calculated values. In this case potholes are detected by acceleration values above specific threshold level. Modified version of the STDEV algorithm includes aggregation of all 3-axis accelerometer data standard deviation values as well as thresholding of acquired aggregate values. In this case vehicle standing mode is detected by acceleration values below specific threshold level. During this proof of the concept experiment sliding window with size 10 samples was used as well as threshold value 0.04 g. Vehicle activity profile created from accelerometer data is shown in Fig. 2 on the right but performance of modified STDEV algorithm in the context of vehicle mode monitoring - in the Table 1.

Table 1 Performance of modified STDEV algorithm in the context of vehicle mode monitoring

Parameter	Value
All vehicle positions fixed during experiment	2043
Standing mode positions fixed using GNSS receiver data	514
Standing mode positions fixed using accelerometer data	634
Standing mode positions fixed using both above mentioned approaches	500

Vehicle mode monitoring using 3-axis accelerometer data as input and modified STDEV algorithm for data processing allows detecting of 500 from 514 (percent equivalent 97%) standing mode vehicle positions fixed by vehicle speed data from GNSS receiver. 123 from 134 (percent equivalent 92%) from other vehicle positions detected by this algorithm were characterized by vehicle speed below 25 km/h as well as location in the close proximity to the positions where vehicle speed according GNSS receiver data was 0 km/h (Fig. 1 - on the right).

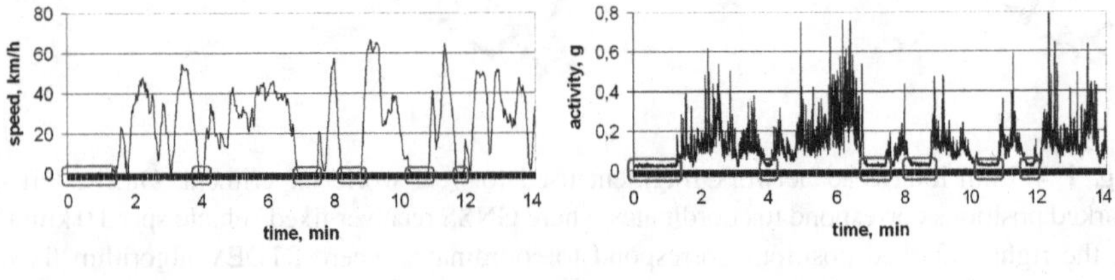

Fig. 2 Fragments of the speed and activity profiles of the passenger car BMW 323 Touring used for real world experiment. On the left - vehicle standing mode corresponds to places where GNSS receiver fixed vehicle speed 0 km/h, on the right - vehicle standing mode corresponds to places where modified STDEV algorithm fixed activity level less than 0.04 g.

Analysis of the created vehicle activity profile revealed a necessity for a mechanism to prevent vehicle mode detection instability in cases when values of vehicle activity profile are changing in a narrow range near the threshold value. Such mechanism could be established due usage of not only one but already two threshold values - T_{low} and T_{high}. Other improvements of the algorithm include increasing of the accelerometer data acquisition rate from 37x to 100x per second as well as increasing of sliding window size from 10 samples (0.25 seconds) to 100 samples (1 second). These changes are related to usage of full potential of accelerometer characteristics.

The goal of the next real world test was to verify this solution using several vehicle types such as passenger cars and buses, several road surface types such as city streets and intercity highways, several vehicle "stop" types such as traffic lights and bus stops. These experiments were carried out using following version of the algorithm:

1. Acquisition of 3-axis accelerometer data 100x per second.
2. Storage of all accelerometer data acquired during last second using appropriate data structure:

$$X[] = \{X_{n-99},...,X_n\}; Y[] = \{Y_{n-99},...,Y_n\}; Z[] = \{Z_{n-99},...,Z_n\} \quad (1)$$

3. Calculation of all 3 axis standard deviation values as well as their aggregation:

$$STDEV(X[]) + STDEV(Y[]) + STDEV(Z[]) \quad (2)$$

4. If aggregation of all 3-axis standard deviation values is
   a. > 0.1g (above T_{high}) then vehicle mode is set to driving;
   b. < 0.05 g (below T_{low}) then vehicle mode is set to standing;
   c. ≥ 0.05 g and ≤ 0.1 g (between T_{low} and T_{high}) then previous set vehicle mode is preserved.

Analysis of the accelerometer data acquired from bus Setra S415 HDH revealed that activity profile of this vehicle type is characterized by relatively lower g values (Fig. 3). In this case typical value for vehicle standing mode was 0.02 g therefore corresponding values for T_{low} and T_{high} could be 0.03 g and 0.06 g respectively. Differences in the activity profile could be explained with different vehicle volumes as well as different distances between data acquisition hardware and vehicle engine as most significant source of the vibrations during vehicle standing mode.

Following real world experiments were carried out using different hardware platform POGA v.1b consisting of WSN mote Tmote Mini and accelerometer ADXL330 similar to previous used ADXL335. This hardware item includes also two LED's used for visual real time control of currently detected vehicle mode. After analysis of data acquired in three different vehicles (Table 2) the conclusion about suitability of improved algorithm for usage in different vehicle types was taken as well as the necessity for self-calibration functionality to adapt corresponding threshold values for each individual vehicle was approved.

Table 2 Performance of vehicle mode detection algorithm in the context of different vehicles using threshold levels T_{low}=0.05 g and T_{high}=0.1 g.

	Volvo V70	VW Passat Variant	Setra S415 HDH
Standing	OK	OK	OK
Driving	OK	OK	NOK[c]
Stopping	OK[a]	OK[a]	OK[d]
Starting	OK[b]	OK[b]	NOK[e]

[a] Switching could be pre-emptive if stopping is performed rolling on flat surface
[b] Switching could be delayed if starting is performed smoothly
[c] There is a tendency switch to standing during rolling on flat surface
[d] Switching could be particularly pre-emptive if stopping is performed rolling on flat surface
[e] Switching could be particularly delayed if starting is performed smoothly (typical for busses)

4.2 Adaptive Functionality

Based on the data acquired during previous real world experiments two assumptions were defined as a basis for development of an adaptive functionality for vehicle mode detection algorithm:

1. There exist vehicle activity levels with certain adherence to main vehicle modes. The first approximation of these levels is 0.1 g and more for driving mode as well as 0.03 g and less for standing mode.
2. Certain vehicle standing mode is represented by at least 5 seconds long period characterized by narrow range of vehicle activity level.

Previous version of the algorithm assumes calculation of vehicle activity level 1x per second. This frequency should be preserved in the adaptive version to maintain vehicle mode detection with granularity 1 second. Additional calculations should be performed for detection of certain vehicle standing mode periods and subsequent main algorithm calibration:

1. Storage of all vehicle activity level values acquired during the last 5 seconds using appropriate data structure:

$$ACTIVITY[] = \{ACTIVITY_{n-4}, ..., ACTIVITY_n\} \quad (3)$$

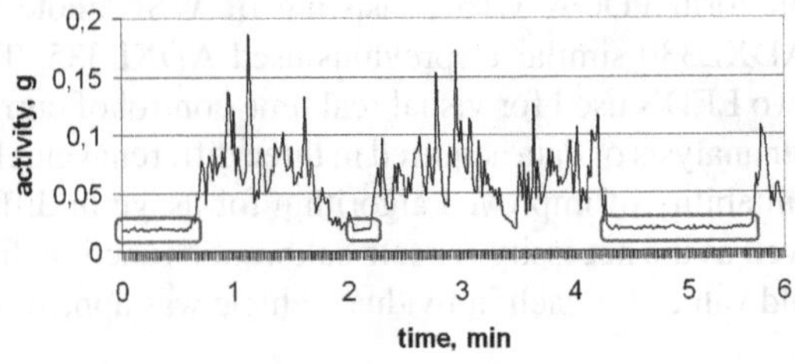

Fig. 3 A fragment of the activity profile of the bus Setra S415 HDH used for real world experiment. Vehicle standing mode corresponds to places where modified STDEV algorithm fixed activity level less than 0.03 g.

2. Calculation of vehicle activity level standard deviation value after each added value:
$$STDEV(ACTIVITY[]) \tag{4}$$
3. If calculated vehicle activity level standard deviation value is under a certain threshold (first approximation of this level is 0.005 g) the counting of consecutive calculations is started or continued.
4. If count of consecutive values under certain threshold reaches 5 and vehicle standing mode period is detected, the arithmetic mean (AM) of corresponding vehicle activity level values is calculated as well as both threshold values T_{low} and T_{high} in the algorithm changed:
$$T_{low} = AM\{ACTIVITY[]\} \times 1.25; T_{high} = T_{low} \times 1.5 \tag{5}$$
5. To avoid miscalibration of the algorithm due vehicle standing without working engine as well as driving with stable acceleration several additional constraints should be set:
$$T_{low} \neq 0g; T_{high} \leq 0.11g \tag{6}$$

Adaptive version of vehicle mode detection algorithm was tested using passenger car Volvo V70 and bus Setra S415 HDH. Results of these test drives are shown in Fig. 4. Standing and driving modes are properly detected despite different vehicle activity profiles.

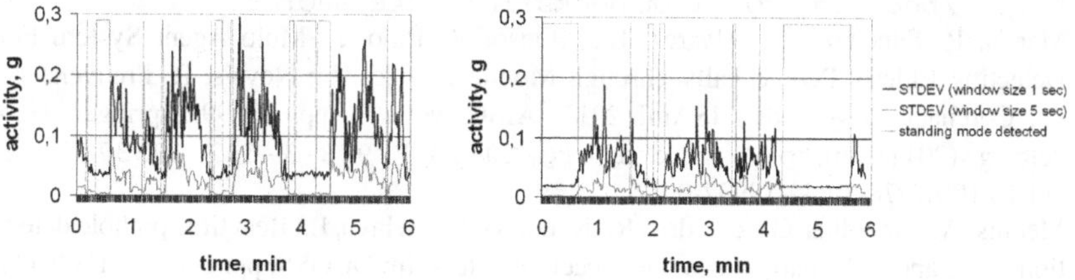

Fig. 4 Vehicle activity profiles - on the left Volvo V70, on the right - Setra S415 HDH. Detected vehicle standing mode corresponds to places where adaptive STDEV algorithm fixed activity level under T_{low}.

5 Conclusion

We have proposed an adaptive vehicle mode monitoring solution that includes embedded device with a 3-axis accelerometer. The solution was evaluated on a particular application - detection of vehicle standing and driving modes with 1 second granularity. The detection was performed by adaptive thresholding of accelerometer data standard deviation values. We performed several test drives on different road surface types using several vehicle types. The experimental results were evaluated by comparison with the corresponding vehicle speed data acquired from GNSS receiver as well as real time observations using the visual interface of embedded device. The results show, that our solution detects vehicle standing mode with 97%

reliability and adaptive functionality allows its usage in different vehicle types characterized by different vehicle activity profiles.

The future work includes evaluating of the developed solution using broader vehicle type set and improvement of vehicle mode switching detection using characteristic activity patterns of particular vehicle and driver.

Acknowledgements. This work was supported by European Social Fund grant Nr. 2009/0219/1DP/ 1.1.1.2.0/APIA/VIAA/020 "R&D Center for Smart Sensors and Networked Embedded Systems".

References

1. Baker, P., Catterson, V., McArthur, S.: Integrating an agent-based wireless sensor network within an existing multi-agent condition monitoring system. In: 15th International Conference on Intelligent System Applications to Power Systems, ISAP 2009, pp. 1–6 (2009), doi:10.1109/ISAP.2009.5352888
2. Chowdhary, M., Zhang, Q., Chansarkar, M., Zhang, G.: Systems and methods for detecting a vehicle static condition (2011), http://patents.com/us-7979207.html
3. Fokum, D.T., Frost, V.S., Depardo, D., Kuehnhausen, M., Oguna, A.N., Searl, L.S., Komp, E., Zeets, M., Deavours, D., Evans, J.B., Minden, G.J.: Experiences from a transportation security sensor network field trial. Tech. rep. (2009)
4. Gleason, S., Gebre-Egziabher, D.: GNSS applications and methods. In: GNSS Technology and Applications. Artech House (2009), http://books.google.com/books?id=juXAE3SHQroC
5. Martín, P., Sánchez, M., Álvarez, L., Alonso, V., Bajo, J.: Multi-Agent System For Detecting Elderly People Falls Through Mobile Devices. In: Novais, P., Preuveneers, D., Corchado, J.M. (eds.) ISAmI 2011. AISC, vol. 92, pp. 93–99. Springer, Heidelberg (2011), http://dx.doi.org/10.1007/978-3-642-19937-0_12, doi:10.1007/978-3-642-19937-0_12
6. Mednis, A., Strazdins, G., Zviedris, R., Kanonirs, G., Selavo, L.: Real time pothole detection using android smartphones with accelerometers. In: DCOSS, pp. 1–6. IEEE (2011)
7. Rammal, A., Trouilhet, S., Singer, N., Pécatte, J.M.: An adaptive system for home monitoring using a multiagent classification of patterns. Int. J. Telemedicine Appl. 2008, 3:1–3:8 (2008), http://dx.doi.org/10.1155/2008/136054, doi:10.1155/2008/136054
8. Sánchez, M., Martín, P., Álvarez, L., Alonso, V., Zato, C., Pedrero, A., Bajo, J.: A New Adaptive Algorithm for Detecting Falls through Mobile Devices. In: Corchado, J.M., Pérez, J.B., Hallenborg, K., Golinska, P., Corchuelo, R. (eds.) Trends in PAAMS. AISC, vol. 90, pp. 17–24. Springer, Heidelberg (2011), http://dx.doi.org/10.1007/978-3-642-19931-8_3, doi:10.1007/978-3-642-19931-83
9. Schwartz, R.: Vehicle state detection (2010), http://ip.com/patapp/US20100204877
10. Wang, J.H., Gao, Y.: Multi-sensor data fusion for land vehicle attitude estimation using a fuzzy expert system. Data Science Journal 4, 127–139 (2005)
11. Zviedris, R., Elsts, A., Strazdins, G., Mednis, A., Selavo, L.: LynxNet: Wild Animal Monitoring Using Sensor Networks. In: Marrón, P.J., Voigt, T., Corke, P.I., Mottola, L. (eds.) REALWSN 2010. LNCS, vol. 6511, pp. 170–173. Springer, Heidelberg (2010)

Using MAS to Detect Retinal Blood Vessels

Carla Pereira, Jason Mahdjoub, Zahia Guessoum,
Luís Gonçalves, and Manuel Ferreira

Abstract. The segmentation of retinal vasculature by color fundus images analysis is crucial for several medical diagnostic systems, such as the diabetic retinopathy early diagnosis. Several interesting approaches have been done in this field but the obtained results need to be improved. We propose therefore a new approach based on an organization of agents. This multi-agent approach is preceded by a preprocessing phase in which the fundamental filter is an improved version of the Kirsch derivative. This first phase allows the construction of an environment where the agents are situated and interact. Then, edges detection emerged from agents' interaction. With this study, competitive results as compared with those present in the literature were achieved and it seems that a very efficient system for the diabetic retinopathy diagnosis could be built using MAS mechanisms.

Keywords: Diabetic retinopathy, Fundus images, Image processing, Kirsch filter; Multi-agent system.

1 Introduction

The segmentation of the retinal vasculature is of major importance for clinical purposes. In fact, by analyzing the vascular structures it is possible to have the

Carla Pereira · Manuel Ferreira
Industrial Electronics Department,
University of Minho,
Campus de Azurém, 4800-058 Guimarães, Portugal

Jason Mahdjoub · Zahia Guessoum
CReSTIC – MODECO, University of Reims,
rue des Crayères, 51100, Reims, France

Luís Gonçalves
Ophthalmology Service,
Centro Hospitalar do Alto Ave, Guimarães, Portugal

J.B. Pérez et al. (Eds.): Highlights on PAAMS, AISC 156, pp. 239–246.
springerlink.com

early diagnosis of several chronic pathologies, such as diabetic retinopathy (DR). In this study, the blood vessels segmentation was performed concerning this pathology that has been revealed as the leading cause of vision impairment in developed countries. In order to prevent loss of vision it is important to detect the pathology as early as possible.

At present, the most common way to analyze vascular modifications is the digital color fundus photograph as it is a non invasive technique. Numerous research efforts have been done in segmenting the blood vessels by image processing techniques applied in the fundus images [1]-[6]. The main difficulties in accurately segment the vessels are due to the presence of pathologies, noise, the low contrast between vasculature and background, the variability of vessels width, brightness and shape. To solve this problem of variability, it is important to locally adapt interpretations on the image instead of applying only one algorithm on the entire image. A multi-agent system (MAS) is thus proposed as a solution since agents allow the cohabitation of several algorithms. In fact, agents can analyze problems they are locally confronted with, and select the most suitable algorithm to their local context [7].

There are some works described in the literature that associate MAS with image processing in medical images [7]-[12]. Different MAS proprieties and mechanisms have been used in these approaches, such as interaction, self-organization, adaptation, communication, cooperation and negotiation [12]. Richard et al. [9] proposed a hierarchical architecture of situated and cooperative agents to segment the cerebral tissue in MRI. Duchesnay et al. [10] present an approach where the agents are organized as an irregular pyramid and cooperate to aggregate regions. Bovenkamp et al. [11] proposed a MAS for the segmentation of the IntraVascular UltraSound images. The main characteristic of this approach is the elaboration of a high-level knowledge-based control over the low-level image processing algorithms.

Therefore, the association of MAS and image processing has been revealed as an expanded area of research. As far as we know, multi-agent approaches have never been applied to the retinal images. In this paper, an approach based on the previous work of Mahdjoub et al. [7] is applied to the digital color fundus images for the blood vessels segmentation. Despite of the wide literature in the segmentation of blood vessels trough fundus images analysis [1]-[6], none of them performs as needed due to the use of centralized mechanisms at the macro level. Since the retinal images are very complex, it is necessary to develop algorithms that can be locally adapted to the image proprieties. This new approach uses some image processing algorithms as concrete perception and action tools for defining autonomous agents that interact among themselves and with the environment (the image). Then the segmentation of the blood vessel emerges as a global behavior.

This paper is organized as follows: Section 2 describes the proposed approach compounded of two main phases: the preprocessing and the MAS model. The results are demonstrated and discussed in Section 3. Finally, conclusion is presented in Section 4.

2 Methodology

The proposed approach aims to use a MAS model to improve the detection of retinal blood vessels edges resulting from a preprocessing phase. This preprocessing phase consists of a group of conventional image processing algorithms and provides the information (environment) for the MAS model reconstructs and reorganizes them.

2.1 Preprocessing

For this step of the approach, the methods used in the preprocessing phase of work developed by Niemeijer et al. [13] were first employed. The resultant image had no intensity variation in the background across the image and the bright structures were eliminated. A Gaussian filter (width 3 pixels; $\sigma = 2$) was then applied to attenuate the high frequency noise.

In order to remove the noise of the fundus image while preserving the edges it was then applied the Kuwahara filter [14]. Finally to the image resulted from the last step a modified version of the Kirsch filter was employed. The Kirsch filtering is a method for enhancing the edges in an image using a basic convolution filter rotated eight times. All eight filters are applied to the image with the maximum being retained for the final image. The improved Kirsch filter [7] enables to detect edges with a two pixels thickness of which the external edge is represented by a positive or negative value, whereas the internal edge has an opposite value (Fig. 1 a)). This facilitates the detection process of the MAS model since the gradient of the blood vessels reveals a specific pattern (Fig. 1 b)) as they can be represented by series of two parallel linear segments. Thus, the agents search the edges of the blood vessels by looking for this specific pattern of the gradient.

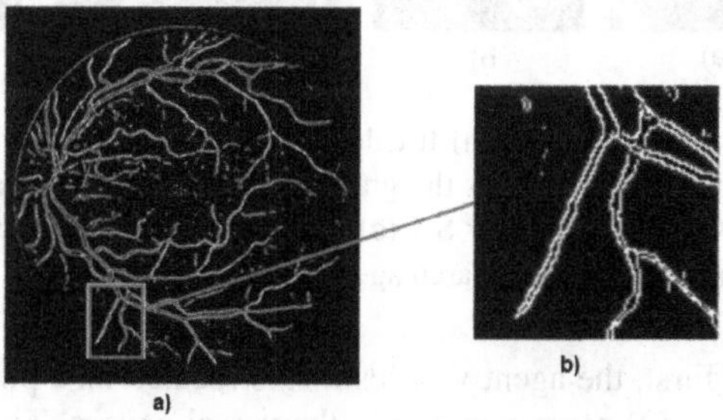

Fig. 1 a) Resultant image of the modified Kirsch filter where the blue and white pixels represent negative and positive gradient values, respectively. b) Expanded version of a section of image a), where is possible to see a characteristic pattern in the vessel gradient values.

2.2 A Multi-Agent Approach for the Edge Detection

Our MAS is composed by a set of agents and their environment. The environment contains the green plane image in which each pixel contains the grey level of intensity and a boolean value defining if the pixel has already been explored by an agent. Moreover, when located in the environment the agents perceive the Kirsch gradient which defines a right visible edge. The agents are of several kinds with different behaviors according to their current state and perception: search agent (SA), following agent (FA), node agent (NA), and end agent (EA). The algorithm initializes with a search agent launched on one of the white points from Fig. 1 a), randomly chosen. At the end of the process the agents have to reconstruct the vessels edge by representing it with a succession of segments.

Agents

The SA initially launched by the system has to find edges and evolves in the environment by following the points with positive gradient. When it finds an edge, it determines the possible directions to follow the contour, creates a NA and moves to another position. The NA creates FAs in the directions given by the SA and establishes segments with them. The FAs follow the edge detected until its segment doesn't correspond with the explored contour anymore. Then they create NAs, give them information about the direction to follow and die. These new NAs create new FAs, and so on. Considering that several agents can follow the same contour, they end up meeting and merging. When there is no direction to follow the NA creates an EA.

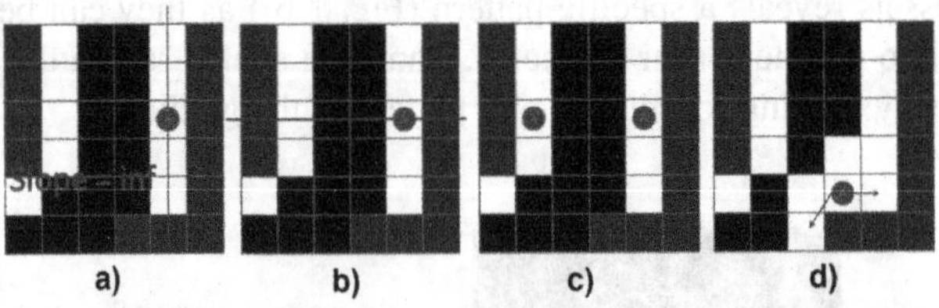

Fig. 2 The SA (red circle) behavior. a) It calculates the slope of the line to determine the perpendicular line (b). Then it verifies the gradient values profile and as the pixel belongs to a vessel pattern it launches another SA (c); d) Possible directions (red arrows) that the agent has to follow according to the search agents' restrictions

Search behavior*:* First, the agent verifies if it is located on a pixel: (1) not visited yet by another agent; (2) corresponding to the vessel pattern. To verify if the pixel belongs to a vessel pattern, the agent calculates the slope of the line constituted by its positions and the neighbors with positive gradient (Fig. 2 a)). Then, it analyses the gradient values profile on the perpendicular line to it (Fig. 2 b)), and if it corresponds to negative-positive-null-positive-negative values, the pixel belongs to a vessel edge. If it verifies the two conditions it determines the possible directions to follow, launches a NA on its position and moves to another white point, also randomly chosen. Moreover, it launches another SA on the parallel line to the one where it is initially located (Fig. 2 c)). For the determination of the directions to follow the agents look for the white points in their 8-neighbooring having a blue point in the 4-neighbooring. This blue point also has to belong to the 8-neighboors of the target pixel (Fig. 2 d)). The initially launched SA stops its behavior and disappears when all the white points were analyzed.

Node behavior*:* The node behavior is executed just one time. The NA launches FA in the directions given by the launcher agent (SA or FA) in the list of the possible directions to follow. If this list is empty it launches an EA, and then it establishes segments with those agents and keeps these segments in its segment list. After that, the NA stops its behavior.

Following behavior*:* At the beginning of the following behavior the agent checks its mail box to verify if there is another FA moving on the same line vessel but in opposite direction. In this case, they ask for a fusion process with each other by linking the respective NA. To construct the edge, the FA moves pixel per pixel and it stores the pixel position on it is located. The objective is to make sure that each position characterizing the edge portion which separates it from its neighbor can be approximated (using a threshold) by the segment connecting the two agents. The FAs move by determining the possible directions to follow as the SA do. If the FA has just one direction to follow, it checks if the segment formed between the position of its neighbor and its position is a valid segment, and if it is not, it launches another NA on its position and disappears. If it has more than one direction to follow it also launches a NA and disappears. After each movement, the FA sends a message to all the FAs in order to attempt a fusion.

End behavior*:* This behavior of the EA is to check if it is located on a blood vessel edge by analyzing the green intensity profile on the perpendicular direction to its segment. It just verifies if the profile is similar to a Gaussian shape as the blood vessels should be. If the profile does not fit a Gaussian curve, the EA disappear with its segment. If the profile fits a Gaussian curve the EA stops its behavior. This process is important to clean the small segments not belonging to the blood vessels, but to some noise and other imperfections of the background that still remains after the preprocessing phase.

3 Results and Discussion

The proposed MAS model was implemented with MadKit [15] that is a generic multi-agent platform written in Java and built upon the AGR (Agent/Group/Role) organizational model. That is, the MadKit agents play roles in groups and thus create artificial societies.

The DRIVE dataset, a publicly available dataset developed by Niemeijer et al. [2], was used to test the proposed approach. It is compound of 40 images, in which 7 present signs of mild early diabetic retinopathy.

The performance of the edge reconstruction by the proposed MAS model depends directly on the image processing algorithms used in the preprocessing phase. It also depends of how the system interprets information resulting from this first phase of the approach. To measure the performance of the overall approach, it is important to compare the resulting image with the information detected by the Kirsch filter. Moreover, the differences between the resulting edge map image and the ground truth vessel map also present in the DRIVE dataset should be evaluated.

Results of the proposed approach applied to normal retinal images are shown in Fig. 3. These illustrate the original and resulting images where the proposed approach had the best and the worst performance respectively, with true positive rates of 88.9% and 71.7%. The overall true positive rate for the DRIVE dataset was 83.1%.

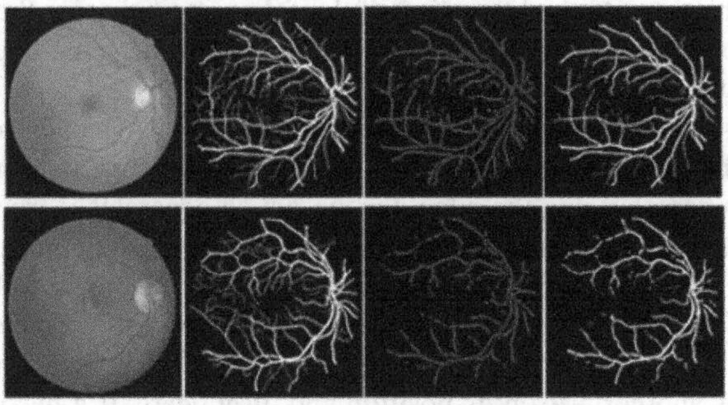

Fig. 3 Resulting images where the proposed approach had the best (above) and worst (below) performance in the DRIVE database. From left to right: original color fundus image; hand labeled image; edge detection by the MAS approach; the same image as the left one, but with a morphological posprocessing to better visualize the performance of the approach.

By analyzing Fig. 1 a) and Fig. 3 (right one, above) it can be observed that the MAS reconstructed the edges of the most part of the vessels, especially of the thickest ones. Furthermore, they eliminated a large part of the pixels detected by the Kirsch filter that do not belong to the vessels. However, the MAS also cleaned some pixels belonging to the thinnest vessels, and some of these vessels were not detected at all, affecting the sensitivity values (Fig. 3). So, improvements have to be made in MAS model to deal with the small vessels.

Nevertheless, the experiments show that the use of a MAS model at the micro level could be an effective way to segment structures in complex images as the retinal images. In fact, through perception of the environment and local interactions, a simple agent organization can have as global behavior the detection of the most part of the retinal vasculature. The use of an improved version of the society of agents with some knowledge a priori about the retina proprieties, complemented with the use of some other traditional image processing algorithms could have the potential to develop a system to detect and differentiate all the anatomic and pathological structures of the fundus images. Actually, this approach is just a preliminary study of a large project where the goal is to develop a computer aided diagnosis system to be applied in regular screening programs to detect DR.

4 Conclusion

In this paper, a MAS approach is proposed where agents enrich a traditional edge detector algorithm allowing local processing adaptation and cooperative behaviors. The system is able to represent edges present on fundus image through segments, mainly the edges belonging to the vasculature.

It seems that a very efficient system for the diabetic retinopathy diagnosis could be built using a micro level approach, applying MAS mechanisms such as stigmergy for instance. Such an approach will overcome the classic image processing algorithms that are limited to macro results which cannot take into account the local characteristics of a complex image as the digital color fundus images are.

References

1. Hoover, A., Kouznetsova, V., Goldbaum, M.: Locating blood vessels in retinal images by piecewise threshold probing of a matched filter response. IEEE Trans. Med. Imag. 19(3), 203–210 (2000)
2. Staal, J.J., Abràmoff, M.D., Niemeijer, M., Viergever, M.A., van Ginneken, B.: Ridge-based vessel segmentation in color images of the retina. IEEE Trans. Med. Imag. 23(4), 501–509 (2004)
3. Soares, J.V.B., Leandro, J.J., Cesar Jr., R.M., Jelinek, H.F., Cree, M.J.: Retinal vessel segmentation using the 2-D Gabor wavelet and supervised classification. IEEE Trans. Med. Imag. 25(9), 1214–1222 (2006)
4. Ricci, E., Perfetti, R.: Retinal blood vessel segmentation using line operators and support vector classification. IEEE Trans. Med. Imag. 26(10), 1357–1365 (2007)
5. Mendonça, A., Campilho, A.: Segmentation of retinal blood vessel by combining the detection of centerlines and morphological reconstruction. IEEE Trans. Med. Imag. 25(9), 1200–1213 (2006)
6. Al-Diri, B., Hunter, A., Steel, D.: An active contour model for segmenting and measuring retinal vessels. IEEE Trans. Med. Imag. 28(9) (2009)
7. Mahdjoub, J., Guessoum, Z., Michel, F., Herbin, M.: A multi-agent approach for the edge detection in image processings. In: 4th European Workshop on Multi-Agent Systems, Lisbon, Portugal (2006)
8. Haroun, R., Boumghar, F., Hassas, S., Hamami, L.: A Massive Multi-agent System for Brain MRI Segmentation. In: Ishida, T., Gasser, L., Nakashima, H. (eds.) MMAS 2005. LNCS (LNAI), vol. 3446, pp. 174–186. Springer, Heidelberg (2005)
9. Richard, N., Dojat, M., Garbay, C.: Automated segmentation of human brain MR images using a multi-agent approach. Artificial Intelligence in Medicine 30, 153–176 (2004)
10. Duchesnay, E., Montois, J.J., Jacquelet, Y.: Cooperative agents society organized as an irregular pyramid: A mammography segmentation application. Pattern Recognition Letters 24, 2435–2445 (2003)
11. Bovenkamp, E.G.P., Dijkstra, J., Bosch, J.G., Reiber, J.H.C.: Multi-agent segmentation of IVUS images. Pattern Recognition 37, 647–663 (2004)
12. Benamrane, N., Nassane, S.: Medical Image Segmentation by a Multi-agent System Approach. In: Petta, P., Müller, J.P., Klusch, M., Georgeff, M. (eds.) MATES 2007. LNCS (LNAI), vol. 4687, pp. 49–60. Springer, Heidelberg (2007)

13. Niemeijer, M., van Ginneken, B., Staal, J., Suttorp-Schulten, M., Abràmoff, M.D.: Automatic detection of red lesions in digital color fundus photographs. IEEE Trans. Med. Imag. 24(5), 584–592 (2005)
14. Kuwahara, M., Hachimura, K., Eiho, S., Kinoshita, M.: Digital Processing of Biomedical Images, pp. 187–203. Plenum Press, New York (1976)
15. Madkit Homepage, `http://www.madkit.org/`

Metalearning in ALBidS: A Strategic Bidding System for Electricity Markets

Tiago Pinto, Tiago M. Sousa, Zita Vale,
Isabel Praça, and Hugo Morais

Abstract. Metalearning is a subfield of machine learning with special propensity for dynamic and complex environments, from which it is difficult to extract predictable knowledge. The field of study of this work is the electricity market, which due to the restructuring that recently took place, became an especially complex and unpredictable environment, involving a large number of different entities, playing in a dynamic scene to obtain the best advantages and profits. This paper presents the development of a metalearner, applied to the decision support of electricity markets' negotiation entities. The proposed metalearner takes advantage on several learning algorithms implemented in ALBidS, an adaptive learning system that provides decision support to electricity markets' participating players. Using the outputs of each different strategy as inputs, the metalearner creates its own output, considering each strategy with a different weight, depending on its individual quality of performance. The results of the proposed method are studied and analyzed using MASCEM - a multi-agent electricity market simulator that models market players and simulates their operation in the market. This simulator provides the chance to test the metalearner in scenarios based on real electricity markets' data.

Keywords: Adaptive Learning, Electricity Markets, Intelligent Agents, Metalearning, Simulation.

1 Introduction

A Metalearner is characterized as a learning algorithm that uses different kinds of meta-data, derived from other learning algorithms to effectively solve a given

Tiago Pinto · Tiago M. Sousa · Zita Vale · Isabel Praça · Hugo Morais
GECAD – Knowledge Engineering and Decision-Support Research Center
Institute of Engineering – Polytechnic of Porto (ISEP/IPP)
Rua Dr. António Bernardino de Almeida, 431
4200-072 Porto
e-mail: {tmp,tmsbs,zav,icp,hgvm}@isep.ipp.pt

J.B. Pérez et al. (Eds.): Highlights on PAAMS, AISC 156, pp. 247–256.
springerlink.com

learning problem [1]. The main goal is to use such meta-data to improve the performance of existing learning algorithms, by understanding how automatic learning can become flexible in solving different kinds of learning problems [2]. This is a particularly useful tool when the problem being dealt with presents a high level of uncertainty, *i.e.* when the characteristics of the problem present high variations depending on different situations and contexts.

An obvious example of a problem with such characteristics is the energy market. This type of market suffered a restructuring with the purpose of increasing the competition in this sector, and consequently leading to a decrease in energy prices. However, the complexity in energy markets operation also suffered an exponential increase, bringing new challenges to the participating entities operation [3]. In order to overcome these challenges, it became essential for professionals to fully understand the principles of the markets, and how to evaluate their investments in such a competitive environment [4]. The necessity for understanding those mechanisms and how the involved players' interaction affects the outcomes of the markets, and thus, the revenues of the investments, contributed to the need of using simulation tools, and machine learning techniques, with the purpose of taking the best possible results out of each market context for each participating entity. Multi-agent based software is particularly well fitted to analyze dynamic and adaptive systems with complex interactions among its constituents [5, 6, 7].

To explore and study such approaches is the main goal of this research, and for that the multi-agent system MASCEM (Multi-Agent System for Competitive Energy Markets) [7, 8] is used. This system simulates the electricity market, considering all the most important entities that take part in such operations. Players in MASCEM are implemented as independent agents, with their own capability to perceive the states and changes of the world, and acting accordingly. MASCEM main entities include: a market operator agent, a system operator agent, a market facilitator agent, buyer agents, seller agents, Virtual Power Player (VPP) agents, and VPP facilitators.

Supporting MASCEM players' actions in what concerns their strategic behavior, an adaptive multiagent system - ALBidS (Adaptive Learning strategic Biding System) [8] is used. ALBidS is presented in section 2.

This paper presents a new methodology to support the negotiating agents' bidding definition. The proposed approach is defined as a metalearner, using the outputs of all ALBidS strategies, to create a new strategic proposal for the MASCEM's supported player to adopt when acting in the market. The metalearner takes into account the performance confidence values that each strategy is presenting, in order to define the weight with which each strategy will affect the metalearner's output. This way the metalearner is able to adapt its output, giving higher influence to the results of the strategies that are proving to be more adequate, while lowering the contribution of the strategies which are presenting worst results.

2 ALBidS System

For each market, the way prices are predicted can be approached in several ways. Through the use of statistical methods, data mining techniques [9], neural networks (NN) [10], support vector machines, or several other methods [5, 11]. There is no method that can be said to be the best for every situation, only the best for one or other particular case.

To take advantage of the best characteristics of each technique, we decided to create a system that integrates several distinct technologies and approaches. The set of algorithms is placed below the main reinforcement learning algorithm, which allows that in each moment and in each circumstance the technique that presents the best results for every actual scenario is chosen as the simulator's response [12]. So, given as many answers to each problem as there are algorithms, the reinforcement learning algorithm will choose the one that is most likely to present the best answer according to the past experience of their responses and to the present characteristics of each situation, such as the week day, the period, and the particular market that the algorithms are being asked to forecast. ALBidS is implemented as a multiagent system itself. There is one agent per distinct algorithm, with only the knowledge of how to perform it. This way the system can be executing all the algorithms in parallel, increasing the performance of the method, and as each agent gets its answer, sends it to the main agent, which chooses the most appropriate answer among all that it received. The forecast algorithms used by the agents of this method are:

- Dynamic Feed Forward **Neural Network** (NN) [10] trained with the historic market prices, with an input layer of eight units, regarding the prices and powers of the same period of the previous day, and the same week days of the previous three weeks. This NN is retrained in each iteration so that the data observed at each moment is considered for the next forecasts, this way constantly adapting the NN forecasting results.
- **Adaptation of the AMES bidding strategy**. This strategy uses the Roth-Erev [5] reinforcement learning algorithm to choose the best among a set of possible bids that are calculated based on the relation cost/profit that the player presents when producing electricity.
- The **SA-QL** strategy uses the Simulated Annealing heuristic to accelerate the process of convergence of the Q-Learning [13] algorithm in choosing the most appropriate from a set of different possible bids to be used by the market negotiating agent whose behaviour is being supported by ALBidS.
- The **Game Theory** strategy is characterized as a scenario analysis algorithm able to support strategic behaviour, based on the application of game theory.
- The **Economic Analysis** strategy implements an analysis based on the two most commonly used approaches of forecasting in a company's scope. These approaches are the internal data analysis of the company, and the external, or sectorial, data analysis [14].
- The **Determinism Theory** strategy executes a strategy based on the principles of the Determinism Theory [15]. A theory that states that due to the laws of cause and effect, which apply to the material universe, all future events are predetermined.
- The **Error Theory** strategy's goal is to analyze the forecasting errors' evolution of a certain forecasting method, to try finding patterns in that error

sequence and provide a prediction on the next error, which will be used to adequate the initial forecast.

The main reinforcement algorithm presents a distinct set of statistics for each context, which means that an algorithm that may be presenting good results for a certain period, with its output chosen more often when bidding for this context, may possibly never be chosen as the answer for another period, since they are completely independent from each other [8]. The tendencies observed when looking at the historic of negotiation periods independently from each other show that they vary much from each other, what suggests that distinct algorithms can present distinct levels of results when dealing with such different tendencies.

The way the statistics are updated, and consequently the best answer chosen, is defined through the use of one of three reinforcement learning algorithms. All the algorithms start with the same value of confidence for each strategy, and then according to their particular performance that value will be updated. The three versions are [12]: a simple reinforcement learning algorithm, which updates the strategies' confidence weights based on the profits each strategy is allowing the supported player to achieve; a learning algorithm based on the Bayesian Theorem of conditional probability [16], in which, each strategy's confidence weights are defined by their probability of success; and a reinforcement learning algorithm based on the Roth-Erev algorithm.

3 Metalearner

Metalearning is one-step higher than ordinary learning and means Learning-About-Learning. As described in [17], metalearning is the state of "*being aware of and taking control of one's own learning*". It can be defined as an awareness and understanding of the phenomenon of learning itself, opposed to subject knowledge.

Taking advantage on the concept of metalearning, and its adequacy to the scope of this work, it has been decided to develop two versions of metalearners, in order to test the advantage of using such type of learning algorithms, which perform a combination of the several different outputs provided by ALBidS' strategy agents.

Simple Metalearner

The idea for the development of the Simple Metalearner has arisen from the analysis of the work of Hans Georg Zimmermann, a Siemens AG, Corporate Technology researcher whose work deals with, among many other issues, forecasting using neural networks (NN) [10]. In his widely recognized work [19], [20], he recurrently uses simple averages of the outputs of the used NN, to originate a final output, in order to overcome the uncertainty that affects the forecasts. In order to overcome such uncertainty, Zimmermann uses different initializations for the used NN, covering a larger space of solutions, and then uses an ensemble averaging to create the final solution. "Obviously the solution of the system identification will depend on the initialization which can be handled by an ensemble averaging." [20].

The scope of the ALBidS system is similar, although many other approaches other than NNs are used. For this reason it has been found interesting to experiment if the ensemble averaging of the outputs of the various ALBidS strategies could lead to interesting results.

Weighted Metalearner

The Weighted Metalearner intends to extend the Simple Metalearner, creating an adaptive tool that can in fact be called a metalearner, using the concept of stacked generalization. This metalearner uses as inputs, the outputs of the various approaches, but presents the additional feature of attributing importance weights to each of those inputs. The weights provide the chance for the metalearner to adapt its output, giving higher focus to the results of the strategies that are proving to be more adequate, while partially or completely ignoring the contribution of the strategies which are presenting worst results.

This procedure allows the Weighted Metalearner to adapt its output according to the observed results of each of its inputs. The weights used for defining the importance that each input has for the metalearner, are based on the confidence values of the main reinforcement learning algorithm used by ALBidS. The reinforcement learning algorithm's confidence values are adapted and updated according to the results each strategy is presenting, hence being exactly what this metalearned requires understanding which are the strategies' outputs that it should consider as most influent to the final output.

The generation of this output Xp is performed through a weighted average, using the reinforcement learning algorithm's confidence values as weights p1, p2, p3, ..., pn for each strategy's output's x1, x2, x3, ..., xn contribution to the final metalearner's solution. The procedure is expressed in (1).

$$\bar{x}_p = \frac{p_1 \times x_1 + p_2 \times x_2 + \ldots + p_n \times x_n}{p_1 + p_2 + \ldots + p_n} = \frac{\sum_{i=1}^{n} p_i \times x_i}{\sum_{i=1}^{n} p_i} \tag{1}$$

The adapted output of the Weighted Metalearner is expected to be able to generate better results than the Simple Metalearner, since it takes higher regard for the inputs that are expected to point the final result towards a better solution.

4 Experimental Findings

This section presents two simulations undertaken using MASCEM, with the purpose of analyzing the performance of the proposed methodology. The data used in this case study has been based on real data extracted from the Iberian market - OMEL [21]. The executed simulations concern 61 consecutive days, starting from the 15th October. In the first simulation Seller 2 uses the Simple Metalearner, and in the second it uses the Weighted Metalearner.

These results purpose is to demonstrate the use of these strategies to support a market player's decisions and to show the influence of the supporting strategies over these metalearners performance.

For these simulations, the main reinforcement learning algorithm used by the Main Agent of ALBidS is the Bayes Theorem algorithm. This algorithm's confidence values for each strategy are used by the Weighted Metalearner, Additionally, as supporting strategies, nine random strategies from all the ones supported by ALBidS, were selected. These strategies are the same for both simulations.

Figure 1 presents the incomes obtained by Seller 2 in the first simulation, and a comparison between the proposed bid and the market price.

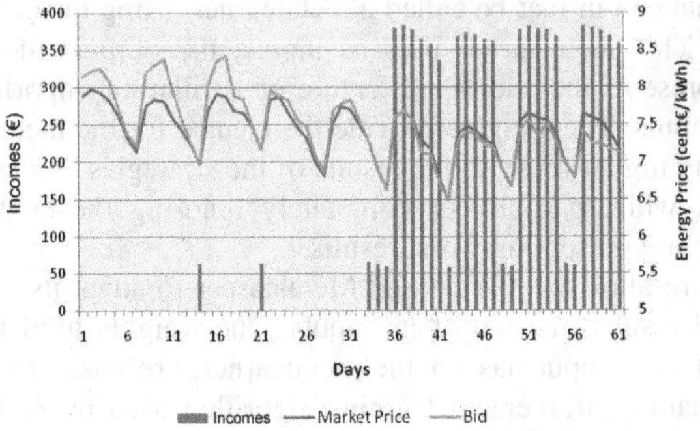

Fig. 1 Results of the Simple Metalearner.

Analyzing Figure 1 it is visible that this strategy starts by achieving bad results, which start to improve with the passing of the days. This is due to some supporting strategies' worst suggestions at the start, while they do not have the experience to learn adequately yet. As this metalearner considers all suggestions in a similar way, the good outputs that some strategies may be presenting are muffled by the bad ones. As the time progresses and the strategies start to learn and provide best individual results, the metalearner's results improve as well. Fig. 2 presents a comparison between the metalearner's bid price and the supporting strategies' proposals.

Figure 2 shows that, in the first days, the strategies' proposals present a high variance among each other, therefore the metalearner's bid follows that variance trend. In the later days, the proposals start to converge and improve their quality,

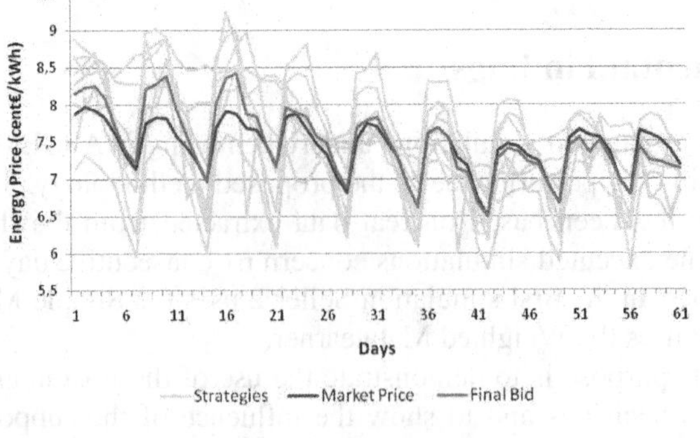

Fig. 2 Simple Metalearner's and supporting strategies' bids.

through their individual learning capabilities, and consequently, the metalearner's bid gets closer to the market price.

The consideration of all suggestions in an equal way results in a bad performance of the Simple Metalearner especially in the first days. This is why the use of the Weighted Metalearner is important. This method considers adequate weights to instigate the attribution of higher importance to the strategies that are presenting the best results at each time. Figure 3 presents the results of Seller 2 using the Weighted Metalearner.

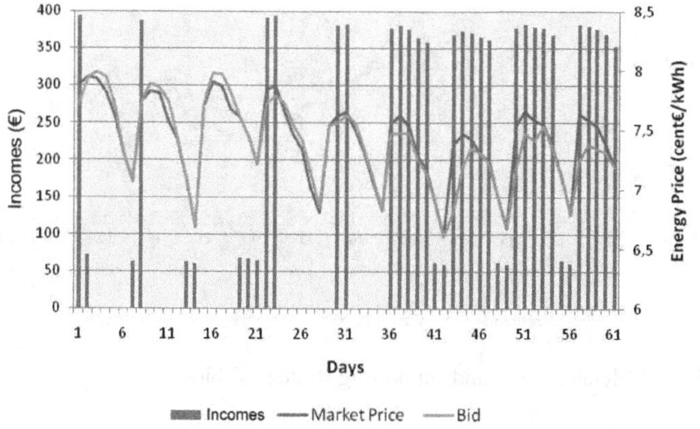

Fig. 3 Results of the Weighted Metalearner.

Figure 3 shows that, in spite of these results following the same tendency as the Simple Metalearner, with bad results at first, and a visible improvement over time, it could achieve good results in some early days, proving the advantage of attributing higher weights to the most appropriate supporting strategies' proposals. Figure 4 presents the development of the supporting strategies' confidence values, which determine the weights each one will be attributed.

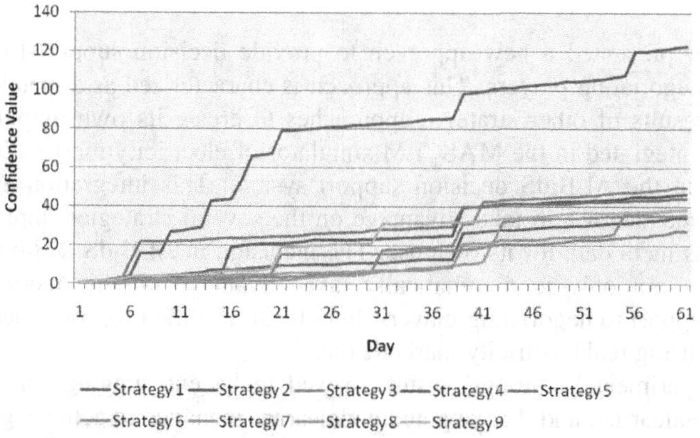

Fig. 4 Supporting strategies' confidence values.

From Figure 4 it is visible that Strategy 4 detaches from the others. This means that this strategy is the one that influences the most the final bid from the Weighted Metalearner. Its influence increases over time. Figure 5 shows how this metalearner's bid follows the Strategy 4's responses, in comparison with the rest of the supporting strategies.

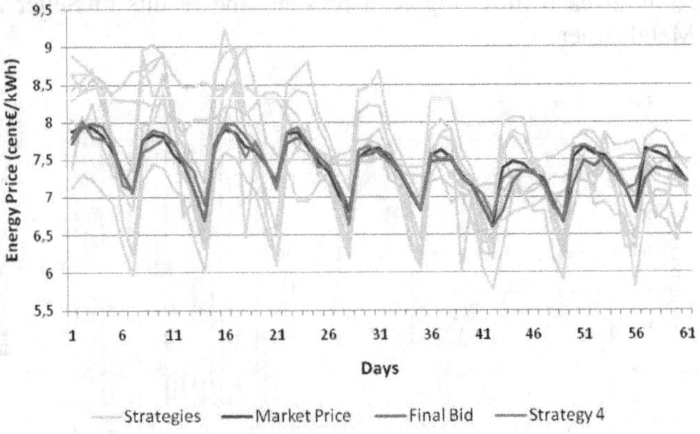

Fig. 5 Weighted Metalearner's and supporting strategies' bids.

Figure 5 shows that, in spite of in the first days the strategies' proposals presenting an high variance among each other, the metalearner's bid stays close to the market price, influenced by Strategy 4. This tendency remains visible throughout the days. Comparing the Weighted Metalearner's performance with the Simple Metalearner, a much more constant behaviour is found. The adjustment of the final bid, taking into account the strategies' results has proven to be an added value in adequating the metalearner's results.

5 Conclusions

This paper presented a new approach to provide decision support to electricity markets' negotiating players. This approach is characterized as a metalearner that uses the results of other strategic approaches to create its own output. This approach is integrated in the MASCEM simulator of electricity markets, through its inclusion in the ALBidS decision support system. This integration allowed the proposed metalearner to take advantage on the several strategies implemented in ALBidS as meta data for its own use. The inclusion in ALBidS additionally gave the chance not only to automatically take advantage on this methodology to provide support to negotiating players, but also to test the proposed method's performance using real electricity market's data

The experimental findings' results proved to be encouraging, since both the simple metalearner, and the weighted metalearner managed to achieve good results on the market. Most importantly, both variations showed a visible improvement over time, as a consequence of the metalearners' capability of taking advantage on the learning abilities of the strategies available to them. Concerning the weighted metalearner, its dependence on the supporting strategies confidence values proved to be a real added value, since the metalearner gained the capability of adapting its outputs, highlighting the strategies that are showing to be able of achieving better results at each time, and for each negotiating context. The metalearners' ability to improve the performance over time is accentuated by the achievement of increasing profits on behalf of the supported market player, as demonstrated by the case studies based on real electricity markets' data. The metalearners' performance is also a valuable asset for the ALBidS system, in what regards this system's purpose.

Improving the metalearners' performance in this scope includes the addition of a neural network, considering as inputs the outputs of the supporting strategies, and originating as output the predicted answer. This way, the metalearner should increase even further its ability to follow trends of the times and contexts each strategy should be considered with the higher influence.

Acknowledgement. The authors would like to acknowledge FCT, FEDER, POCTI, POSI, POCI, POSC, POTDC and COMPETE for their support to R&D Projects and GECAD Unit.

References

[1] Goldkuhl, G., et al.: Method integration: the need for a learning perspective. IEEE Proceedings on Software 145(4), 113–118 (1998)

[2] Bruha, I.: Meta-Learner for Unknown Attribute Values Processing: Dealing with Inconsistency of Meta-Databases. Journal of Intelligent Information Systems 22(1) (2004)

[3] Shahidehpour, M., et al.: Market Operations in Electric Power Systems: Forecasting, Scheduling, and Risk Management, pp. 233–274. Wiley-IEEE Press (2002)

[4] Meeus, L., et al.: Development of the Internal Electricity Market in Europe. The Electricity Journal 18(6), 25–35 (2005)

[5] Li, H., Tesfatsion, L.: Development of Open Source Software for Power Market Research: The AMES Test Bed. Journal of Energy Markets 2(2) (2009)

[6] Koritarov, V.: Real-World Market Representation with Agents: Modeling the Electricity Market as a Complex Adaptive System with an Agent-Based Approach. IEEE Power & Energy Magazine, 39–46 (2004)

[7] Praça, I., et al.: MASCEM: A Multi-Agent System that Simulates Competitive Electricity Markets. IEEE Intelligent Systems 18(6), 54–60 (2003); Special Issue on Agents and Markets

[8] Vale, Z., et al.: MASCEM - Electricity markets simulation with strategically acting players. IEEE Intelligent Systems 26(2), 54–60 (2011)

[9] Azevedo, F., et al.: A Decision-Support System Based on Particle Swarm Optimization for Multi-Period Hedging in Electricity Markets. IEEE Transactions on Power Systems 22(3), 995–1003 (2007)

[10] Amjady, N., et al.: Day-ahead electricity price forecasting by modified relief algorithm and hybrid neural network. IET Generation, Transmission & Distribution 4(3), 432–444 (2010)

[11] Greenwald, A., et al.: Shopbots and Pricebots. In: Proceedings of the Sixteenth International Joint Conference on Artificial Intelligence - IJCAI, Stockholm (1999)
[12] Pinto, T., Vale, Z., Rodrigues, F., Morais, H., Praça, I.: Strategic Bidding Methodology for Electricity Markets Using Adaptive Learning. In: Mehrotra, K.G., Mohan, C.K., Oh, J.C., Varshney, P.K., Ali, M. (eds.) IEA/AIE 2011, Part II. LNCS, vol. 6704, pp. 490–500. Springer, Heidelberg (2011)
[13] Juang, C., Lu, C.: Ant Colony Optimization Incorporated With Fuzzy Q-Learning for Reinforcement Fuzzy Control. IEEE Transactions on Systems, Man and Cybernetics, Part A: Systems and Humans 39(3), 597–608 (2009)
[14] Porter, M.: How Competitive Forces Shape Strategy. Harvard Business Review (1979)
[15] Berofsky, B.: Determinism. Princeton University Press (1971)
[16] Korb, K., Nicholson, A.: Bayesian Artificial Intelligence. Chapman & Hall/CRC (2003)
[17] Biggs, J.: The role of meta-learning in study process. British Journal of Educational Psychology 55, 185–212 (1985)
[18] Boyarshinov, V., et al.: Efficient Optimal Linear Boosting of a Pair of Classifiers. IEEE Transactions on Neural Networks 18(2), 317–328 (2007)
[19] Zimmermann, H., et al.: Multi-agent modeling of multiple FX-markets by neural networks. IEEE Transactions on Neural Networks 12(4), 735–743 (2001)
[20] Zimmermann, H., et al.: Dynamical consistent recurrent neural networks. In: IEEE International Joint Conference on Neural Networks, vol. 3, pp. 1537–1541 (2005)
[21] Operador del Mercado Ibérico de Energia – homepage, http://www.omel.es (accessed on August 2011)

An Agent-Based Prototype for Enhancing Sustainability Behavior at an Academic Environment

N. Sánchez-Maroño, A. Alonso-Betanzos, O. Fontenla-Romero, V. Bolón-Canedo, N.M. Gotts, J.G. Polhill, T. Craig, and R. García-Mira

Abstract. A prototype of an Agent-based Model (ABM) for the LOCAW (LOw Carbon At Work) projet is presented. The main goal of LOCAW is *foresight to enhance behavioral and societal changes enabling the transition towards sustainable paths in Europe*. It will involve examining large employer-organizations in six different countries. This paper presents a shared ontology created to ensure that maximum comparability of the case-studies is maintained. Next, a preliminary prototype, using one of the six organizations, the University of A Coruña, is shown.

1 Introduction

In this paper, we present a prototype of an Agent-based Model (ABM) that constitutes an important part of the LOCAW (LOw Carbon At Work) project. This project (*www.locaw-fp7.com*) will address the challenge of *Foresight to enhance behavioral and societal changes enabling the transition towards sustainable paths in Europe*, to conduct "an analysis of barriers and drivers for engaging on sustainable, low-carbon paths individually, on the level of individual organizations, and collectively" and to contribute to "scenario development and a back-casting exercise in order to identify potential paths to engaging on an integrated effort to support the transition to a sustainable Europe". The LOCAW project will examine six large employer-organizations in six different countries, occupying different positions in the sustainability debate.

N. Sánchez-Maroño · A. Alonso-Betanzos · O. Fontenla-Romero ·
V. Bolón-Canedo · R. García-Mira
University of A Coruña, A Coruña, 15071, Spain
e-mail: {noelia.sanchez,amparo.alonso.betanzos}@udc.es,
{oscar.fontenla,vbolon,ricardo.garcia.mira}@udc.es

N.M. Gotts · J.G. Polhill · T. Craig
The James Hutton Institute, Craigiebuckler, Aberdeen. AB15 8QH
e-mail: {nick.gotts,gary.polhill,tony.craig}@hutton.ac.uk

J.B. Pérez et al. (Eds.): Highlights on PAAMS, AISC 156, pp. 257–264.
springerlink.com

Agent computational models are powerful analytical tools as well as an effective presentation technology, therefore they have been applied to solve different types of problems [1]. They have been extensively used in the study of social dilemmas: situations in which each of a group of actors will be better off if all cooperate, for example to limit use of some resource, but each has an incentive to cheat [2]. Moreover, they are becoming very popular for dealing with sustainable development, in fact, in 2009 a joint workshop of Global System Dynamics and Policies and the European Climate Forum (www.european-climate-forum.net) with the title "Agent-based modeling for sustainable development" was held. Agent-based modeling tries to establish a direct correspondence between the entities that are identified as participants in an observed system and the agents constituting the model. In addition, not only this identification and entity correspondence exists, but also a direct correspondence between the interactions of the aimed system and the agents of the model. So, at the time of developing an ABM applied to economic, political and social sciences, there are some considerations that need to be taken into account [3]:

1. One issue is common to all modeling techniques: a model has to serve a purpose. A general-purpose model cannot work. The model has to be built at the right level of description, with just the right amount of detail to serve its purpose.
2. Another issue has to do with the very nature of the systems being modeled with ABM in the social sciences: they most often involve human agents, with potentially unpredictable behavior, subjective choices, and complex psychology-in other words, soft factors, difficult to quantify, calibrate, and sometimes justify.
3. The last major issue in ABM is a practical issue that must not be overlooked. By definition, ABM looks at a system not at the aggregate level but at the level of its constituent units. Although the aggregate level could perhaps be described with just a few equations of motion, the lower-level description involves describing the individual behavior of potentially many constituent units. Simulating the behavior of all of the units can be extremely computation intensive and therefore time consuming. Although computing power is still increasing at an impressive pace, the high computational requirements of ABM remain a problem when it comes to modeling large systems.

Agent-based modeling will be used within LOCAW in a synthesis role, bringing together evidence from different methods of data acquisition to construct "back-casting" scenarios for the organizations studied. In a back-casting exercise, members from each organization will start from a desirable endpoint for the organization within a sustainable, low-carbon Europe and will then construct several narrative paths to that outcome and imagine ways in which the organization could fail to adapt successfully. In the LOCAW context, and related to consideration (1), the main purpose of the ABM is clear: *Modeling the effectiveness of the organization's policies in changing CO2 with respect to 3 main practices*:

1. Energy and materials consumption
2. Management and generation of waste
3. Organization-related mobility

2 The LOCAW Project

The LOCAW project (LOw Carbon At Work, *www.locaw-fp7.com*) will advance understanding of the drivers of and barriers to sustainable lifestyles by an integrative investigation of the determinants of everyday practices and behaviors within large scale organizations on different levels:

- Analyzing the patterns of production and consumption in the workplace with their resulting greenhouse gas (GHG) emissions.
- Analyzing organizational strategies to reduce emissions and implement EU regulations regarding the "greening" of their production processes.
- Analyzing everyday practices and behaviors at work of employees on different levels of decision-making within the organization.
- Analyzing the relationship between behaviors and practices at work and behaviors and practices outside work.
- Analyzing the patterns of interaction between relevant agents and stakeholders in the organization's environment and the resulting barriers and drivers for implementing sustainable practices and behaviors in the workplace.

The project will provide a theoretical and empirically grounded analysis of everyday practices in the workplace, and of the factors promoting or hindering the transition to more sustainable behaviors and practices in the workplace. It will do so *by examining large employer-organizations in six different countries*, occupying different positions in the sustainability debate, specifically this project involves:

- Two state organizations
  - University of A Coruña (UDC), Spain
  - The municipality in Groningen, The Netherlands
- Two private service providers in the field of natural resources/energy
  - A company of public water and wastewater services: Aquatim, Romania
  - A company dedicated to developing and managing energy generation from renewable sources: Enel Green Power, Italy
- Two heavy industry companies
  - Vehicle production, the Volvo plant in Umeå, Sweden
  - Oil production, Shell company in United Kingdom

In four of the cases, the focus will be on the everyday practices of the organization itself and the interactions between structural/organizational conditions and individual factors in generating barriers to and drivers for a sustainable transition to a low carbon Europe (the first four organizations). While the interactions of the organizations with relevant outside agents will form part of the study of structural conditions in these cases, two case studies will be ethnographic studies for two companies, including their management, their trade unions and their workforce (in the United Kingdom and in Sweden).

3 Development of the ABM

The modelers' priority will be to create the shared core, based on the construction of a shared ontology, of which a first draft has been developed. The common ontology will ensure that maximum comparability of the case-studies is maintained. Once the core ontology and model exist, they can be extended to include case-study-specific aspects. In each case study, the agent-based model for that case will focus on a key part of the organization to be decided (e.g., a particular process in manufacturing for a vehicle manufacturing firm or staff and student use of lighting and electrical equipment in a university). Figure 1 shows the initial ontology that, as explained before, should not be seen as a definitive ontology for the project. The importance of this ontology is to ensure that, where possible, the data collected within the different case studies is in some way 'representable' within it. Besides, it will help to find a way to communicate between the modeling and non-modeling teams.

Apart from this ontology, modelers and non-modelers are working together in order to obtain quality data that enrich the ABM. From this point of view, although all the acquisition phases of the LOCAW project have not been done yet, modelers have started to do a preliminary model that "illustrates non-modelers about data requirements and how policies could be implemented". Bearing in mind that students and teachers of UDC are working on LOCAW project, this organization turns out to be a good candidate as a first case study in order to do a scale model or prototype. This simple model can help computer scientist and sociologist to identify the "drawbacks of the model and lack of data" in order to obtain more informative data for all the

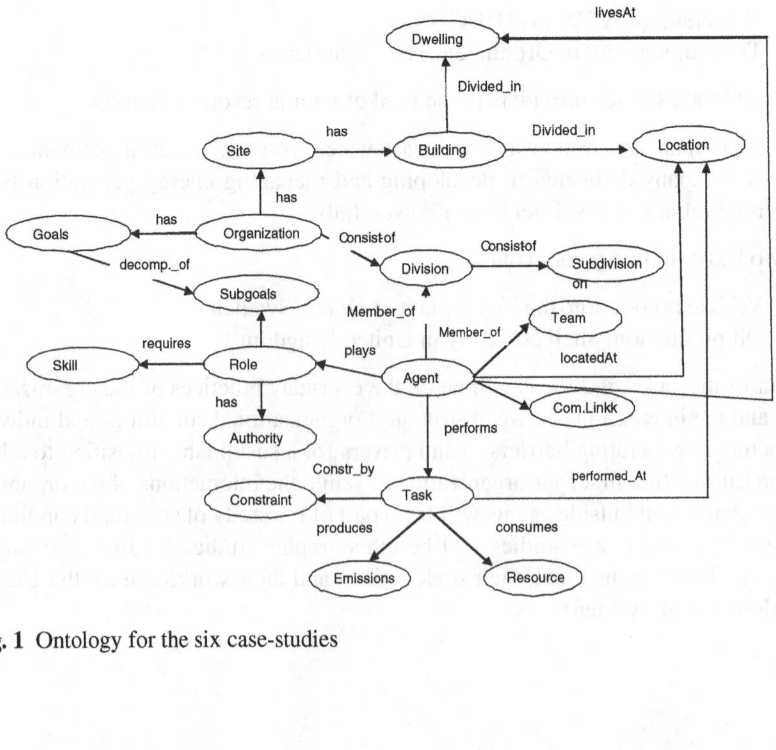

Fig. 1 Ontology for the six case-studies

organizations in the subsequent steps of the project. For doing this prototype, some technical considerations have been taken and some information have been gathered. Next subsections gives details about both, apart from showing the prototype.

3.1 Analysis at the University of a Coruña

The University of A Coruña is territorially situated in the Campus of A Coruña and the Campus of Ferrol, each one placed at a different city, A Coruña and Ferrol, at a distance of 52 kilometers. Besides there are different sites at each one of both campus, at present 6 in A Coruña and 2 in Ferrol. The university has around 20000 students distributed into 40 different degrees. The number of teachers is around 1500 and 750 persons take charge of the administration matters. Therefore, students represent more than 90% of the university community, whereas teachers suppose around 6% and administration staff is only 3%.

Table 1 Transport use (%) at UDC by community in 2008.

	Car	Bike	Urban bus	Interurb. bus	Moto	Walk	Train
Students (90.56%)	47.4	0.22	29.5	8.1	0.12	13.9	0.78
Teachers (6.16%)	80.8	2.16	4.7	3.8	0.60	7.6	0.34
Admin. (3.28%)	73.3	0	8.6	4.8	0	11.1	2.27
Total	50.2	0.33	27.3	7.8	0.15	13.5	0.80

In 2008, the Office for the Environment (OFE) of the UDC presented an study related to the ecological footprint of the organization that turns to be 3475 hectares(ha) of forest in that year. The ecological footprint was evaluated by considering 8 components: mobility, energy (electricity and gas), building, territory and consumption (water, paper and waste management) [4]. The footprint per capita is 0,15ha/person, obtaining a global value that is 50 times greater than the territory occupied by the UDC campus. Energy consumption and mobility comprised 80% of the footprint, so both areas are preferential for future interventions. There are data available at [4] showing the differences between energy consumption from one campus to another. Related to communities, Table 1 indicates the type of transport used in percentage, it can be noted that *Car* is preferred by members of any community, specially by teachers; it is worth to notice the reduced used of environmental friendly transports such as bikes.

3.2 ABM Toolkit Election

There are several agent based modeling and simulation tools [5][6][7]. Popular ABM toolkits are SWARM[8], Repast[9], MASON[10] and NetLogo[11]. MASON should be used where speed an/or sophisticated batch runs are required. Netlogo is a easy-to-use platform and language that is slower than Repast and also difficult to extend. In [6], general agent based modeling (ABM) are evaluated, and MASON and Repast are clear winners and are strongly recommended as ABM Platforms, although MASON is suggested as an overall best choice. On the contrary, authors in [7] advise Repast when the complexity of the system grows up and simulation speed is required. In both cases, thanks to its easy-to-use properties, Netlogo is highly recommended for not too complicated models. For that reason, Netlogo was elected for developing the initial prototype, however Repast toolkit would be used in further steps of the project.

3.3 Data Acquisition

The first part of the LOCAW project will provide a diagnosis of existing practices at work, relevant for achieving sustainable working behaviors. This part has been already done and has been addressed through a mix of qualitative methods: document analysis, direct observations, and interviews with workers situated at different levels of decision-making within the organizations under study. These data acquisition process provide an assessment of existing everyday practices and behaviors in the organizations under study, especially of those most related to the emissions of greenhouse gases. However, it does not provide individual factors promoting or hindering sustainable behaviors at work, i.e., there are not data related to individuals. In order to solve this inconvenient for the initial prototype, questionnaires covering every day practices of UDC personnel were distributed. The questionnaire focused into areas such as mobility (number of trips per week, transport used, ...), cafeteria used (lunch,drinks,...) and so on. Twenty individual answers were collected from teachers and students.

Undoubtedly, it will be necessary to have a higher number of agents reflecting the existing heterogeneity in the University of A Coruña, tying this heterogeneity to the intention of the project, that is to say, to determine which students exhibit a more or less ecological behavior, who have a major influence over their counterparts, who are more sociable, which are their preferred timetables to work (day-night), etc. In fact, this matter is considered in the project, but in a further step. However, the preliminary questionnaire helps to identify some issues:

- Students noticed that some posible answers were absent.
- Modelers realized that some questions were erroneously formulated, in the sense that the information provided was not completely useful.

Identifying these problems in an initial stage is very important because the arduous data acquisition phase is being done in 6 different companies, each one at a different country. Therefore, it is not possible to repeat this costly process. It must be taken into account that any policy that was not considered cannot be implemented, and thus its effects cannot be predicted with the model. Therefore, it is essential for the modelers to know all the possible policies and action lines, so as to be included in the model.

3.4 Development of the Prototype

The prototype includes two types of agents (students and teachers), the main difference between both types is the tasks they can develop, for example, students attend to class whereas a professor teaches. However, they have similar properties such as environmental degree, name, livesAt, etc. The model has a repeating daily time step consisting of the different tasks those agents do:

- Going to work and going back home
- Attending to a class / Teaching
- Break-going for a coffee/snack
- Having lunch
- Studying / Researching

For each task, the agent selects from a set of choices, for example, the transports available to go to work or switching off/on the heating in a classroom. Each choice has a environmental cost and the environment reflects this impact (increases the green house emission properly). In this initial prototype this impact has no effect over the agent, however, in further steps of development, the agent will get reinforcement from its neighbors and the environment, varying its elections consequently. Policies are represented by the set of choices available, for example, if car is not allowed at the campus, this transport is not included in the choice set for the agent.

Figure 2 shows a screenshot of the prototype. The left panel allows to modify the use-of-car and use-of-plastic and to indicate the data to load (different files are provided). The central panel shows different locations (blue squares) and agents are moving from one location to another to develop their tasks. The right panel shows plots associated with the GHG emissions related to transport used and plastic consumption of the different agents.

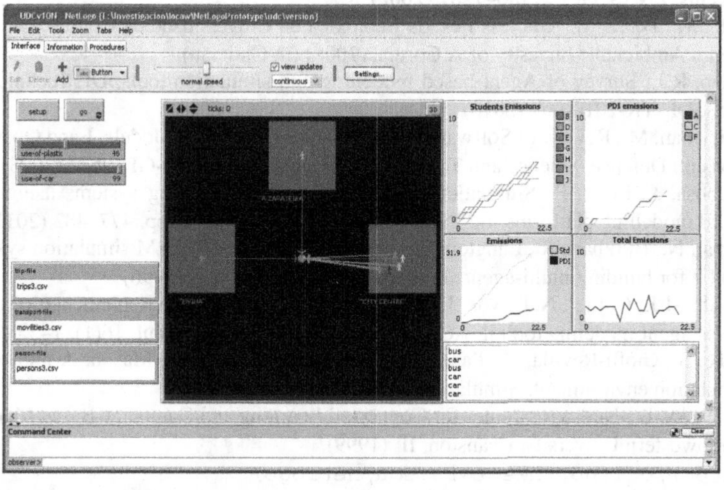

Fig. 2 Screenshot of the prototype

4 Conclusions

A prototype for the LOCAW projet is presented. This project (*www.locaw-fp7.com*) will address the challenge of *Foresight to enhance behavioral and societal changes enabling the transition towards sustainable paths in Europe* by examining 6 large employer-organizations in different countries. Data collected in these organizations, will then be synthesized and fed into the ABM that will allow to derive policy solutions. A preliminar ontology and prototype was done by modelers' in order to guide the next acquisition phase of the project and to assist sociologists to obtain quality data from the organizations. Not missing any informative data is crucial in the initial stages, because repeating any acquisition phase, bearing in mind the magnitude of the companies, would be unapproachable. The prototype is continuously evolving to reflect more aspects of the university community. As more agents and tasks are going to be included, Repast would be considered in further versions that will also reflect other cases of study.

Acknowledgements. This work was supported by the European Commission under project "LOw Carbon At Work: Modelling agents and organization to achieve transition to a low carbon Europe", 7th Framework Programme, ENV.2010.4.3.4-1 Grant Agreement 265155.

References

1. Axtell, R.: Why Agents? On the Varied Motivations for Agent Computing in the Social Sciences. In: Macal, C.M., Sallach, D. (eds.) Proceedings of the Workshop on Agent Simulation: Applications, Models, and Tools, pp. 3–24. University of Chicago: Argonne National Laboratory (1999)
2. Gotts, N.M., Polhill, J.G., Law, A.N.R.: Agent-based simulation in the study of social dilemmas. Artificial Intelligence Review 19, 3–92 (2003)
3. Bonabeau, E.: Agent-based modeling: Methods and techniques for simulating human systems. PNAS 99(3), 7280–7287 (2002)
4. Soto, M., Pérez, M. (eds.): A Pegada Ecolóxica da Universidade da Coruña. Oficina de Medio Ambiente, Univesity of A Coruña (2008) (in Galician)
5. Allan, R.J.: Survey of Agent based modelling and simulation tools, DL Technical Reports, DL-TR-2010-007 (2010)
6. Berryman, M.: Review of Software Platforms for Agent Based Models, Land Operations Division Defence, Science and Technology Organisation DSTO-GD-0532 (2008)
7. Barbosa, J., Leito, P.: Simulation of multi-agent manufacturing systems using agent-based modelling platforms. In: INDIN 2011, Lisbon, Portugal, pp. 477–482 (2011)
8. Minar, N., Burkhart, R., Langton, C., Askenazi, M.: The SWARM simulation system: a toolkit for building multi-agent simulations. Technical report (1996)
9. North, M.J., Collier, N.T., Vos, J.R.: Experiences creating three implementations of the repast agent modeling toolkit. ACM Trans. Model. Comput. Simul. 16(1), 1–25 (2006)
10. Luke, S., Cioffi-Revilla, C., Panait, L., Sullivan, K., Balan, G.: Mason: A mul- tiagent simulation environment. Simulation 81(7), 517–527 (2005)
11. Wilensky, U.: NetLogo. Center for Connected Learning and Computer-Based Modeling. Northwestern University, Evanston, IL (1999), `http://ccl.northwestern.edu/netlogo/`
12. Janssen, M.A., Ostrom, E.: Empirically based, agent-based models. Ecology and Society 11(2), 37 (2006)

Mobile Virtual Platform to Develop Multiplatform Mobile Applications

R. Berjón*, M.E. Beato, and G. Villarrubia

Abstract. This article presents a multiplatform framework PMV (Mobile Virtual Platform). This framework allows obtain a native application in the main mobile systems (Windows Phone, Android and iOS). PMV needs a XML specification of our application (virtual application) and automatically it returns three different native applications to the different operating systems. The idea is to facilitate multiplatform mobile application development. Using this platform the developer can obtain the application skeleton easily.

Keywords: Smart mobility, Multiplatform, framework.

1 Introduction

Nowadays the smartphones are like small computers: with operating system, storage capacity, Internet access and it is possible to develop applications running in them. In fact, the applications are now the most valuated part by the users. Nowadays using the mobile only for calling or sending messages it is not usual. The mobiles are used to carry in your pocket a set of application to play, buy, read the email, obtain information in real time, …

This affirmation is present in DISTIMO's [5] inform on April 2011. In this inform are presented the applications number available in the different markets. The reality is more impressive if you see downloads number. For example it exists applications like Google maps of Android [2] with 50 millions of downloads and in

R. Berjón · M.E. Beato · G. Villarrubia
Universidad Pontificia de Salamanca,
c/ Compañía 5, 37002 Salamanca, Spain
e-mail: {rberjonga,ebeatogu}@upsa.es
gabri@grupodeltron.com

* This work has been partially supported by the Junta de Castilla y León, project PON002B10-2.

J.B. Pérez et al. (Eds.): Highlights on PAAMS, AISC 156, pp. 265–272.
springerlink.com

Apple AppStore [1] reached in 2010 100 millions of downloads since July the 2008 when it was born.

In addition to the smartphone, the tablets have entered the market with great force. The first was iPad in 2008 and in 2011-2012 is expected to be its final takeoff, mainly because the strong emergence of various Android tablets [3] and the BlackBerry tablet [4]. So there will so many applications targeting the tablets, the tablets characteristics (mainly screen size and use of them) make these applications different of smartphone applications.

So nowadays there is great demand for applications by users of smartphones and tablets and in the next years this demand is going to grow. From the user (of smartphones and tablets) point of view the applications utility is clear but which is the point developer of view?

The developer is faced with different operating systems, the main are Windows Phone, iOS, Android, BlackBerry OS, ... these operating systems are incompatible, an application designed for a concrete operating system does not work in other operating systems. This fact it is a great inconvenient to the developers. The developer need to learn different platforms with different language, structure and they have to develop different applications one for each platform mobile. Moreover they have to develop for smartphone and tablets.

To resolve this inconvenient we present a multiplatform solution (Mobile Virtual Platform PMV) with which it is possible obtain code for different operating systems, specifically Windows Phone, Android and iOS.

Recently some multiplatform solutions have been born; these solutions are classified in two types:

- **For Web applications.** These solutions are based in HTML5 [6]. The solution is simple but it is limited to Web applications. In this category nowadays PhoneGap[11], Sencha Touch [16], WAC [18] and jQuery [9].
- **For native applications.** Applications similar to applications developer using the native framework. In this category nowadays Titanium [13], and Rhodes [12]. Our application is in this category.

2 PMV Architecture

PMV Architecture is showed in the figure 1. PVM works as follows: first we need application specifications and it is necessary obtain a virtual application using XML. After that we have different machines (NAG Native Application Generator) for each operating system. These NAG transform the virtual application in a native application. The native application is not a final application else it is the skeleton, and we have to add the logic to get a final application. We have studied the native application from different point of view: application structure, life cycle, user interface and event handle (data access is not implemented yet).

In the following section we describe how to get a virtual application using XML. The virtual application is the base to get native code in all platforms.

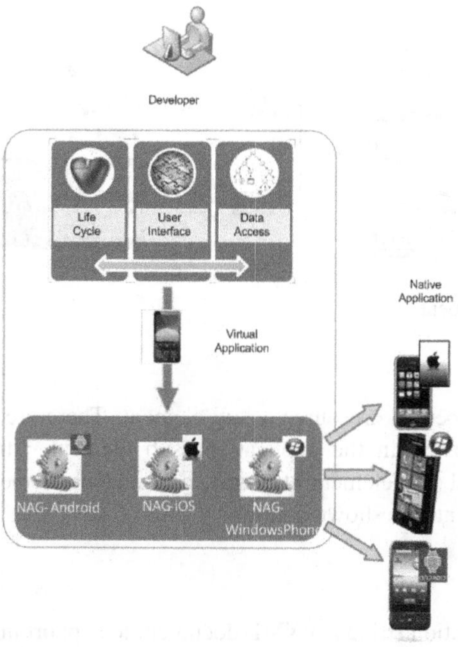

Fig. 1 PMV Architecture

3 Virtual Application

Applications are designed using Model View Controller (MVC) pattern. The view is where user acts; it contains the controllers to different screens. The controller is closely related to the view and it is distributed around all screens.

On the other hand, the model is the most complicated part because the model information is obtained from different sources: databases, web services, sensor data, ... Because this is a problem, we have created a intermediate layer, in this layer are defined types and data to the model. So the data source to the view is this intermediate layer, the view does not directly access to original source.

Now we are going to describe in more detail the XML elements of our virtual application

Application Description

The root element in XML document represents the mobile application. In this element is described the application name, its prefix (the packet in Android or namespace in Windows phone) and the languages preferences, for this propose we use name, prefix, language and country attributes (see figure 2).

To describe the application we need specify the resources and the different screens, we put this information in resources and forms sub elements.

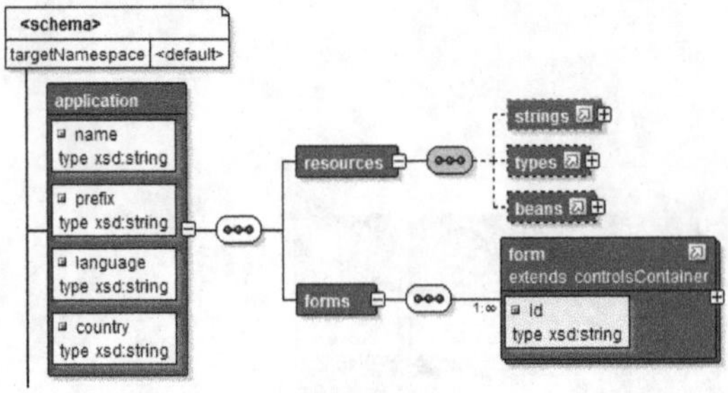

Fig. 2 Application element

The Model

The model is described in <resources> sub element. There are three types of resources: strings showed in the application (sub element <strings>), datatypes presents in the model (sub element <types>) and data (sub element <beans>). How describe these elements are shown below.

<strings>

We represent application strings in XML document to support internationalization. For each string we can specify different values to different languages (see figure 3).

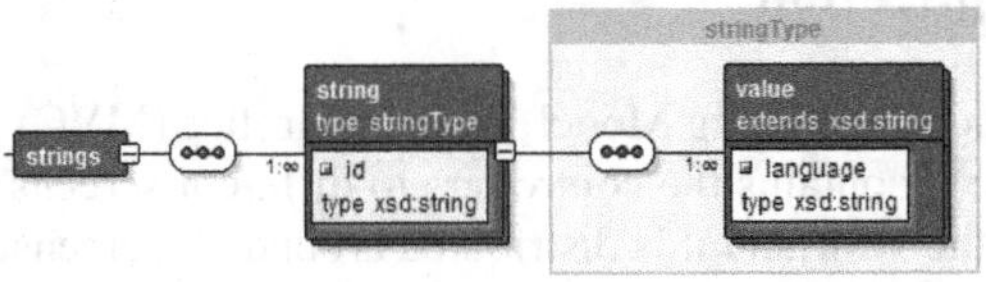

Fig. 3 Strings element

<types>

As we have explained above, the model of MVC pattern is integrated by objects which represent the information showed in application screens in each moment. So, It is necessary to know the objects type, this information will be in <types> element (see figure 4).

<types> element has a sub element <type> for each data type defined by the user. And <type> element has a sub element <property> for each type property.

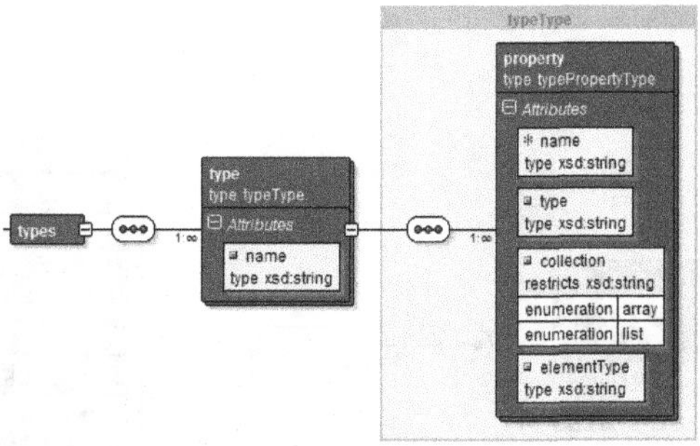

Fig. 4 Types element

<beans>

In this element we specify the objects for the application model. <beans> element has a sub element <bean> for each object represented. For each object we need its type (any <type>) and optionally an initial value. You can see the XML structure for <beans> element in the figure 5. Moreover <bean> element has two attributes name and type, name is used to identify to bean and type describes its datatype.

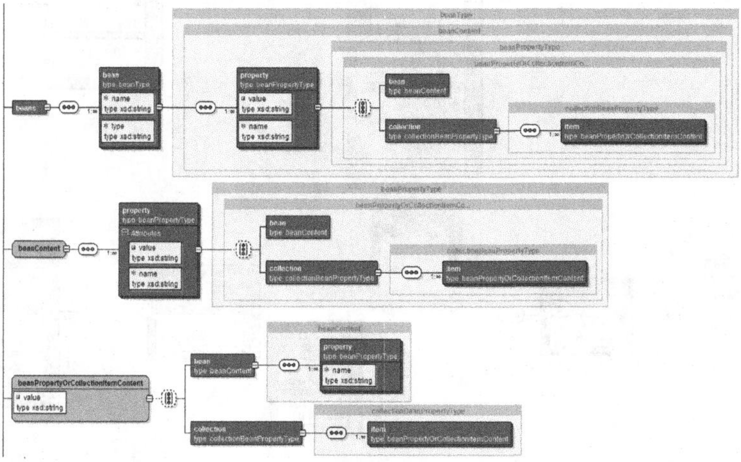

Fig. 5 Beans element

View / Controller

These components are defined together (see figure 6). In user interface there are controls to be included: text (TextView in Andoid, UILabel in iOS), button (Button, UIButton), edit (EditText, UITextField), radiogroup (RadioGroup, UISegmentedControl) and check (CheckBox, UISwich).

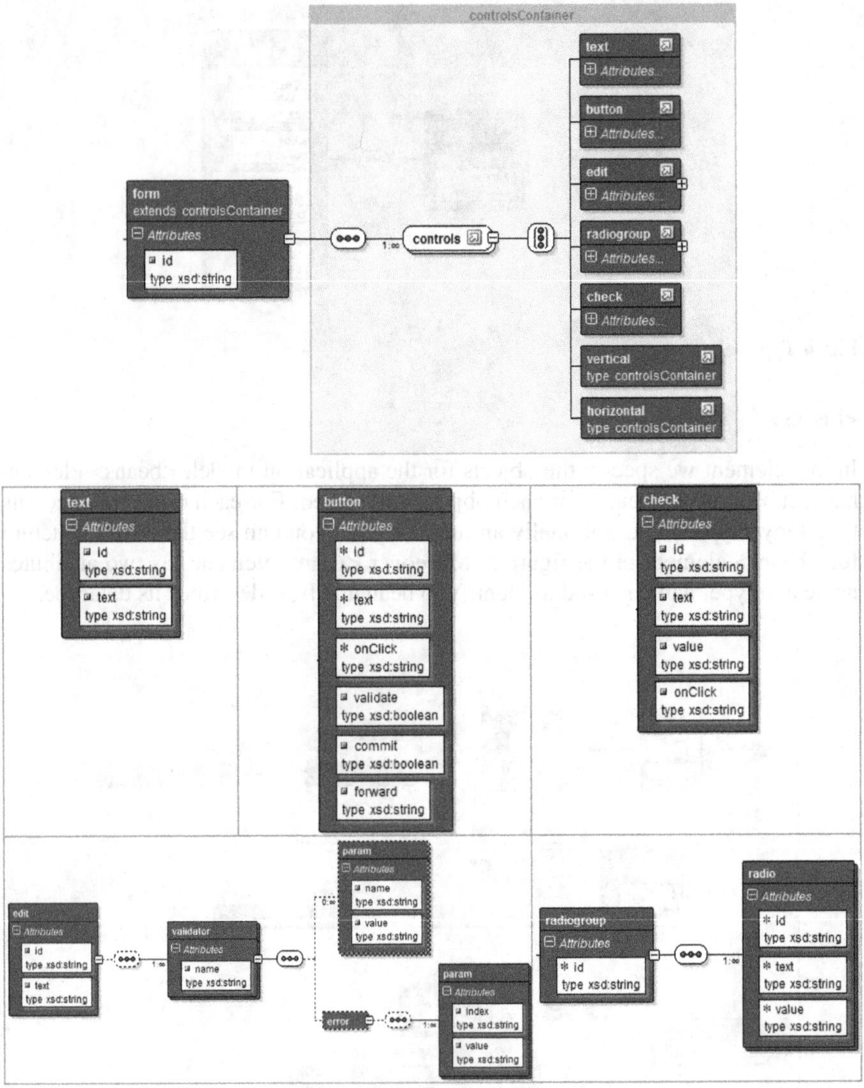

Fig. 7 UI controls

For edit is possible define validators. A validator is used to check if the user text meets certain conditions. A validator can receive parameters to initialization or to customize error message.

A virtual application specification example is shown below. This example includes one form with an edit and validators.

```
<application … >
 <resources>
  <strings>
   : : :
   <string id="label_age">
    <value language="es">Edad</value>
    <value language="en">Age</value>
   </string>
   <string id="label_ok">
    <value language="es">OK</value>
    <value language="en">OK</value>
   </string>
  </strings>
  <types>
   <type name="Contact">
    <property name="name"  type="string"/>
    <property name="email" type="string"/>
    <property name="age"   type="int8"/>
   </type>
  </types>
  <beans>
    <bean name="charles" type="Contact">
      <property name="name"
                value="Charles"/>
      <property name="email"
                value="charles@gmail.com"/>
    </bean>
  </beans>
 </resources>
```

```
<forms>
 <form id="edits">
  <edit id="edit_age"
        text="@bean/charles.age">
   <validator name="required">
    <error>
     <param index="0"
            value="@string/label_age"/>
    </error>
   </validator>
   <validator name="int8">
    <error>
     <param index="0"
            value="@string/label_age"/>
    </error>
   </validator>
   <validator name="numberInRange">
    <param name="minValue" "0"/>
    <param name="maxValue" "127"/>
    <error>
     <param index="0"
            value="@string/label_age"/>
     <param index="1" value="0"/>
     <param index="2" value="127"/>
    </error>
   </validator>
  </edit>
  <button text="@string/label_ok"
          onclick="onClickOk"
          validate="true"/>
 </form>
</forms>
</application>
```

The result of this virtual application in Android is shown in figure 8.

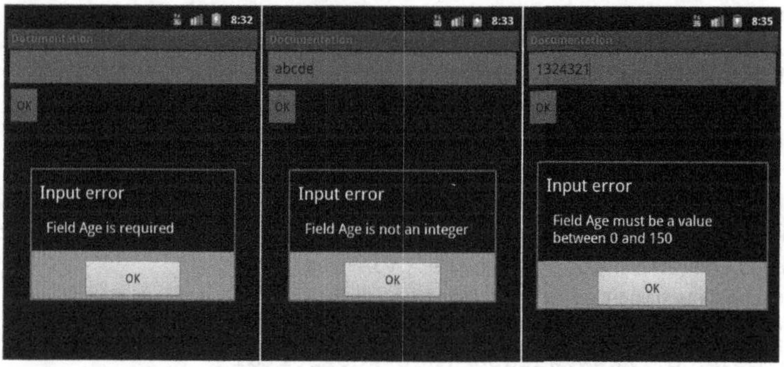

Fig. 8 Android UI

4 Conclusions and Future Work

In this report we have presented a main part to our multiplatform solution, with it you can obtain application for Android, Windows Phone and iOS. This work is not finished; our idea is following with it in the future. To the future, the main goals are the followings:

- Work with tablets. In this moment the user have to specify the interface characteristics using XML. In the future the native application obtained will be prepared as smartphone as tablets.
- Comparative between similar frameworks. The idea of multiplatform framework is new but it is not only our. Very recently different multiplatform frameworks have been born with similar objectives to PMV. So our goal is perform a comparative study of them. This study will show us interesting information with which improves our solution.
- Extend PMV to other operating system. Concretely we think extend to blackberry due to its market penetration

Acknowledgements. This work has been partially supported by the Junta de Castilla y León, project PON002B10-2.

References

1. AppStore, `http://www.apple.com/iphone/appstore/`
2. Android Market, `http://www.android.com/market/`
3. Android, `http://developer.android.com/`
4. BlackBerry Market, `http://es.blackberry.com/services/appworld/?`
5. Distimo report (April 2011), `http://www.distimo.com`
6. HTML5. W3C Working Draft (May 2011), `http://www.w3.org/TR/html5`
7. iOS, `http://developer.apple.com/`
8. Java, `http://www.java.com`
9. jQuery, `http://jquerymobile.com/products/touch`
10. Microsoft Windows Mobile (2008), `http://www.microsoft.com/spain/windowsmobile/default.mspx`
11. PhoneGap, `http://www.phonegap.com`
12. Rhodes, `http://rhomobile.com/products/rhodes`
13. Titanium, `http://www.appcelerator.com`
14. Windows Marketplace (2009), `http://marketplace.windowsphone.com`
15. Nokia Ovi (2009), `http://store.ovi.com`
16. Sencha Touch, `http://www.sencha.com`
17. Symbian (2008), `http://www.symbian.com`
18. WAC, `http://www.wacapps.net/web/portal`
19. Beato, M.E., et al.: Mobile devices: Practical cases on this new applications platform. In: Intl. Workshop on Practical Applications of Agents and Multiagent Systems, Salamanca, pp. 231–240 (2007)
20. Fraile, J.A., Delgado, M., Sánchez, M.A., Beato Gutiérrez, M.E.: The UPSA Mobile Information System –MoviUPSA. In: Intl. Workshop on Practical Applications of Agents and Multiagent Systems, pp. 221–230 (2007)

Automatic Route Playback for Powered Wheelchairs

R. Berjón, M. Mateos, I. Muriel, and G. Villarrubia

Abstract. This paper presents a system to make a wheelchair go along recorded routes automatically, without user intervention. Also introduces several interfaces to control an electric wheelchair, with the head, voice and a mobile device. The system aims at making the daily routes as easily as possible for the person on the wheelchair, and also at covering the need of independence of severe disabled people that cannot use a common wheelchair joystick. The structure of the system as well as the preliminary results obtained is presented in this paper.

Keywords: Head control, disabled people, mobile devices, motion detection, RFID, automatic routes.

1 Introduction

Many times, people on wheelchairs need to repeat the same movements over and over. For example, when a person inside his, or her, house goes from the living room to the kitchen, is always doing the same route.

Also, in a public building, like a Hospital, or a Ministry, departments are always on the same place, so there are also different common routes one can follow to reach them.

Aescolapius[1] presents a solution to this situation, by automating the routes inside the home of the person, and inside those public buildings.

Automating those routes also means less stress, and less fatigue, as the person does not have to know, or guess, where to go and how to get there, just needs to select the destination and let the wheelchair take control.

R. Berjón · M. Mateos · I. Muriel · G. Villarrubia
Universidad Pontificia de Salamanca, Compañía 5,
37002, Salamanca, Spain
e-mail: {rberjonga,mmateossa,imurielnu.inf}@upsa.es,
gabri@grupodeltron.com

J.B. Pérez et al. (Eds.): Highlights on PAAMS, AISC 156, pp. 273–280.
springerlink.com

There are several other situations where Aescolapius can be useful: In museums it can guide the visitors and show them multimedia content to explain the art, or at airports, showing travellers where their plane is, and even in industrial environments, supplying components at the right time.

The automatic route playback system uses lines on the ground and RFID tags to guide itself along a route. The use of lines on the ground provides a cheap and efficient way to guide the wheelchair and, what's more, many new Hospitals in Spain already have deployed lines all over their floors to guide their visitors, so it does not exist any need to re-deploy more lines as Aescolapius could use the existing ones. Figure 1 shows the architecture of the automatic routes system.

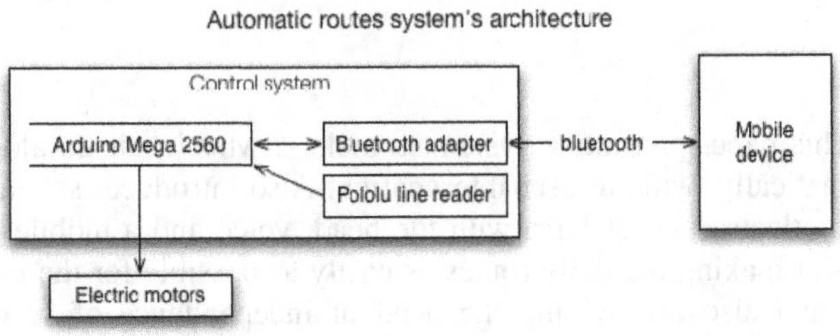

Fig. 1 Shows automatic routes system's architecture.

On the other hand, when an automatic route is not available and the wheelchair needs to be controlled manually, Aescolapius can also serve as a Human Machine Interface (HMI) system that aims at those severe disabled people, as well as any other person, trying to help them and make them as independent as possible.

There are several other HMI systems being developed to control wheelchairs, many of them are based on the SIAMO [2] project, and some use infrared sensors [3] to detect head movements.

Aescolapius system can be adapted to work with most of the modular based wheelchair systems out there, like SIAMO [2].

Like [3], our system is able to detect a person's face, but using an RGB camera instead of infrared light, and like [2], Aescolapius is able to interpret voice commands and move the wheelchair accordingly. But what really makes it different is that an external person also can control it, through a mobile device.

The rest of the paper is structured as follows: Section 2 presents the alternative ways of detecting the user face and describes the behaviour of the head tracking system. Section 3 shows and describes the voice recognition module, which is part of the system's core. Section 4 and 5 describe the HMI implemented to be used on mobile devices. Section 6 shows and describes the automatic route playback system. Section 7 shows the Aescolapius prototype made for testing the HMI and the prototype created for testing the route playback system. Finally, section 8 presents the preliminary results obtained and the conclusion.

2 Head Based Control

We manage to move the chair by using a camera connected to a portable computer to track the face of the person sitting on the wheelchair. An algorithm called Optical Flow [4] is used to track the face's movements in real time.

Haar-like features [5], used to find the human face in the image, consider adjacent rectangular regions at a specific location in a detection window; it sums up the pixel intensities in these regions and calculates the difference between them. This difference is then used to categorize subsections of an image. For example, let us say we have an image database with human faces. It is a common observation that among all faces the region of the eyes is darker than the region of the cheeks. Therefore a common haar feature for face detection is a set of two adjacent rectangles that lie above the eye and the cheek region. The position of these rectangles is defined relative to a detection window that acts like a bounding box to the target object (the face in this case).

In our system we use a Haar-like features combined with an Optical Flow algorithm (both supplied by the Open Computer Vision libraries created by Intel [6]). Optical flow allows us to get the pattern of apparent motion of objects between an observer (the camera) and the scene (the subject).

The threshold of the system can be adjusted by setting the minimum distance to exist between the first centroid and the current centroid.

By doing this, the system can be adapted to the person sitting on the chair, so it is always sure that the person will be able to move the chair properly.

3 Voice Based Control

Using an Android platform and a Windows 7 computer there are two different ways of getting the voice translated to commands. One way is using Google Speech Recognition API and the other is to use Microsoft SAPI [7] (Speech API) 5, bundled inside Windows 7.

Both ways have been tested and it is been found that Microsoft's one is better for our purpose. It doesn't only allows to continuously get sounds from the audio stream from the microphone, avoiding any command losses, it is also faster because the audio stream is converted locally, and not in a server. And when we are talking about a wheelchair, and that someone could get harmed, fastness becomes a critical factor.

We use the Microsoft Speech API to get the voice commands in plain text. A dictionary of 'allowed commands' is defined and everything that is not in that dictionary is just ignored.

Once a valid voice command is obtained, the action attached to that command is performed. The action can be one of this: turning the lights on, making the wheelchair go forward, backwards, left or right, turning on the face tracking system, reading the news, reading the weather forecast or stopping the wheelchair.

Once the voice recognition system was implemented, we found out that any person just talking by the wheelchair could fire a voice action and make the wheelchair start moving. To prevent this, a special word that makes no sense in a normal conversation has been added before any command, so there is no way to make it work by just talking to it.

4 Mobile-Device Based Control

When we were designing the system, we thought about the person sitting on it. But many times another person goes along with him, or her. And we asked ourselves "What about the person walking by the chair?"

It is difficult to control a normal electric wheelchair if you walk along with it by using its joystick, because as you walk you move it and make the chair go faster or slower that your walking speed. And if we are talking about the speech recognition system, or the face tracking system, it is even more difficult (you cannot put your face in front of the camera as you walk).

So, we thought about it and realized that nearly every person nowadays has a mobile phone in his or her pocket, and the number of smart phones is rising exponentially. Why not use that smart phone as an interface to the system, removing this way the physical link between the wheelchair and the person next to it?

Most of the smart phones nowadays, even the cheapest ones, have an acceleration sensor bundled inside. That sensor allows us to measure the inclination of the device with respect to the ground, in other words the gravity vector direction. With this, we are able to simulate a kind of joystick using the phone, by tilting the phone.

If the person tilts the phone left, or right, forward, or backwards, the wheelchair moves accordingly.

This has been the favourite interaction way. People who have tested the system find that tilting the phone is very usable and comfortable, as they walk beside the wheelchair.

Another input method available on many smart phones is the touch screen. The system has the ability to receive input events from a mobile device provided with a touch screen.

Using a finger, slide it from the bottom to the top of the screen to make the system move the wheelchair forward. The opposite movement, while moving forward, will make the chair stop.

Sliding the finger from left to right will move the wheelchair to the right, and vice versa.

There are also severe disabled people that have very little movement in their hands, but not enough strength or freedom to use the joystick.

With the mobile device under the hand, on the chair, the touch screen worked smooth and succeeded moving the chair properly.

5 Automatic Route Playback System

Aescolapius provides a system that allows a wheelchair to run along a route without user intervention [8][9].

To achieve this, the system consists of several components:

- Black lines: painted on the floor, they guide the wheelchair along the route.
- Mobile device: Controls all of the decisions the system can take. Calculates the shortest route between the origin and the destination.
- Robot prototype: Has the motors, and receives commands from the mobile device through a Bluetooth connection. It has to deal with the line following algorithms and, when it founds an RFID tag it has to transmit the ID of the tag to the mobile device.

The process of going from origin to destination is as follows:

When a destination is selected on the mobile device, a signal is sent to the robot to make it read the RFID tag that it has below, all of the communication between the robot and the mobile device is achieved using Bluetooth. The robot sends the ID of the tag back to the device, and then it calculates the entire route.

When the route is calculated, the mobile device tells the robot to follow the line until the next line cross. Once the robot gets to the next line cross, it reads again the RFID tag below the cross, sends the ID to the device and the device answers with the action to take (turn left/right or continue forward).

Figure 2 shows a line cross, with the RFID tag inside.

Fig. 2 Detail of lines cross with RFID tag below

This process continues on and on, with the robot following the line, reading tags and sending them, and the device answering with the appropriate actions until the destination is reached.

Once the route is finished, the mobile device goes back to the main screen in order to allow another person to start a new route.

The robot prototype is made up from several different components; each of them is described below.

- Arduino Mega 2560: Manages and processes signals from the hardware (wheels, sensors, and communications). Turns the actions sent from the mobile device into electrical signals.

- Bluesmirf Bluetooth adapter: Creates a communication link between the device and the Arduino, so they can exchange data.
- LCD display: Shows status messages coming from the system.
- ID-12 RFID reader: It's an RFID antenna that allows us to read the ID of the tags.
- 8-sensor pololu: It is an array of pairs of infrared emitters and sensors. It detects the reflectance of the ground, so it can tell if the floor is black or white, allowing the robot to follow a line.
- H-Bridge: An integrated chip to control the motors easily, allowing them to roll forward or backwards quickly and safely (without risk of short circuits).

Figure 3 shows a scheme of how all of these components are connected to each other.

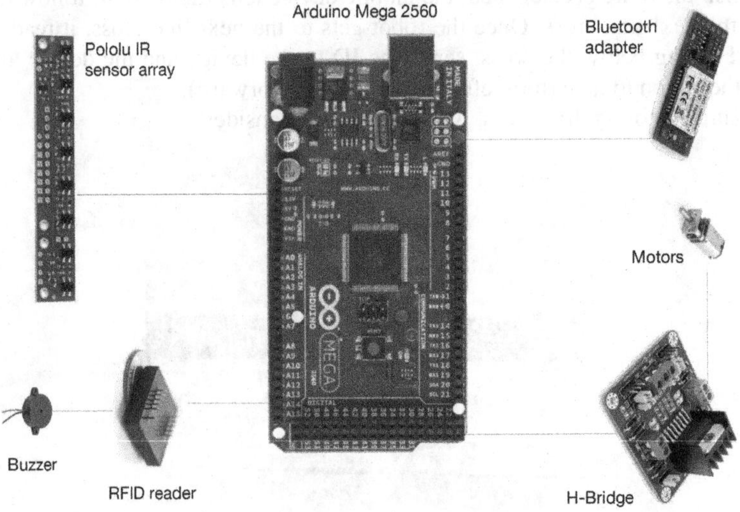

Fig. 3 Components of the robot and relations between them

6 Case Study: Aescolapius 2.0

We wanted to build an indoor location system to allow a wheelchair to freely move between rooms. We built a fixed routes system which simplified the development and allowed us to spend more time with the route calculation algorithm, and some others functionalities which took a lot of time.

Aescolapius 2.0 is the second prototype of the project, a small robot.

The circuitry has been enhanced since the previous prototype, making it simpler and more powerful. Now the robot uses an Arduino board, instead of an 8-relay controller, which is more powerful because it has its own processor and can make decisions on its own.

The communications have been moved from Wi-Fi to Bluetooth, now there is no need of a Wi-Fi router.

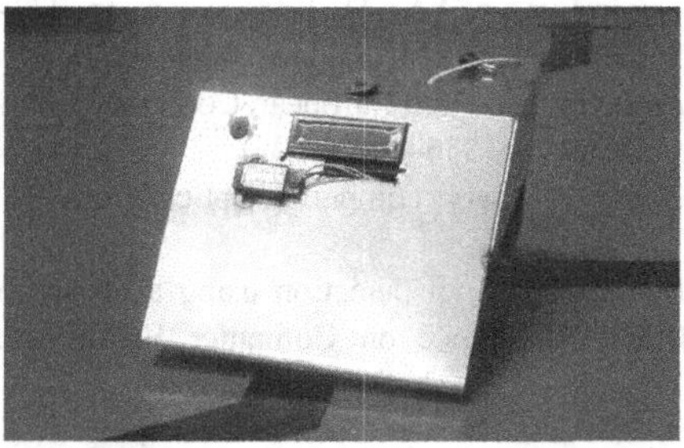

Fig. 4 Aescolapius 2.0 robot prototype

7 Conclusions

Aescolapius can be adapted to virtually any wheelchair system. The robot has been built so you only need to change the motors for some powerful ones, wheelchair motors, and everything will run as expected.

The HMI provides the following interfaces: head tracking, speech recognition and two more on the mobile device: accelerometer and touch screen.

While testing, the system behaved properly. It was able to follow the lines on the floor properly. The routes calculated with the algorithm were always the shortest routes, and it worked quite well.

The HMI also reacted properly to user commands, whatever voice, face or mobile device commands were sent. Reaction times were short enough.

The system also showed robustness against different kinds on faces, with and without glasses, long and short hair or different hair colours and styles. Light conditions have to be good for the head-tracking module to work properly, although the Aescolapius prototype has its own lighting system.

Future work involves obstacle detection using infrared and ultrasonic sensors, so the system can react when it encounters an object, or other people, on its route.

“Shared control” [10] is another goal for the project. Sometimes, while the user controls the wheelchair, he, or she, can make mistakes and, for example, can hit a wall. Shared control avoids these risk situations, when the system detects that the wheelchair is going to hit something, or fall down a cliff, it takes control over the user actions and can solve the situation in the best possible way, for example stopping the motors o turning to the opposite side.

References

1. Berjón, R., Mateos, M., Barriuso, A., Muriel, I., Villarrubia, G.: Head Tracking System for Wheelchair Movement Control. In: Pérez, J.B., Corchado, J.M., Moreno, M.N., Julián, V., Mathieu, P., Canada-Bago, J., Ortega, A., Caballero, A.F. (eds.) Highlights in PAAMS. AISC, vol. 89, pp. 307–315. Springer, Heidelberg (2011)

2. Garcia, J.C., Mazo, M., Bergasa, L.M., Ureña, J., Lázaro, J.L., Escudero, M.S., Marrón-Romera, M., Sebastian, E.: Human-machine interfaces and sensory system for autonomous wheelchair. In: Proc. 5th European Conference on the Advancement of Assistive Technology in Europe, AAATE 1999, vol. 6, pp. 272–277 (1999)
3. Christensen, H.V., Garcia, J.C.: Infrared non-contact head sensor, for control of wheelchair movements. In: Proc. 8th European Conference for Advancement of Assistive Technology in Europe, pp. 336–340 (2005)
4. Stavens, D.: The OpenCV library: computing optical flow. Stanford Artificial Intelligence Lab
5. Viola, P., Jones, M.: Rapid object detection using a boosted cascade of simple features. In: Proc. IEEE Conference on Computer Vision and Pattern Recognition, pp. 511–518 (2001)
6. Bradski, G., Kaehler, A.: Learning OpenCV: computer vision with the OpenCV library. O'Reilly (2008)
7. Microsoft Tellme, `http://www.microsoft.com/enus/Tellme/developers/default.aspx`
8. Vicente Diaz, S., Amaya Rodriguez, C., Diaz Del Rio, F., Civit Balcells, A., Cagigas Muniz, D.: TetraNauta: a intelligent wheelchair for users with very severe mobility restrictions. In: Proc. of the IEEE International Conference on Control Applications, Glasgow, Schotland, pp. 778–783. IEEE Press (2002)
9. Kurozumi, R., Fujisawa, S., Yamamoto, T., Suita, Y.: Development of an automatic travel system for electric wheelchairs using reinforcement learning systems and CMACs. In: International Joint Conference on Neutral Networks, Honolulu, Hawai, pp. 1690–1695 (2002)
10. Levine, S.P., Bell, D.A., Jaros, L.A., Simpson, R.C., Koren, Y., Borenstein, J.: The NavChair Assistive Wheelchair Navigation System. IEEE Transactions on Rehabilitation Engineering 7, 443–451 (1999)

Evaluation of Labor Units of Competency: Facilitating Integration of Disabled People

Amparo Jiménez, Amparo Casado, Fernando de la Prieta,
Sara Rodríguez, Juan F. De Paz, and Javier Bajo

Abstract. Education and training for disabled people has acquired a growing relevance during the last decade, especially for labour integration. Disabled people represent a considerable percentage of the current population and require special education. This paper presents a competencies model to evaluate the skills of disabled people previous to their integration to working scenarios.

Keywords: units of competency, disabled, active learning.

1 Introduction

Remote training methods acquire an important role nowadays, as they allow the users to interact with the workplace before facing reality. Therefore, one of the objectives to be achieved is to obtain an active learning methodology via internet or TV proposing innovative learning strategies that enable people with visual, hearing and motor disabilities to perform training and learning prior to their employment. This paper proposes a novel learning method that combines the advantages of active learning with the technological capabilities of adapted devices. Active learning allows the users to learn for themselves and acquire knowledge through their own experiences. In addition, the introduction of interactive channels is made available to all the users as a mechanism for communication and interaction with great audio-visual possibilities.

Amparo Jiménez · Amparo Casado · Javier Bajo
Universidad Pontificia de Salamanca, Salamanca, Spain
e-mail: {ajimenezvi,acasadome,jbajope}@usal.es

Fernando de la Prieta · Sara Rodríguez · Juan F. De Paz
Departamento Informática y Automática,
Universidad de Salamanca, Salamanca, Spain
e-mail: {fer,srg,fcofds,corchado}@usal.es

J.B. Pérez et al. (Eds.): Highlights on PAAMS, AISC 156, pp. 281–288.
springerlink.com

Without going into specific definitions of what skills are or what their different levels of specification, evaluation and training, an essential step in the career counseling of people with disabilities, is to clearly determine the job profile to develop. Depending on the definition of the job position and the desired profile, it is possible to propose a model for assessing competence. It is necessary to consider, very specifically, which are the functions of the worker and, from that description, provide the technological resources needed to carry out the work.

The aim of this paper is to propose a teaching method for designing a new panel of execution of exercises based on active learning methodologies that can adapt to the characteristics of the disabled user that runs them. The approach will provide a new technological solution for training and distance learning, and opens a wide range of research into new methods of training disabled people. We propose an adaptive mechanism incorporating new algorithms at each stage of a case-based reasoning. We propose the use of cooperative Hebbian learning [12] in the recovery phase and SOM neural networks [11] incorporating new modifications for reuse phase.

The rest of the paper is organized as follows: Section 2 revises the related work. Section 3 describes the proposed approach. Finally, section 4 presents the initial conclusions obtained.

2 Related Work

There is an ever growing need to supply constant care and support to the disabled and the drive to find more effective ways to provide such care has become a major challenge for the scientific community [5]. Education is the cornerstone of any society and it is the base of most of the values and characteristics of that society. The new knowledge society offers significant opportunities for AmI applications, especially in the fields of education and learning [7]. The new communication technologies propose a new paradigm focused on integrating learning techniques based on active learning (learning by doing things, exchange of information with other users and the sharing of resources), with techniques based on passive learning (learning by seeing and hearing, Montessori, etc.) [6]. While the traditional paradigm, based on a model focused on face to face education, sets as fundamental teaching method the role of the teachers and their knowledge, the paradigm based on a learning model highlights the role of the students. In this second paradigm the students play an active role, and build, according to a personalized action plan, their own knowledge. Moreover, they can establish their own work rhythm and style. The active methodology proposes learning with all senses (sight, hearing, touch, smell and taste), learn through all possible methods (school, networking, etc.), and have access to knowledge without space or time restrictions (anywhere and at any time).

More and more learning platforms are designed for use by disabled people, given the ease of using and the increase of technology in our daily lives. Moreover, in the case of young children, an electronic device as a tablet or smartphone as a learning tool can be especially appropriated, given the attraction of these devices. Currently there are several types of platforms for education for the disabled. There are games based platforms as PICAA [4]. It is an application for smartphones (iPhone or Apple iPod Touch). Other platforms are based on exercises like PROAF (Program Articulation and Phonation) [2]. This program is conducted from the Office suite that offers Microsoft, PowerPoint tool, which allows us to create presentations. The program P.R.O.A.F. is for students with poor auditory skills and language disorders [2]. In 2009, a student of the UNEF-Maracay, developed by the end of grade one braille learning platform for the blind and deaf blind [3]. The O.N.C.E. (National Organization of Spanish Blind) makes available to its members a range of applications for global development, employment, education and social development. These resources are open to anyone who wants to use them [1]. Among them, we can distinguish from braille printer drivers to games and stories for children through voice synthesizer applications.

3 Proposed Mechanism Overview

Without going into specific definitions of what skills are or what their different levels of specification, evaluation and training, an essential step in the career counseling of people with disabilities, is to clearly determine the job profile to develop. Depending on the definition of the job position and the desired profile, it is possible to propose a model for assessing competence. It is necessary to consider, very specifically, which are the functions of the worker and, from that description, provide the technological resources needed to carry out the work.

To this end, an essential step in this description is to place ourselves in a particular model of time and labor needs. In Spain, as in other countries, there exists a framework document that determines the qualifications of individual workers, and that unifies and determines what "to do" in the various work environments. Therefore, we understand that when addressing the issue of training for employment, we must take as reference the proposal presented in the Spanish document of INCUAL (National Institute of Professional Qualifications) [14]. This document allows us to indicate the structure of the professional qualifications that will be used to design the programs, resources, methodologies and educational interventions in labor scenarios. Every qualification is associated with a general competence, which briefly defines the duties and essential functions of the worker. We also need to describe the professional environment in which the workers can develop their skills, the corresponding productive sectors and occupations or relevant job positions to which they will be able to gain access. Each qualification is divided into units of competency. The unit of competency is the minimum aggregate of professional skills, capable of being recognized, being partially accredited and, in training environments, to training and assessment. Each unit of competency has associated a training design (no training module as in the SNCP), which describes the necessary training to acquire this unit of competency. In this case, each unit of competency will be associated with different technology-based educational resources that allow individuals to learn or verify their appropriateness in the performance taking into account the special characteristics of the target persons, who have disabilities in this case. The unit of competency is divided into professional achievements. The personal achievements describe the expected behaviors of the person, objectified by its consequences or outcomes, so that it can be considered competent in that unit.

Thus, we propose as valid a proposal following the guidelines of the Spanish National Catalogue of Professional Qualifications (CNCP), since it is the instrument used by the Spanish National System of Qualifications and Vocational Training (SNCFP) that regulates the professional qualifications that can be recognized and accredited, identified in the production system based on appropriate competencies for professional practice. Taking this proposal as starting point, the next step is to determine, from the units of competencies described, the professional achievements and performance criteria selected from those identified in the CNCP. This selection allows us to determine which strategies and training resources any company can include in the training and development plans for their workers, or for the selection of new candidates. We must emphasize the current interest in technology-based design tools (current, new and technically valid) to help people with certain disabilities to effectively integrate in labor environments and job profiles.

3.1 A Specific Competencies Model

We have selected a professional qualification: Auxiliary operations and general administrative services, associated with the following occupations or relevant jobs: Office Assistant, General Services Assistant, Archives Assistant, Classifier and / or delivery of mail, Ordinance, Information Assistant, Telephonist, Box office. Within this first professional qualification the following units of competence are identified, which are associated to a certain performance criteria:

- UC0969_1: Conduct and integrate basic administrative support operations
- UC0970_1: Transmit and receive routine operational information with external efforts of the organization
- UC0971_1: Perform additional reproduction and archiving on conventional or compute

By means of an initial analysis work we select the most relevant professional achievements related to the particular disability. We identify as relevant those requiring greater effort in their level of training, learning and achievement in relation to the particular characteristics of the professionals who must perform these tasks. These are:

RP 1: Register periodically updates on information of the organization, its departments and areas, and staff assigned to them, as instructed, to have the information necessary to provide good service.

RP 2: Collate and record, business, administrative, financial, or other documents in the organization, as instructed, using specific or generic software, to ensure the reliability of the information.

RP 3: Draw up notices, alerts, informational signs or other internal documents used in the communication process, following the instructions and using conventional means or computers.

RP 4: Maintain optimum operating conditions of equipment and multimedia resources, detecting and correcting incidents or managing repairs, according to the instructions received, in order to support the normal functioning of the office.

RP 5: Archiving and accessing both traditional and digital documentation, introducing specific codes, and following the procedures of the organization and the instructions received, so as to allow a quick search.

RP 6: Archivar y acceder a la documentación, en soporte convencional o digital, introduciendo los códigos precisos y respetando los procedimientos de la organización y las instrucciones recibidas, de manera que se permita una búsqueda rápida.

PR 7: Transfer and purge of documents according to the rules and deadlines and instructions, to maintain updated and functional the file system.

RP 8: Update and extract different types of information, using computer applications or databases, according to the rules and instructions, to display and query the information stored, and to facilitate upon request.

RP 9: Get the necessary copies of working documents on conventional or computational supports, in the quantity and quality required by the characteristics of the document itself, and according to the instructions given in accordance with applicable environmental protection.

RP 10: Scan or digitally reproducing documents, ensuring the highest quality of service according to the manuals and received instructions, in order to preserve and archive.

3.2 A Reasoning Mechanism

Case-based Reasoning is a type of reasoning based on the use of past experiences [9]. The fundamental concept when working with case-based reasoning is the concept of case. A case can be defined as a past experience, and is composed of three elements: A problem description which describes the initial problem, a solution which provides the sequence of actions carried out in order to solve the problem, and the final state which describes the state achieved once the solution was applied. The way in which cases are managed is known as the case-based reasoning cycle. This CBR cycle consists of four sequential steps: retrieve, reuse, revise and retain. The retrieve phase starts when a new problem description is received. Similarity algorithms are applied in order to retrieve from the case's memory the cases with a problem description more similar to the current one. Once the most similar cases have been retrieved, in the reuse phase the solutions of the cases retrieved are adapted to obtain the best solution for the current case. The revise phase consists of an expert revision of the solution proposed. Finally, the retain phase allows the system to learn from the experiences obtained in the previous phases and updates the case memory in consequence.

We propose an adaptive mechanism that incorporates new algorithms at each stage of a case-based reasoning system. We propose the use of cooperative Hebbian learning [12] in the recovery stage and SOM neural networks [10, 11] incorporating new modifications in the reuse stage, and CART [13] for knowledge extraction. Thus, we propose an active learning method for disabled people, proposing an adaptable learning model with great customization capabilities. The proposed mechanism uses the user profile and the services available and requested, it is possible to provide the best learning plan to facilitate their labor integration. The mechanism is mainly composed of a case-based reasoning system, which can be integrated into CBR-BDI agents [8], specifically designed to provide advanced reasoning mechanisms.

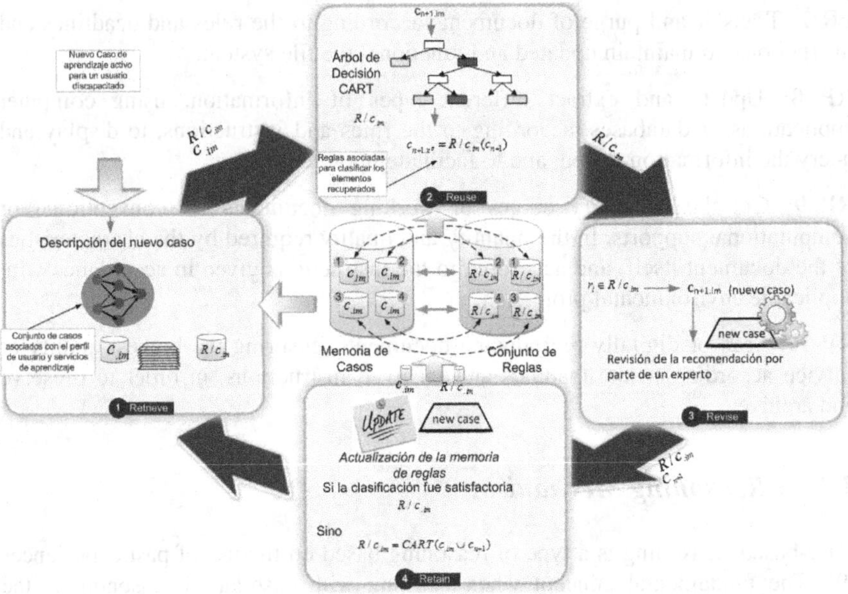

Fig. 1 Proposed mechanism to recommend training tasks.

A case is composed of the following elements: User profile (based on dissabilities, etc.), Available resources, Units of Competency, Professional achievements. The problem description consists of the user profile and units of competency taken into account in the workplace. Thus, a new problem seeks to find the solution that best suits the user's and work environment characteristics (described by the units of competency to be taken into account). The professional achievements are also taken into consideration. The solution consists of the most appropriate performance criteria for the user and the tools that were more appropriated for the user. The solution describes the professional achievements more appropriated for the user and the tools considered most suitable for carrying out the professional activities. The efficiency of the solution evaluates the degree of satisfaction and is provided by an expert. Thus, a case is defined as:

Table 1 Case Description

Fields	Type	Variable
IDCase	Int	*i*
Profile_Type	String	*m*
Disability_Type	Int	*s*
Units_of_Competency	Array of Boolean	*n*
Professional_Achievements	Array of Boolean	*l*
Tools	Array of Boolean	*u*
Achievement_Criteria	Array of Integer	*c*
Efficiency	Double	*k*

4 Conclusions

This paper has proposed a teaching method for designing a new panel of execution of exercises based on active learning methodologies that can adapt to the characteristics of the disabled user that runs them. The approach provides a new technological solution to support training and distance learning of disabled people. We have revised the related work and it was not possible to find existing approaches solving the problem of providing a generic solution for training of disabled people. In this paper we have focused on analyzing the units of competency and personal achievements for a specific professional qualification. The proposed model has been used to design a technological solution, that makes use of a case-based reasoning system, Hebbian learning, SOM artificial neural networks and CART to provide recommendations about panels of exercises taking into account previous experiences. We propose an adaptive mechanism incorporating new algorithms at each stage of a case-based reasoning.

Our future work focuses on evaluating the proposed approach in a case study scenario. This will help to have a quantitative evaluation of the proposal. Moreover, we will focus on tuning and improving the proposed model, as well as its possible extension to additional professional qualifications.

Acknowledgements. This project has been supported by the Spanish CDTI. Proyecto de Cooperación Interempresas. IDI-20110343, IDI-20110344, IDI-20110345. Project supported by FEDER funds.

References

1. Once (2011), ftp://ftp.once.es/pub/utt/tiflosoftware/
2. Proaf (2011), http://www.educarm.es/templates/portal/images/ficheros/diversidad/10/secciones/11/contenidos/625/proaf.pps

3. Rondon, R.: (2009), http://tecnologiasdeapoyodiscapacidad.blogspot.com/2009/05/plataforma-electronica-de-aprendizaje.html
4. Rodríguez-Fórtiz, M.J., Fernández-López, A., Rodríguez, M.L.: Mobile Communication and Learning Applications for Autistic People. In: Williams, T. (ed.) Autism Spectrum Disorders - From Genes to Environment. InTech (2011) ISBN: 978-953-307-558-7
5. Bajo, J., Molina, J.M., Corchado, J.M.: Ubiquitous computing for mobile environments. In: Issues in Multi-Agent Systems: The AgentCities. ES Experience. Whitestein Series in Software Agent Technologies and Autonomic Computing, pp. 33–58. Birkhäuser, Basel (2007)
6. Brown, T.H.: Beyond constructivism: Exploring future learning paradigms. Education Today 2005(2) (2005)
7. Friedewald, M., Da Costa, O.: Science and Technology Roadmapping: Ambient Intelligence in Everyday Life (AmI@Life). In: Working Paper. Institute for Prospective Technology Studies IPTS, Seville (2003)
8. Laza, R., Pavón, R., Corchado, J.M.: A Reasoning Model for CBR_BDI Agents Using an Adaptable Fuzzy Inference System. In: Conejo, R., Urretavizcaya, M., Pérez-de-la-Cruz, J.-L. (eds.) CAEPIA/TTIA 2003. LNCS (LNAI), vol. 3040, pp. 96–106. Springer, Heidelberg (2004)
9. Kolodner, J.: Case-Based Reasoning. Morgan Kaufmann, San Francisco (1983)
10. Kohonen, T.: Self-Organization and Associative Memory. Springer, Berlin (1984)
11. Baruque, B., Corchado, E.: A weighted voting summarization of SOM ensembles. Data Mining and Knowledge Discovery 21(3), 398–426 (2010)
12. Corchado, E., Baruque, B., Yin, H.: Boosting Unsupervised Competitive Learning Ensembles. In: de Sá, J.M., Alexandre, L.A., Duch, W., Mandic, D.P. (eds.) ICANN 2007, Part I. LNCS, vol. 4668, pp. 339–348. Springer, Heidelberg (2007)
13. Breiman, L., Friedman, J., Olshen, A., Stone, C.: Classification and regression trees. Wadsworth International Group, Belmont (1984)
14. INCUAL (National Institute of Professional Qualifications), https://www.educacion.es/iceextranet/accesoExtranetAction.do

A Study Case for Facilitating Communication Using NFC

Montserrat Mateos, Juan A. Fraile, Roberto Berjón,
Miguel A. Sánchez, and Pablo Serrano

Abstract. This article proposes a novel solution that facilitates the use of mobile technology for the elderly. The senior population is rapidly growing, representing 16-17% of the total population in developed countries. The use of a mobile device presents two large obstacles for this group: problems from disabilities due to age, and problems associated with the lack of familiarity with mobile technology.The combined use of contactless technologies such as NFC and MiFARE eliminates the barriers encountered by older individuals. These technologies make it possible to use basic services of a mobile telephone, such as placing a call or sending a message, in a way that is easy, user friendly and intuitive. The ComunicaME system makes it possible to place a call or send a sms on a mobile by simply moving it closer to a contactless tag.

Keywords: NFC, accessibility elderly, mobile usability.

1 Introduction

The use of mobile telephony and mobile terminals has been widely integrated into modern society; recent years have seen an exponential growth that has surpassed 100 lines for every 100 inhabitants. In 2010 Spain had 109 lines for every 100 inhabitants [1]. Subgroups of the population, such as the elderly, are often excluded from accessing technological advances, primarily because of the barriers encountered in their use [2, 3].

The difficulties experienced by these individuals are primarily derived by the actual characteristics of the terminals (small screens, shortcuts, touchscreens, etc.)

Montserrat Mateos · Juan A. Fraile · Roberto Berjón ·
Miguel A. Sánchez · Pablo Serrano
Pontificia University of Salamanca,
c/ Compañía 5, 37002 Salamanca, Spain
e-mail: {mmateossa,jafraileni,rberjonga,masanchevi}@upsa.es,
pablo1984@gmail.com

J.B. Pérez et al. (Eds.): Highlights on PAAMS, AISC 156, pp. 289–297.
springerlink.com

and by problems associated with the lack of familiarity with the mobile technology itself [5].

This supposes a serious problem if we consider that the subgroup of seniors is very high in all developed countries. In a few years the number of people older than 65 in the United States will reach nearly 16% of the population; in Asia, and specifically Singapore, it will reach nearly 18% within just over a decade; while in Spain that group is already nearing 20% of the total population [6, 2, 7, 8]. Consequently, a significant number of individuals do not have access to the most basic services offered by the mobile telephone, such as placing a call or sending a message. The lack of adaptation to technological advances among seniors has not gone unnoticed in our current society. For example, the Advancement Plan in Spain[1] contains a proposal that would encourage the use of TICs (Information and Communication Technologies) among citizens aged 55 years and older. In Europe, this group is addressed directly by the AGE[2] platform whose ambit of operation includes a component specific to TICs.

This article presents the ComunicaME system, which is intended to facilitate communication and make it more accessible through the use of a mobile telephone for senior citizens. For these individuals, simply writing a short message or searching the contact list for a telephone number is a complicated task [5]. The objective of this system it to provide a simple solution to this problem, and ensure that sub-groups of the population with difficulty adapting to mobile technology can place a call or send messages in a simple and intuitive way. This can be achieved by using contactless technologies such as NFC (Near Field Communication) [9, 10] and MiFare[3]. The combination of these technologies allows for the identification of objects and actions associated with these objects. A mobile device with NFC can identify an action that needs to be performed, such as sending a message or placing a call, by moving the mobile closer to an object containing an associated MiFare tag that represents these actions. Once the action has been identified, the same mobile device uses another MiFare tag to identify the person with whom one wishes to communicate.

This article first reviews the problems seniors have in accessing TIC. It then describes the contactless technologies used in this work as well as techniques for voice synthesis and recognition used to improve the accessibility of the application. Part 4 describes the proposed solution, while part 5 provides a detailed presentation of the experiments that were carried out with the ComunicaME system and the results obtained. Finally, the conclusions are presented, with an emphasis on the innovative approach of this proposal.

2 Accessibility and Dependent Environments for the Elderly

When speaking of technology, it is generally more common to think of young people; rarely does one associate technology with senior citizens. Nevertheless, this is a group of individuals that, given the proper training and, in some cases, specific

[1] http://www.planavanza.es/Paginas/Inicio.aspx

[2] http://www.age-platform.org/

[3] http://www.mifare.net/products/mifare-smartcard-ic-s/

adaptations to lessen their limitations, can benefit from the numerous advantages that technology offers to improve their quality of life [8]. To do so, however, it is necessary to overcome certain obstacles [11] and find a way to point out how technology can improve their daily lives. This could increase their interest in working with different technological resources. The resources most used by seniors are the personal computer, landline telephone, mobile telephone and television. They use these resources to a lesser degree than the remainder of the general population, but according to statistics from the INE (National Institute of Statistics) [12], the number of older people using advanced technological resources is ever increasing. Among seniors, the mobile phone is the most prized resource, along with the television, as they consider these items to be necessary in their daily lives.

However, there are also a number of barriers that complicate and in some cases impede the use of TIC among the elderly. The aging process involves a progressive deterioration of both physical and cognitive capabilities, which assumes some limitations in accessing TIC [6]. Neither personal computers nor mobile telephones in particular are designed to ensure easy access for those with limitations. In some cases, the size of the device or the amount of information displayed, the lack of a standard design or function, the volume, placement of the keys, etc., may create an obstacle impossible for many of these individuals to overcome [13]. The limitations that most commonly affect the elderly and prevent them from accessing or using TICs include: poor vision, loss of hearing, or reduced motor skills due to problems with articulation or coordination.

The majority of works or projects related to accessibility of mobile telephones focus on improving the services and content of mobile devices. One example is the WebA Mobile, an assistance tool for the design and evaluation of websites for mobile devices. This line includes the W3C mobileOK Checker validation tool or the Mobile Web Best Practices 1.0 practical recommendations. These tools and recommendations serve as a guideline for future developments aimed at improving the accessibility of the elderly to specific technological elements.

There are also numerous studies, such as that of Yelmo-Garcia et al. [15], that focus on the accessibility of services from mobile devices, particularly those related to Public Administration, Public institutions have tried to lead the way in ensuring that their services are accessible. Other authors, such as Lindholm et al. [16] have studied the general usability of mobile devices. The usability of mobile devices is directly related to their ability to provide accessible services.

Among projects with a broader scope, there are studies that deal with the interoperability of mobile devices, such as the INREDIS project (Interfaces for Relations between people with Disabilities). The lines of investigation for this project include their objective of facilitating interaction between users and their environments by using mobile devices.

Other works, such as Abascal and Civit [13], analyze the novelties that have become part of the daily lives of senior citizens with mobile technology, and point out the goals that stem from using the technology and usage guidelines for this group of individuals. The objective of other studies has been to understand the best use of technology for the elderly, with particular emphasis on mobile games [8]. Keating et al [6] studied the needs of the elderly and those with visual or audio impairment, and how these needs can be included in the design and development of future products.

Additionally, Thomson and Cupples [17] have analyzed the use of SMS, highlighting that its use, in addition to being a means of communication between people, is also a relationship or association between a mobile and a person. This relationship places great importance on the ease of use of the mobile device.

Notable among all the described research is the absence of specific systems aimed at simplifying the basic services offered by mobile telephones.

The ComunicaME system is a specific solution that aims to facilitate the interaction between the user and the mobile device through the use of NFC technology. Additionally, this project is innovative for its use of wireless NFC technology, which is integrated in the most innovative mobile devices.

3 The ComunicaME System

This system facilitates and makes communication accessible to older persons. Using a NFC mobile telephone and tags that identify both people and actions, the user need only move the mobile close to the tag that represents the person or corresponding action.

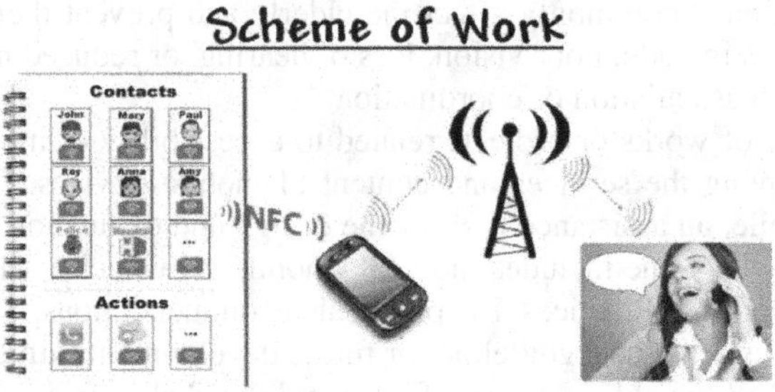

Fig. 1 Flowchart.

The flowchart is represented in Fig. 1. It includes a panel that displays a graphical representation of the list of contacts in a mobile telephone and the actions that can be performed. The contacts are displayed as photos and the actions as an image with which the user can very easily identify an action to be performed. Each of the objects (photo and image) are then associated with a NFC or MiFare tag containing the contact information or the action to be performed. The user must have a mobile telephone with NFC technology in order to read the tags.

This system is composed of two applications: ComunicaME and ComunicaME Writer. With ComunicaME Writer, the panel is configured with the contacts and the actions by writing the information onto the contactless tags. This module is used by persons who do not have any problems using and handling mobile telephones.

In order to write the action as shows Fig. 2, the user need only select the action and move the mobile close to the tag they want to record. To write a contact, the user can either select the list of contacts or input the telephone number of the contact. Once this is done, the user simply moves the mobile close to the tag that represents that contact.

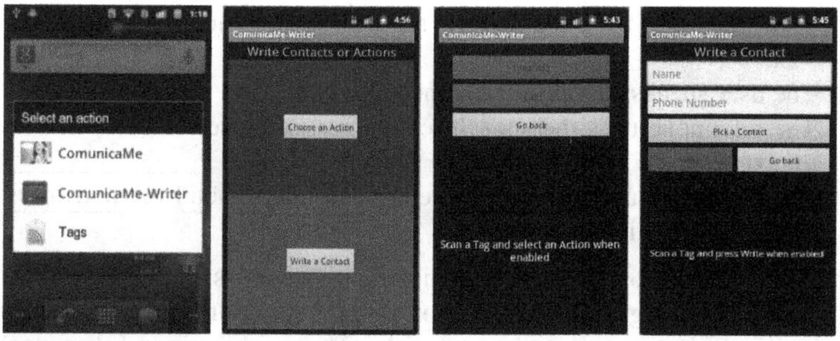

Fig. 2 Process to write a tag.

ComunicaME is the application used by seniors to communicate in a way that is simply, easy and intuitive.

To place a call as shows Fig. 3, the user performs the following steps:

1. The user selects the person to call by moving the mobile to the photo of that person.
2. The mobile telephone uses NFC to read the NFC tag and obtain the telephone number of contact information for the person to be called.
3. The user selects the *call* action by moving the mobile near the image that represents the action of calling.
4. The telephone uses the information to establish communication with the telephone number retrieved in step 2.

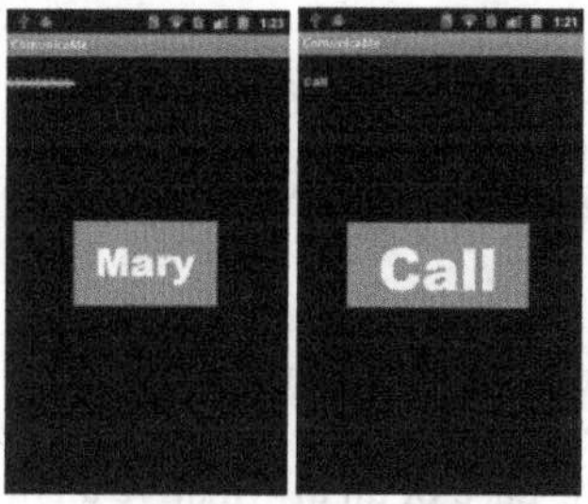

Fig. 3 Making a call.

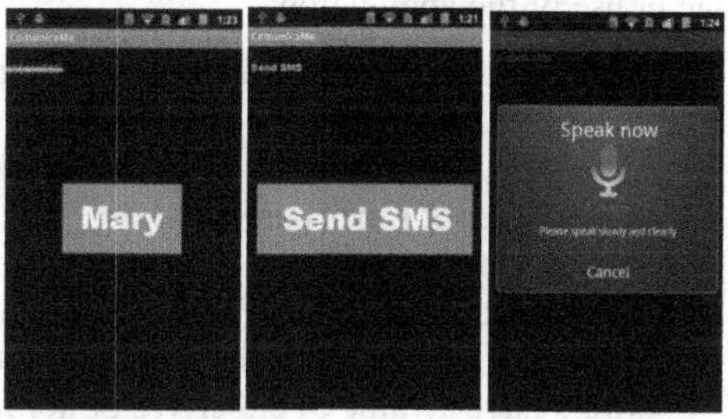

Fig. 4 Sending a SMS.

The user can send a short message as show Fig. 4, by following these steps:

1. The user selects the person to receive the message by moving the mobile phone near the photograph of that person.
2. The mobile telephone uses NFC to read the NFC tag and obtain the telephone number of the contact information for the person to receive the message.
3. The user selects the *send message* action by moving the mobile near the image that represents the action of sending a short message.

4. The telephone uses the speech recognizer to request the message to be sent.
5. The user speaks to transmit the message to be sent.
6. The telephone uses the speech recognizer to transcribe the voice message to text.
7. The telephone uses the mobile telephone line to send the text message to the number retrieved in step 2.

The primary innovation of the ComunicaME systems is in the combination of NFC mobile technology, MiFare technology and the connection between mobile devices and mobile communication networks. The combination of these technologies and means of communication ensure that mobile communication is simple and easy for seniors. Additionally, it does not require a complex technical infrastructure, and the development is simple. Because NFC technology is free, easy and secure, routine tasks can be carried out very easily by individuals of certain sectors of the population such as the elderly, who are normally quite reluctant to use new technologies.

4 Evaluation

In order to demonstrate the usability of the ComunicaME system, we developed a study involving 50 seniors of various ages. They were each provided with a panel as shown in Fig. 1, which includes the tags that were correctly written with the photos of each person's contacts and the images for each action.

The following user variables were evaluated prior to the user having any knowledge or use of the application:

1. Previous experience of the user with mobile terminals. Rated on a scale of 0-10 with 0 being no experience and 10 much experience.
2. Frequency with which the user uses a mobile telephone. Rated according to the number of times per week.
3. Age of the user.
4. Does the user live alone? The answers have a value of either yes or no.
5. Does the user utilize other tele-assistance mechanisms? Which? The answers have a value of either yes or no. In the case of an affirmative answer, the mechanisms are identified.

The user first indicates previous experience with the ComunicaME system and then uses the system with the correctly configured panel of contacts. To evaluate the efficiency of the system, the following application variables were evaluated:

1. The user's rating of the system.
2. Rating the use of the call.
3. Rating the use of SMS messages sent by speech recognition.
4. Have you used the ComunicaME system?
5. Do you think the ComunicaME system has helped you in performing daily tasks with your mobile telephone?

The responses were ranked on a scale of 0-10, with 0 being the lowest score and 10 the highest.

After the study was concluded, we noted that, in general terms, in 90% of the cases, the users believe that the ComunicaME system is a good idea and that its use helps them to perform daily tasks. After evaluating the system, we could appreciate the distinction between older (older than 75 years) and younger (between 65 and 74 years) seniors. Fig. 5 shows that the younger seniors rate the application higher and use the system more than their older counterparts. We can also see that the automatic call service is ranked higher than the SMS service by the users.

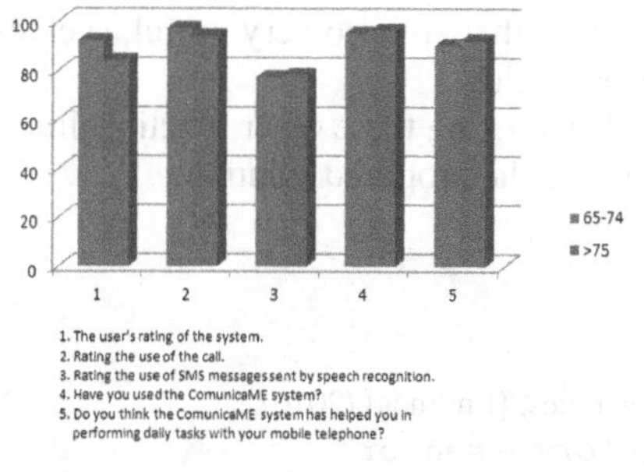
100
80
60
40
20
0
1
2
3
4
5
65-74
>75
1. The user's rating of the system.
2. Rating the use of the call.
3. Rating the use of SMS messages sent by speech recognition.
4. Have you used the ComunicaME system?
5. Do you think the ComunicaME system has helped you in performing daily tasks with your mobile telephone?

Fig. 5 Results based on the age.

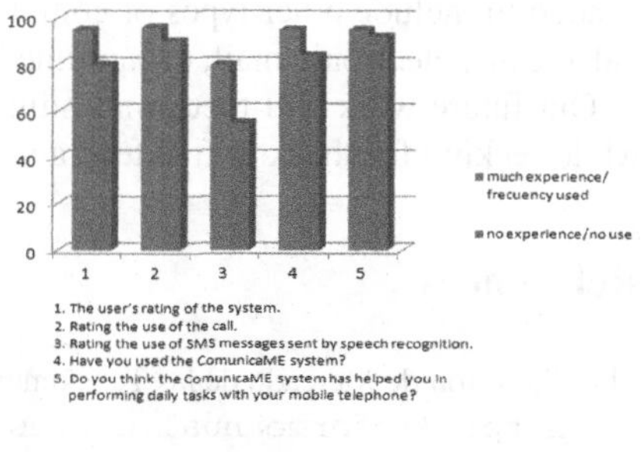
100
80
60
40
20
0
1
2
3
4
5
much experience/ frecuency used
no experience/no use
1. The user's rating of the system.
2. Rating the use of the call.
3. Rating the use of SMS messages sent by speech recognition.
4. Have you used the ComunicaME system?
5. Do you think the ComunicaME system has helped you in performing daily tasks with your mobile telephone?

Fig. 6 Results based-on experience/use.

Another significant point that was observed during the system evaluation involves the disparity among the data gathered: on the one hand, there are persons with experience in using mobile devices who use them frequently, on the other hand are those with little experience who use the mobile devices sporadically. As observed in Fig. 6, the application is ranked higher by those with more experience using a mobile phone, while those with less experience using mobile technology, although they believe that it is a good idea, rate it lower. All users, however, are convinced that the ComunicaME system facilitates the daily tasks with the use of a mobile phone. A significant finding is that the SMS is not highly valued by users with little experience with mobile technologies.

In addition to these evaluations, we were also able to note during the study that 95% of individuals living alone use the ComunicaME system.

We will need to define more case studies that allow us to further optimize and improve the mobile interface of the ComunicaME system. Additionally, we must bear in mind the different problems of communication and the errors that can emerge in real scenarios between mobile devices, the wireless tags and the wireless communication networks.

5 Conclusions and Future Work

Mobile technology provides great advantages that improve quality of life within our society; nevertheless, the elderly use these resources to a lesser degree than the greater population. This low level of use is due to the deficiencies that come with age and represent, in many cases, obstacles to access new technologies. The obstacles faced by this segment of the population when using mobile telephones can be overcome with the ComunicaME system. This system facilitates the use of mobile telephones and mobile communication networks, creating a more user-friendly experience. The ComunicaME system described in this article presents a technological solution that is innovative and facilitates the implementation of basic mobile services. This project is oriented primarily to the elderly, but it may be used by any type of user.

It is also important to note that while the system currently uses the two types of actions that represent the most widely used forms of communication, it can be expanded to include other types of communication that are also very useful, such as video calls, electronic mail, or multimedia messaging.

Our future work will focus on testing and evaluating these other functionalities, while seeking feedback to fine tune and improve the proposed solution.

References

1. Comisión de Mercado de las Telecomunicaciones. [En línea] (2010), `http://informeanual.cmt.es/informe-sector/comunicaciones-moviles` (Citado el: 2 de Octubre de 2011)
2. Del Arcos, C., San Segundo, J., Encinar, J.M.: Los Mayores ante las TIC. Accesibilidad y Accesiquibilidad. Fundacion Vodafone España (2010)
3. Al Mahmud, A., et al.: Desigining and Evaluating the Tabletop Game Experience for Seniors Citizens. In: 5th Nordic Conference on Human-computer Interaction: Building Bridges. ACM (2008)
4. Pavon, F., Casanova, J.: Telefonía móvil y personas mayores: la accesibilidad como derecho. Latinoamericana de Tecnología Educativa 5, 385–395 (2006)
5. Keating, E., et al.: The Role of the Mobile Phone in the Welfare of Aged and Disabled People. University of Texas/NTT DoCoMo (2007)
6. Hix, B.: Seniors, Mobile Communication, and Health Care. 3CInteractive, LLC, Florida (2011)
7. Been-Lirn Duh, H., et al.: Senior-Friendly Technologies: Interaction Design for Seniors Users. ACM, Atlanta (2010)
8. Roebuck, K.: Near Field Communication (NFC): Hight-impact Strategies - What you need to know: definition, adoptions, impact, benefits, maturity, vendors. Emereo Pty Limited (2011)
9. Finkenzeller, K.: RFID Handbook: Fundamentals and Applications in contactless smart cards, radio frecuency identificacion and NFC (2010)
10. Huber, J.T., Walsh, T.J., Varman, B.: Camp For All Connection: a community health information outreach project. Journal Med. Libr. Assoc. 93, 348–352 (2005)
11. INE. Encuesta de Tecnologías de la Información en los hogares, [En línea] (2010), `http://www.ine.es`
12. Abascal, J., Civit, A.: Mobile Communication for Older People: New Oportunities for Autonomous Life. In: Worshop on Universal Accessibility of Ubiquitous Computind: Providing for the Elderly, pp. 255–268. European Research Consortium for Informatics and Matematics, Firenze (2000)

13. Yelmo García, J.C., San Miguel González, B., Martín García, Y.S.: Análisis del potencial de las comunicaciones móviles para la mejora de los servicios proporcionados por la Administración Pública a las personas con discapacidad. Tecnimap (2007), http://www.tecnimap.es
14. Lindholm, C., Keinonen, T.: Mobile usability: How Nokia changed the FACE of the Mobile Phone. McGraw Hill (2003)
15. Thompson, L., Cupples, J.: Seen and not heard? Text messagind and digital sociality. Routledge, Social & Cultural Geography 9, 95–108 (2008)

Enabling Intelligence on a Wireless Sensor Network Platform

M.J. O'Grady, P. Angove, W. Magnin, G.M.P. O'Hare, B. O'Flynn, J. Barton, and C. O'Mathuna

Abstract. Conventional Wireless Sensor Networks (WSNs) usually adopt a centralised approach to data processing and interpretation primarily due to the limited computation and energy resources available on sensor nodes. These constraints limits the potential of intelligent techniques to data analysis and such activities on the centralised host. In contrast, Intelligent WSNs (iWSNs) will be significantly more powerful thus enabling the harnessing of intelligent techniques for diverse purposes. One such purpose is the practical realisation of smart environments, and facilitating mobility and interaction with the inhabitants of such environments. As a step in this direction, this paper presents the design of an iWSN sensor node platform that enables the hosting of lightweight Artificial Intelligence (AI) frameworks whilst enabling the ubiquitous energy constraints be quantified, mitigated and managed.

1 Introduction

From a network topology perspective, networked sensors may be regarded as leaf nodes at the fringes of the network. Historically, many service providers would have regarded the mobile phone as the archetypical leaf node; however, current genres of smart phones are sophisticated computational platforms within their own right. This evolution from relatively dumb mobile terminals to highly sophisticated and robust platforms may be replicated in the coming years in the sensor world. Such an evolution is fundamental if the objectives of Ambient Intelligence(AmI) [8], namely achieving intuitive

Michael O'Grady
CLARITY: Centre for Sensor Web Technologies, University College Dublin
e-mail: michael.j.ogrady@ucd.ie

Philip Angove
Tyndall National Institute, University College Cork
e-mail: philip.angove@tyndall.ie

J.B. Pérez et al. (Eds.): Highlights on PAAMS, AISC 156, pp. 299–306.
springerlink.com

interaction between smart environments and their inhabitants, are to be realised in practice. Thus the question arises - to what degree can intelligent techniques be harnessed by sensor platforms and what particular category of problems may be solved (or mitigated) by adopting such techniques? A prerequisite for exploring these fundamental questions is the availability of robust intelligent Wireless Sensor Network (iWSN) platform that supports a variety of sensing modalities while supporting the practical operation of a lightweight intelligent frameworks.

2 Background

Sensors act in a reactive stimulus/report manner for the most part. On detecting some measurable phenomena, they report this to a central server or, more likely, a nearby base station in the WSN configuration. There, the data is processed and interpreted, according to the criteria specified by the network designers. The net result of this is significant data traffic on the network, which has considerable implications for power consumption, perceived network performance and ultimately, the operational lifespan of the WSN.

From this description, it is useful to classify networks using two schemes. In the first instance, the network connection adopted may serve as a classifier:

1. A wired sensor network uses a fixed network infrastructure, for example, Ethernet, to communicate either between nodes on the network, or with a centralised gateway node and server.
2. A Wireless Sensor Network (WSN) uses some wireless technology, for example, Zigbee, to communicate with other nodes.

Secondly the issue of power may be considered:

1. Many sensor networks can be connected to the electricity grid; others may be powered by significant power packs that can support the operation of the network for many months. Indeed, such networks can be reinforced with mechanisms for recharging, for example small solar panels.
2. Most sensors are powered by conventional batteries, meaning that their practical operation may be limited to days rather than months.

Traditionally, legacy sensor networks have fixed connectionss to the grid, support significant storage facilities. If real-time data feeds are required, the data may be routed via an fixed internet connection to a base station or web server. In industrial environments, a SCADA system would be a typical example.

Recent research in sensors is motivated in many instances by the potential of sensor platforms as enabling technologies for pervasive computing and AmI. Thus WSNs, due to their flexibility and lightweight nature are of particular interest. Such networks are inherently resource constrained from a computational and power supply perspective, in the later case due to the finite power supply of their batteries. The potential of such networks and the practical constraints under which their operate make them a particularly interesting study and a promising platform for harnessing intelligent solutions to manage these constraints.

It should be noted that classifications of sensor and sensor network exist, for example static and dynamic, mobile and fixed, and so on. These have their own particular characteristics and pose many interesting research challenges. For the purposes of this discussion, focus remains on the typical wireless sensor node and its key operational constraint, namely energy supply.

A number of sensor mote platforms are available commercially. Typical examples include the TelosB [4] and the WASP mote [3]. TelosB motes are based on TinyOS and are programmed in NesC. The WASP mote is programmed in C. It can already be seen that the issues of heterogeneity and lack of interoperability will be become significant problems, should WSNs gain traction in the manner envisaged. From a software engineering perspective, the middleware construct has been proposed by many researchers as a solution to this problem [2] [11]. Though relatively simple computational artefacts, developing software for motes can be time consuming in all but the most trivial cases.

Within the WSN mote category, a more powerful platform is that of the Oracle SunSPOT [10]. This prototype mote platform supports a number of standard sensors. More importantly, it supports a J2ME compatible JVM - SQUAWK [9]. This makes application development easier, given the popularity of Java. It also suggests the possibility of interoperability with other embedded devices that support a compatible software framework, including mobile phones.

In summary: many developments have taken place in WSN-enabling technologies, and a number of prototype deployments have been described in the literature. Yet it must be stated unequivocally that WSNs remain theoretical entities for the most part, and a number of fundamental problems must still be overcome before their potential in smart environments can be fully exploited.

3 iWSN Requirements

Designing a node platform for iWSNs demands the consideration of three key issues

1. Power Monitoring - As discussed, power a key determinant of the network operational longevity. Thus its use must be understood and quantified in all scenarios. This exercise is more akin in many regards to embedded systems design rather than conventional software engineering. However, conventional approaches using battery voltage are inherently inaccurate; thus a highly accurate method of measuring power consumption in real-time is required.

2. Intelligent Reasoning It is necessary that the platform supports embedded reasoning. A facility for meta-reasoning is also required, again due the available power being at a premium and its usage in all circumstances must be measured and justified. The applicability of embedded in-situ reasoning are many but one generic issue is that of minimising the need for data traffic. Thus rather than a simple *stimulus/response* behaviour pattern, a *stimulus/deliberation/response* pattern may be realised. In essence, this is a continuation of a trend that increasingly enables the deployment of intelligence in a distributed fashion and on the peripheral nodes of communications networks [7].
3. Scalability - Though scalability can be designed for, and simulation tools can be used to validate these designs, it is only in deploying a sensor test bed in the physical environment that scalability can ultimately be validated. Thus practical economic constraints limit the efficacy of scalability testing. This demands that unit price per node be kept to a minimum.

4 Design of Intelligent Sensor Node

A sensor mote platform has been commissioned by Tyndall National Institute in line with the requirements listed above. Conceptually, it may be regarded

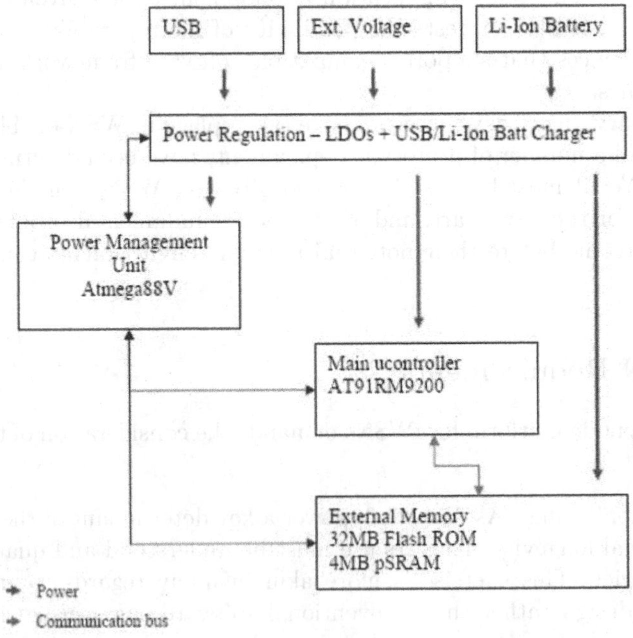

Fig. 1 Architecture of the iWSN Java board.

as constituting four layers - each of which is now described. Figure 2 illustrates the iWSN java board; a schematic may be viewed in Figure 1.

4.1 Power Monitoring Layer

This layer enables real-time system power usage measurement. This incorporates a battery gauge monitor utilizing a Coulomb counter which has a unique feature in that it continues to accumulate (integrate) current flow in either direction even as the rest of the internal microcontroller is placed in a very low power state, further lowering power consumption without compromising system accuracy. This achieves independence of current waveform shape, thus simplifying the required DSP calculations that are required due to the duty cycling & pulse current profiles of typical sensor mote platforms. This approach differs from conventional approaches that rely on a snapshot of the battery voltage - something that is highly susceptible to environmental operating conditions.

4.2 Java Board Layer

The Java board is based on the eSPOT board, an open source design of the SUNSPOT project. Compatible with J2ME, this allows the deployment of intelligent techniques such as intelligent agents, for example Agent Factory Micro Edition [5] or even expert system technology [1]. A software stack that can accommodate intelligent reasoning within a WSN has been successfully deployed on the Intelligent Sensor Node. The ability to host agents, specifically Belief Desire Intention (BDI) agents, on sensor nodes facilitates the opportunistic collaboration of sensing nodes in the collection, routing, filtering and analysis of data. It also enables the sensor network to operate in concert in the collection of data this is particularly important where the sensing devices are computationally challenged and where power depletion is critical to the continued operation of the network as a whole while preserving Quality of Service (QoS) constraints. It is often claimed that such sensor nodes ought to be denuded of in-network processing. Furthermore intelligent reasoning in situ has, historically, been viewed as computationally inappropriate. The iWSN power monitoring layer provides a mechanism by which we can directly compute the cost of intelligent computation and develop utility-based intelligent reasoning where the net benefit of a deduction in terms of potential power savings resulting from reduced data sensing, filtering or simply discarding of certain data.

4.3 Communications Layer

This consists of a 25x25mm Zigbee Layer, and enables interaction with legacy sensor boards. This design could easily be adapted to other wireless protocols technologies developed as a 25*25mm layer [6], for example, a Bluetooth layer has already been demonstrated.

4.4 Sensor Layer

This layer comprises a generic suite of sensors which can be configured as per domain requirements. A wide range of additional sensors can be incorporated if required.

Fig. 2 The prototype iWSN Java Board.

5 Future Work

As an initial step towards validating the platform, the AFME framework has been successfully deployed on a node. Both agents and sensor networks are inherently distributed constructs, thus to fully validate the node, it is necessary to deploy iWSN testbed, and explore its operation under a variety of conditions. Plans for such a testbed are well advanced. It is hoped that this testbed will lead to new insights regarding the practical deployment of agents on WSN nodes, how best to harness their reasoning and social capabilities, as well as identifying scenarios where iWSNs may offer a more apt approach. It may be envisaged that in practice, WSNs will incorporate a heterogeneous suite of nodes. How this will manifest itself in practices remains to be seen; however as an initial step, the potential of supporting the SunSPOT eDemo board is being investigated. Finally, a range of additional power management approaches are also being considered.

6 Conclusions

iWSNs will constitute one of the cornerstones of the next generation of computing paradigms, including AmI solutions as well as services based on the sensor web. A deeper understanding of iWSNs and role they may fulfil is required, particularly in achieving intuitive interaction with a heterogeneous, static or mobile end-user population. Fundamental to this is the availability of a robust node hardware platform that support intelligent decision making in-situ. In this paper, the design of such a node has been presented.

Acknowledgements. The support of the National Access Program (NAP) provided by the Tyndall National Institute, and that of Science Foundation Ireland under grant 07/CE/I1147 is acknowledged.

References

1. Hall, L., Gordon, A., Newall, L., James, R.: A development environment for intelligent applications on mobile devices. Expert Systems with Applications 27(3), 481–492 (2004)
2. Henricksen, K., Robinson, R.: A survey of middleware for sensor networks: state-of-the-art and future directions. In: Proceedings of the International Workshop on Middleware for Sensor Networks, MidSens 2006, pp. 60–65. ACM, New York (2006)
3. Libelium: Waspmote datasheet (2011), `http://www.libelium.com/support/waspmote`
4. MEMIC: Telosb mote platform (2011), `http://www.memsic.com/products/wireless-sensor-networks/wireless-modules.html`
5. Muldoon, C., O'Hare, G.M.P., Collier, R.W., O'Grady, M.J.: Towards pervasive intelligence: Reflections on the evolution of the agent factory framework. In: El Fallah Seghrouchni, A., Dix, J., Dastani, M., Bordini, R.H. (eds.) Multi-Agent Programming, pp. 187–212. Springer, US (2009)
6. O'Flynn, B., Lynch, A., Aherne, K., Angove, P., Barton, J., Harte, S., O'Mathuna, C., Diamond, D., Regan, F.: The tyndall mote. enabling wireless research and practical sensor application development. In: Adjunct Proceedings of the 4th International Conference on Pervasive Computing, pp. 21–26 (2006)
7. O'Grady, M.J., O'Hare, G.M.P., Chen, J., Phelan, D.: Distributed network intelligence: A prerequisite for adaptive and personalised service delivery. Information Systems Frontiers 11, 61–73 (2009)
8. Sadri, F.: Ambient intelligence: A survey. ACM Comput. Surv. 43, 36:1–36:66 (2011)

9. Simon, D., Cifuentes, C., Cleal, D., Daniels, J., White, D.: Java on the bare metal of wireless sensor devices: the squawk java virtual machine. In: Proceedings of the 2nd International Conference on Virtual Execution Environments, VEE 2006, pp. 78–88. ACM, New York (2006)
10. Smith, R.: Spotworld and the sun spot. In: 6th International Symposium on Information Processing in Sensor Networks, IPSN 2007, pp. 565–566 (2007)
11. Wang, M.-M., Cao, J.-N., Li, J., Dasi, S.: Middleware for wireless sensor networks: A survey. Journal of Computer Science and Technology 23, 305–326 (2008)

Touch Me: A New and Easier Way for Accessibility Using Smartphones and NFC

Miguel A. Sánchez, Montserrat Mateos,
Juan A. Fraile, and David Pizarro

Abstract. In this article we describe a real and practical mobile solution called Touch Me. This smart mobility project is based on NFC (Near Field Communication) technology. It is implemented for Android phones and has been tested by older and blind people. After a few tests we believe that this kind of software is a new and easier way for accessibility. It is especially useful for identifying and describing objects. The user simply has to touch the object with the smartphone and then listen an audio description of its content. To make this possible it is necessary to attach NFC smart tags to the object and to write the description on it using another mobile application, called Write Me Description. In addition, Touch Me can be used to manage basic functions of a smartphone like making a call or launching an application. In this case we need to combine this software with a personalized pad that must include smart tags. To configure these tags we need another two applications, called Write Me Contact and Write Me App.

Keywords: NFC, accessibility, smart mobility.

1 Introduction

Near Field Communication (NFC) [9] is a specific protocol that allows data interchange between very near devices. This technology can be used with several proposes like mobile payment or mobile marketing. But we are only interested in using it to read information from smart tags with NFC mobile phones. This capability makes possible a new way of accessing information. In this moment there are only a few smartphones that incorporate this technology. However, the prediction described on the NFC roadmap [6] will change the mobile environment in only two or three years.

Miguel A. Sánchez · Montserrat Mateos,
Juan A. Fraile · David Pizarro
Pontifical University of Salamanca, c/ Compañía 5, 37002 Salamanca, Spain
e-mail: {masanchevi,mmateossa,jafraileni}@upsa.es,
dpizarro89@gmail.com

J.B. Pérez et al. (Eds.): Highlights on PAAMS, AISC 156, pp. 307–314.
springerlink.com

Since 2008 we are using NFC and other mobile technologies to design and develop mobile applications. We are looking for new and better ways of interaction for older [3][7] or disabled people. Some of these projects are:

- **DIAMI** [1][2], a distributed architecture for facilitating the integration of blind musicians in symphonic orchestras.
- **MASEL** [10], a mobile assistant for the elder, specially designed for this people, in collaboration with ESGRA Asistencia, that have many residences like ESGRA Mozarbez and ESGRA *Residencial La Vega*. Thanks to this collaboration we can test our applications with many old persons in this residences.
- ***PharmaFabula*** [8][11], an NFC application that was developed in collaboration with ONCE, the main Spanish blind. It is a mobile application that recognizes medicines and then reports a description of it in audio format. It only requires an NFC mobile and to incorporate an RFID tag [5] in the medicament box. In addition, *PharmaFabula* can provide personal information like dosage or treatment duration of each patient.

We made mistakes in previous projects because we focused too much in technology. After many tests we realised that we need to focus on persons, on what they really demand and how they feel comfortable with technology. *Touch Me* is a completely new project, based on NFC technology because the older and the blind people want what it means: a new and easier way for accessing to useful functions like making a phone call or identifying an object.

For this reason we tell a story to explain how it works and how it is used. This story is based on a real person, Elvira, who helped us to define, test and improve the application.

Fig. 1 Elvira *touches* a medicine to remember how to use it, the dosage and expiration date

Elvira is 70 years old. She has one daughter, Mercedes, who is blind. Elvira doesn't want to use a mobile phone because she doesn't know how to use it. For us, another problem is the barrier that both have got for recognizing objects, meals and medicines (see Fig. 1). Elvira took a wrong pill some years ago and had a serious health problem. For this reason we defined, designed and developed *Touch Me*, a new solution based on our good results with other NFC projects like *PharmaFabula*. They made several tests with older and blind people and finally they realise that NFC technology is the best solution for these persons.

This project was developed using Android and it is complemented with a pad with photos, smart labels and braille texts to make easier to access to some functionalities.

In the next sections we describe the main elements necessary to use this mobile solution, the main features of each mobile application and how they work.

2 Main Elements and Mobile Applications

Touch Me is a mobile solution composed by several elements and applications. The components are:

- **NFC or RFID tags.** Something basic to store an object description, the telephone number of a contact or an application link. Many kinds of tags are available (figure 3). For this project we used stickers with *Mifare Classic 1Kb* tags [4] but other options are possible. These tags can be written many times so we can modify them for different uses.
- **Android phone with NFC.** It is the main element. With it we can touch objects, photos or icons to read the information in tags associated to them. The phone can also be used to write this information in the tags. For this reason we will use different applications to read or write the tags.
- **Personalized pad.** This component can simplify the use of the phone. It is better to customize it for each person with photos or icons. For this, we only need to attach photos in the front, and NFC tags under the pad, right under the photo or the icon. As we see in Fig. 2, Elvira's pad has been personalized with her preferences. In addition, we left enough space to put braille text near each photo or icon. This way, blind people like Mercedes can localize easily each element that can be touched.

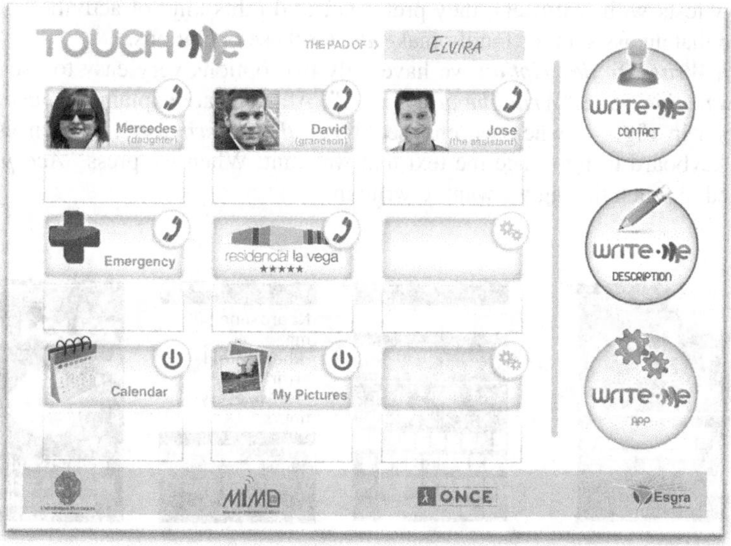

Fig. 2 Elvira's pad, customized with her preferred contacts and applications

- **Screen cover with relief points.** This component is optional but very useful for the blind to localize the main touch buttons and zones of the phone. This cover is easy to make and there are companies that can produce little or big units for each device.

Once we have these elements available, we can use four different applications. The most important is *Touch Me*, which is used to read all NFC tags, with descriptions, contacts or application links. In the other hand, we have three different *Write Me* applications, once for each type of content to write in tags. Now we describe briefly these applications:

1. ***Write Me description.*** It is used to write a text description of an object in a tag. It is possible to use voice recognition or the keyboard to write the information.
2. ***Write Me contact.*** It is used to write a telephone number in a tag. To do this you can choose a contact of the phone or write directly the number. It is possible to use voice recognition or a big numeric keyboard.
3. ***Write Me app.*** Used to write a link to a phone application in a tag. To do this you can choose the phone application from a list. Another option is to use voice recognition to select the application.
4. ***Touch Me.*** It is designed to read the tags and is executed each time the phone recognizes a NFC tag.

3 Administration Use Cases: Writing Tags with *Write Me*

To write information on NFC tags we have to use the *Write Me* applications described in the previous section. The elder or the blind can execute these actions because this software can be managed using voice recognition. However, after only a few tests with real users, they prefer not to do this kind of activities. So, we realised that an assistant or family makes better these operations.

With ***Write Me description*** we have only two options very easy to use: *"Say the description"* and *"Write the description"*. All steps are explained in audio. As described in Fig. 3, when we choose *"Write the description"*, we can use the phone keyboard to introduce the text that we want. When we press *"Accept"* we just need to touch the tag we want to write in.

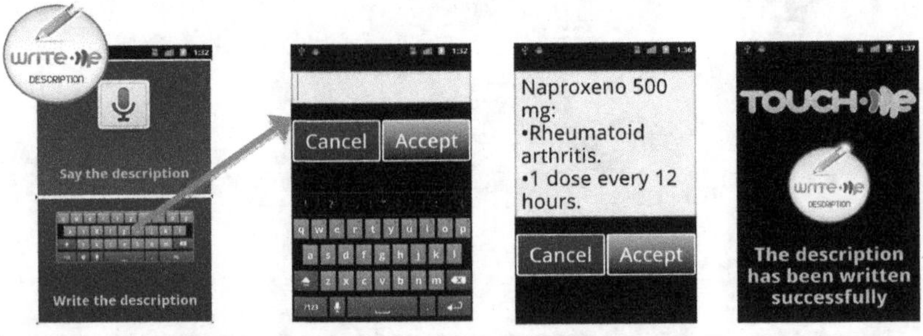

Fig. 3 Steps to write a text description in a tag using *Write Me description*

Write Me contact is other administrative application that we can use with two options: *"Say the name of the person"* and *"Indicate the contact or number"*. In this last case, we can write a phone number or choose it from the phonebook (Fig. 4). When we finish with any of these options, we just need to touch the tag we want to write in.

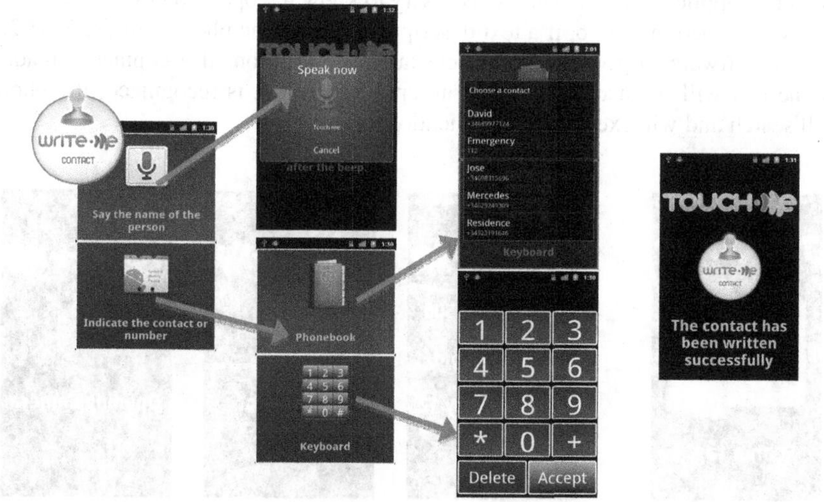

Fig. 4 Options to write a phone number in a tag with *Write Me contact* application

The third application is ***Write Me app***, with two options: *"Say the name of the application"* and *"Select an installed application"*. As described in Fig. 5, it is possible to use voice recognition or to select an application from a list of installed applications. After that, we just need to touch the tag we want to write in.

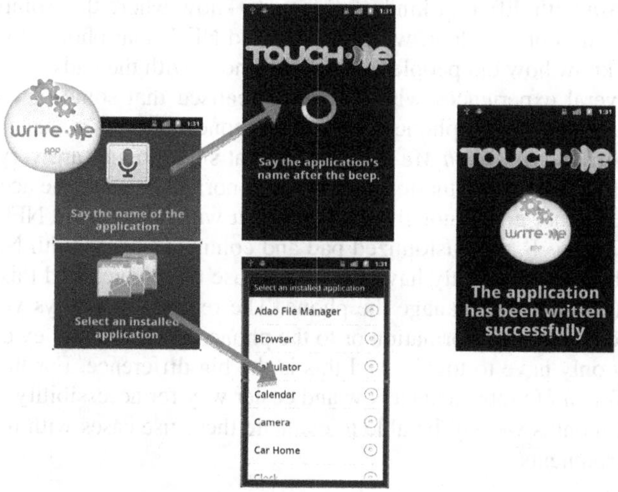

Fig. 5 Options to write an application link in a tag with *Write Me app*

4 User Use Cases: Reading Tags with *Touch Me*

To simplify the use of the phone *Touch Me* is the default NFC application in the phone. For this reason, *Touch Me* is executed quickly when a NFC tag is read.

It can recognize the kind of information that is reading: a contact, a text description or an application link. This is necessary to know the operation to execute.

As described in Fig. 6, if a text description is read, the phone will use *Text-To-Speech* software to give the equivalent audio description. If a contact is read, a phone call will be made. Finally, if an application link is recognized, the phone will search and will execute that application.

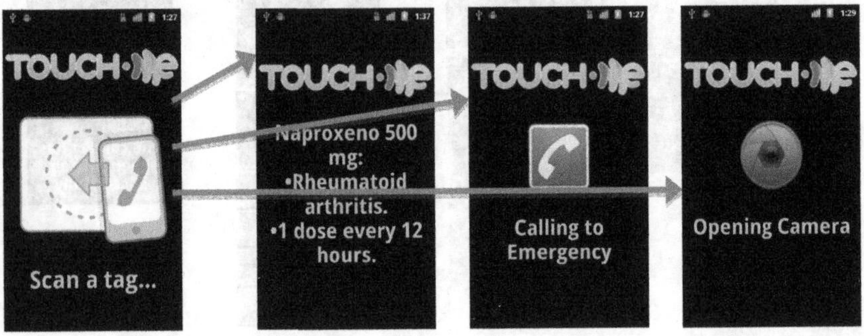

Fig. 6 *Touch Me* can read different types of tags

We have tested this solution with several users but we have not enough results yet. We presented it to Tifloinnova[1] 2011 (Fig. 7), an international showcase of assistive technology for blind and visually impaired people. In three days several user groups tested it. In addition, bind people from ONCE Salamanca is now testing *Touch Me* with different kind of objects to know where this solution is really useful for them. For the older, we gave pads and NFC smartphones to *Residencial La Vega* to know how old people manage the phone with the pads.

After several experiences with users, we realised that some of them want to change the Android smartphone for another phone easier to manage, but always before the firs use of *Touch Me*. They think that smartphones are very difficult to use. We believe that this situation is something normal because the accessibility of these phones is not good enough right now. But with an Android NFC phone like a Samsung Nexus S, the customized pad and common objects with NFC tags, the elder and the blind now only have to touch to use the phone. And this is a radical change in the way they manage the phone. The reaction is always very good because the access to the information or to the phone is something they don't have to learn. They only have to touch. And this is the big difference. For this reason we think that *Touch Me* represents a new and easier way for accessibility.

In a few months we will be able to evaluate these use cases with real users and in real environments.

---

[1] `http://tifloinnova.once.es/index.cfm?idi=ing`

Fig. 7 *Touch Me* can read different types of tags

5 Conclusions and Future Work

For many people, NFC is *the eternal promise* and many manufacturers and developers are waiting for its explosion, always associated to mobile payments. In an opposite side, we believe that is a reality. In only a few months, there will be more NFC terminals on market. The support of Google is very important and, in the last Mobile World Congress, NFC was probably one of the main points for companies and visitors.

As *Touch Me* reveals, NFC is a key technology to break some important barriers for older people and the blind. In addition it is useful for everybody. As in the case of Elvira, who previously did not want to use a smartphone, we think that many people will find this very useful. This is what Elvira told us after a few days using the phone and her personalized pad.

For the blind this is may be the only way they have to access to important information, as we could test with this and previous projects. They support us because they need this kind of software. And we want to give it to all the people who want it.

Our experiences with NFC are so good that we will work on it to incorporate many other applications for these persons: treatments associated to medicines and controlled by doctors, identification, some kind of payments, security access, etc.

For us, this solution is very important because it can give a clear message to society: NFC can help people and it is useful for much more that only mobile payments. *Touch Me* is one of the steps in our strategy of trying to improve the quality of life of older people and the blind.

References

1. Bajo, J., et al.: A distributed architecture for facilitating the integration of blind musicians in symphonic orchestras. Expert Syst. Appl. 37(12), 8508–8515 (2010)
2. Bajo, J., et al.: DIAMI. Director basado en Inteligencia Ambiental para Músicos Invidentes (2009), http://www.upsa.es/diami
3. Been-Lirn Duh, H., et al.: Senior-Friendly Technologies: Interaction Design for Seniors Users. ACM, Atlanta (2010)
4. Dayal, G.: How they hacked it: The MiFare RFID crack explained; A look at the research behind the chip compromise. Computerworld (2008), http://www.computerworld.com/s/article/9069558/How_they_hacked_it_The_MiFare_RFID_crack_explained
5. Finkenzeller, K.: RFID Handbook: Fundamentals and Applications in contactless smart cards, radio frecuency identificacion and NFC (2010)
6. Gamma Solutions, NFC Roadmap (2010), http://www.gammasolutions.es/index.php?option=com_content&view=article&id=93&Itemid=228
7. Hix, B.: Seniors, Mobile Communication, and Health Care. 3CInteractive, LLC (2011)
8. Pérez, J.M., Fernández, F., Fraile, J.A., Mateos, M., Sánchez, M.A.: Multi-Agent System (GerMAS) Used to Identify Medicines in Geriatric Residences. In: Pérez, J.B., Corchado, J.M., Moreno, M.N., Julián, V., Mathieu, P., Canada-Bago, J., Ortega, A., Caballero, A.F., et al. (eds.) Highlights in PAAMS. AISC, vol. 89, pp. 299–306. Springer, Heidelberg (2011), doi:10.1007/978-3-642-19917-2_36
9. Roebuck, K.: Near Field Communication (NFC): Hight-impact Strategies - What you need to know: definition, adoptions, impact, benefits, maturity, vendors. Emereo Pty Limited (2011)
10. Sánchez, M.A., Beato, E., Salvador, D., Martín, A.: Mobile Assistant for the Elder (MASEL): A Practical Application of Smart Mobility. In: Pérez, J.B., Corchado, J.M., Moreno, M.N., Julián, V., Mathieu, P., Canada-Bago, J., Ortega, A., Caballero, A.F. (eds.) Highlights in PAAMS. AISC, vol. 89, pp. 325–331. Springer, Heidelberg (2011), doi:10.1007/978-3-642-19917-2_39
11. Sánchez, M.A., et al.: Pharmafabula. NFC mobile technology and RFID for identification of medicines for the Blind (2009), http://www.upsa.es/pharmafabula

Earn More, Stay Legal: Novel Multi-agent Support for Islamic Banking

Amer Alzaidi and Dimitar Kazakov

Abstract. Multi-Agent Systems can play a critical role in Islamic banking development to create new or reform existing financial products in order to promote profit, reduce risk and to solve a number of legal issues related to Islamic Sharia law. This paper introduces a novel multi-agent platform for commodity trading in support of Islamic banking. The article focuses on the most popular (and debated) Islamic banking product, Tawarruq, and describes the design and implementation of a heterogeneous multi-agent platform assisting three types of participants: banks, retailers and individuals in need of cash. The benefits of our approach are demonstrated experimentally through simulations.

Keywords: Islamic Banking, Market, Multi-agent, Tawarruq.

1 Introduction

The paper introduces a multi-agent platform for a specific form of commodity trading, which is used in Tawarruq, an Islamic banking product providing individual and corporate funds in compliance with the Sharia law. This novel multi-agent platform serves as an e-market to address the issues that this most popular Islamic banking product has at present. Tawarruq provides the bank customer with funds through buying and selling commodities under certain legal constraints. Platform described here is used to assist efficient trading and Islamic law compliance.

2 Islamic Banking

A vital feature of Islamic banking is its entrepreneurial aspect. Risk and equity sharing are characteristic, as is the prohibition of interest (riba). Islamic law

Amer Alzaidi · Dimitar Kazakov
Department of Computer Science,
University of York, York YO10 5GH, UK
e-mail: {amer,kazakov}@cs.york.ac.uk

J.B. Pérez et al. (Eds.): Highlights on PAAMS, AISC 156, pp. 315–322.
springerlink.com

accepts a capital reward for loan providers solely on a profit/loss-sharing basis [3]. This means that the profit or loss margin will depend on how well or how poorly the investment project has performed. Islamic banks are profit-driven institutions just like non-Islamic banks, but their financial activities are managed under strict Islamic rules and principles.

Consider a person or institution in need of approximately £1000, which turns to a bank offering *Tawarruq*. The bank will purchase a commodity (e.g., rice) for, say, £1010, and sell it to the customer for a deferred payment of £1030 (thus making a profit). The customer then has the right to sell the commodity and release the cash they need [2]. In practice, this can be done in a number of ways. To assist their customers, banks offer what is known as *organised Tawarruq,* where the bank is in full control of the transaction and is in charge of both the purchase and the reselling of the commodity in question [4]. Under Islamic law, the following principles direct the legitimacy of the Tawarruq process:

1. The seller must own the product.
2. The product cannot be re-sold back to the original seller or its supplier. If this happens, the entire transaction will be considered void.

When these conditions are met, the financial product is known as *personal Tawarruq*. Organised Tawarruq alters the usual steps: to assist the customer, the bank takes the customer's authorisation to sell that commodity on their behalf to a third party, which may often be the same supplier from which the bank bought it. The funds are then deposited in the customer's account. This, and the fact that reselling prices may be pre-arranged for the customers even before they buy the commodities from the bank, are two ways in which organized Tawarruq may be seen as contradicting Sharia law. Nevertheless, financing based on these principles is still in use in a number of banks, which is perceived as a problem.

3 Analysis and Design

The first goal of the proposed system is to implement Tawarruq in a way that combines the original principles of personal Tawarruq and the smooth processing of organised Tawarruq, but removes the related legal issues. The workflow starts with the customer browsing the commodities offered by the bank as the basis for trading. In our platform, the customer can make use of a novel forecasting tool [1] to guide their choice and chose a commodity that can be resold with a maximum profit or minimum risk. Once this choice is made, the customer sends an offer request to the bank. Different banks use different strategies to determine the customer's expected ability to make the monthly payments. The bank then sends an official offer to the approved customer, and when they both agree, a contract is signed Fig. 1.

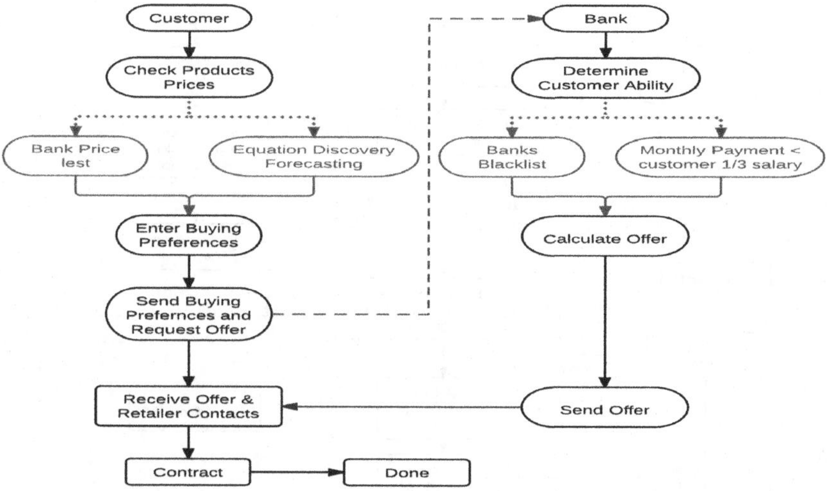

Fig. 1 Tawarruq Order Workflow.

The second goal is for the customer to resell what they have bought from the bank to a retailer. Customers start by entering their selling preferences (which, in our system, are the amount of commodity sold, minimum price and deadline for the sale) and choosing their selling strategy ('best offer' or 'auction'). *Best offer* is a fast and simple strategy for customers to sell, where the CMA finds the highest offer for the customer product. In the best offer strategy, the CMA sends product details to the relevant retailers and requests an offer from each retailer. The CMA determines the highest offer and sends the details of the retailer with the highest offer to the customer. If selling by *auction*, the CMA creates an auction for that customer.

Once all relevant retailers are ready to send bids, the CMA send the reserved price and request bids from retailers. In our implementation, this step is repeated for three rounds and after each round the CMA announces the winner and sends the highest bid to the retailers requesting higher bids. Some retailers drop out because the highest bid or reserved price is higher than their maximum. Otherwise, retailers choose a strategy to calculate their next offer ether through user input, by contacting the retailer representative or using an equation for a more automated process. The CMA sends the details of the winning retailer to the customer and they sign a contract. Retailers send their final offers to the CMA and wait for the customer to contact them or a new offer request from the CMA fig. 2.

4 Experimental Setup

The design is composed of five systems:

1. The customer system is where the product is chosen by the user to order from the bank as part of Tawarruq. This includes all the buying specifications.

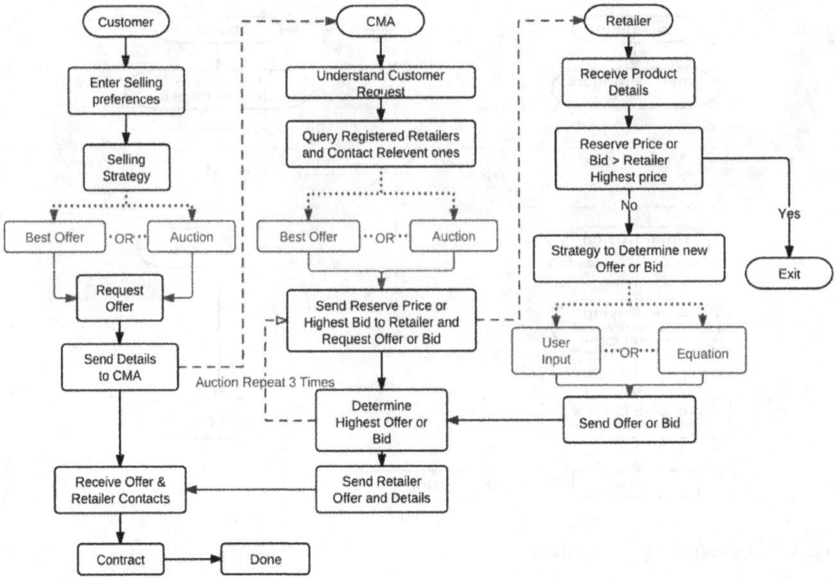

Fig. 2 Workflow for a customer reselling a product to a retailer.

2. The bank system is where the conditions for providing Tawarruq are set and the offer of Tawarruq is made.
3. The retailer system supports the end buyer in the process; the retailer is only allowed to buy from the customer. This is where the retailer chooses a buying strategy (best offer or auction) and calculates offers.
4. The market system is where deals are made and selling offers matched to buying offers.
5. The Sharia system is where deals and products are checked.

The goals of the simulation are to evaluate the market performance and insure that the market workflow is Shria compliance. The main goal is to monitor the customer output for each transaction based on profit & lost percentage then compare customers buying commodities based on Equation discovery forecasting and other methodologies.

The experiment simulates Tawarruq in Saudi Arabia and all the data and scenarios are designed to reflect the Saudi market. The experiment represents a multi-agent commodity market with one bank, which offers a choice of 4 commodities for its Tawarruq product, 480 daily customers and 100 retailers authorised to buy commodities from the customers.

The simulation uses four data generators to simulate real world data, which are described later in the chapter. The experiment uses external data sources to generate random scenarios from real data. Groups were defined as:

- Group 1: customers choose the commodity that will make the most profit based on the results of Equation Discovery forecasting using LAGRAMGE [1].
- Group 2: customers choose the commodity that will make the most profit based on the Neural Network forecasting results.
- Groups 3-6: customers always choose the shares of the AlRajhi Company, Alinma Company, Jabal Company, Emmar Company.

For Dataset we have used the prices of four stock market companies including daily highest, lowest and closing prices over 176 days. The bank uses the closing price as the selling price; the customer uses the minimum as the minimum price to resell and retailers use the highest and lowest price to generate minimum and maximum prices for each retailer. The companies chosen are: Al Rajhi Bank, Alinma Bank, Jabal Omar and Emaar, which were selected from the bank and real estate sector.

5 Results

The market opened for 8 hours a day, 5 days a week. The simulation ran for 176 market days, with 480 new customers and 100 retailers every day. A total of 28,336 Tawarruq deals were made including the reselling. The simulation shows that the market can handle a large volume of transactions. Customers made on average 0.03% profit after reselling, which shows that it can be a profitable market for customers. Total Tawarruq requested is 5,824,617,216 SAR and total reselling is 5,826,400,889 SAR, which give customers total profit of 1,783,673 SAR. Fig. 3 shows the six types of customer illustrated earlier and the number of customers of each type, which were accepted. The left graph shows the daily average accepted by the bank and the right graph shows the total accepted for each type. Customers could be rejected if their income was lower than the bank required for the requested loan. The background check also affected whether the request was accepted or rejected by the bank. The focus is on the type of customer that will buy commodities based on the equation discovery forecasting and comparing this with

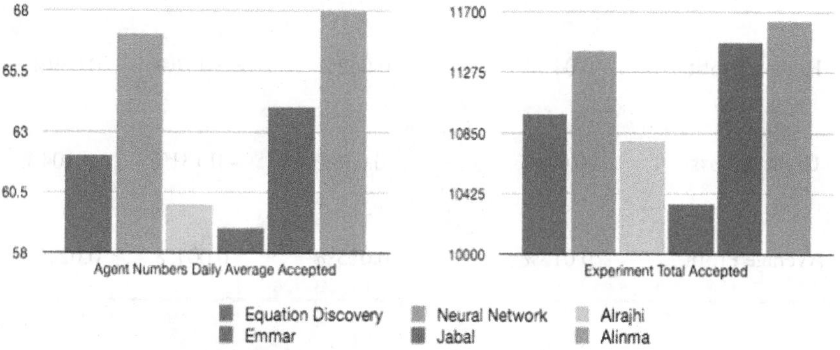

Fig. 3 Final accepted customers by the bank based on their buying strategy.

customer buying on the basis of based on the neural network forecasting. There were 66,666 accepted customers out of 84,480 customers, with an average of 380 customers accepted every day.

Customers resell their commodities using one of two strategies, best offer or auction. For each customer type, 80 customers enter the market every day. Half of the customers use best offer and the other half use an auction. Fig. 4 shows the daily average accepted and the total accepted for each strategy, for each customer type. As expected, similar numbers enter the market for each strategy, proportionally more customers choosing auction get accepted.

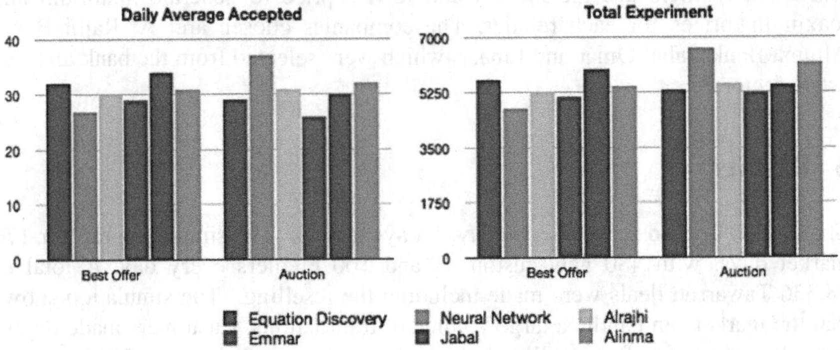

Fig. 4 Customer vs Reselling strategy.

Table 1 Market Results

	Equation Discovery		Neural Network	
	Best Offer	Auction	Best Offer	Auction
Highest Profit	0.041%	0.062%	0.027%	0.039%
Greatest Loss	- 0.045%	-0.036%	- 0.131%	- 0.104%
Average Profit	0.013%	0.025%	0.007%	0.024%

After 176 days of trading buying and reselling, the table below summarises the customer profit from the reselling. The equation discovery customers are compared to the neural network customers. In general customers of both types made a profit overall. When re-selling, auction is more profitable + 0.062%. From the table, profit and loss is a small percentage, which shows a stable market. The highest profit ever made, 0.062%, was achieved through an auction using equation discovery. The greatest loss ever incurred, -0131%, was registered when using neural network and best offer strategy. Even this extreme figure is many times smaller than the typical amount of -3% that a customer is expected to lose in organised Tawarruq based on observations and interviews with some Islamic banks.

Equation discovery customers performed better in the market. The average profit for the experiment for equation discovery customers is 0.033% and for Neural network is 0.017%

6 Conclusion

In this section the simulation and results are studied from economic and Islamic finance perspectives. The market provides two strategies for customers to resell their commodities to retailers, best offer and a three-round auction. Tawarruq customers need cash in a short time and also hope to generate profit in the process of obtaining it, or at least not lose too much. The best offer strategy ensures customers resell quickly for the best offer in the market at that specific time and the three round auction strategy maximizes the customer's return. In the future more strategies can be used in the market, and customers can coevolve their own strategy together with the retailers.

Respected Islamic finance experts declared organized Tawarruq to be illegal, because it is an unfair market to customers, which is against the principles of Islamic finance concerning fair trading, and not exploiting people's needs to force them to do something that do not want to, because it is the only way to obtain the cash that they need. Our simulation shows that customers can obtain profits or they may make a loss in the process depending on the market movements. However, this is a fair market which provides a platform for Sharia-compliant Tawarruq to take place.

The simulation proves that giving the customer a forecasting tool which uses equation discovery can increase the customer's chances of choosing the most profitable commodity in the market, without interfering with the principles of Islamic finance, under which a person should be free to choose the commodity they want. The results show that the market is generally profitable and equation discovery forecasting gives a better average profit, with 0.033% compared to neural networks with 0.017%. The combination of a multi-agent platform and equation forecasting provides the most profitable way to do Tawarruq in the modern day banking systems.

Customers have concerns about whether a banking product is Sharia compliant. Therefore, when Tawarruq was first introduced in banks, people used it heavily and bank promoted it as their number one product; however, after a number of Sharia concerns were raised related to the way banks were practicing Tawarruq, customers were discouraged from using this product. Banks started to change the product name and the way it is described. The two main concerns regarding organised Tawarruq are that customers cannot resell the commodities they have bought by themselves, and customers are almost guaranteed to lose money in the reselling. The results from the simulation show that customers can have full control of the reselling process and that on average they generate profit with our approach.

References

1. Alzaidi, A., Kazakov, D.: Equation Discovery for Financial Forecasting in the Context of Islamic Banking. In: The Proceedings of the Eleventh IASTED International Conference on Artificial Intelligence and Applications, Innsbruck, Austria (2011)
2. Roy, O.: The Failure of Political Islam. Harvard University Press (1994)
3. Sait, S., Lim, H.: Land, Law and Islam. UN-HABITAT, New York (2006)
4. Venardos, A.M.: Islamic Banking & Finance in South-East Asia: Its Development & Future. World Scientific Publishing, Singapore (2011)

Optimal Portfolio Diversification? A Multi-agent Ecological Competition Analysis

Olivier Brandouy, Philippe Mathieu, and Iryna Veryzhenko

Abstract. In this research we study the relative performance of investment strategies scrutinizing their behaviour in an ecological competition where populations of artificial investors co-evolve. We test different variations around the canonical modern portfolio theory of Markowitz, strategies based on the naive diversification principles and the combination of several strategies. We show, among others, that the best possible strategy over the long run always relies on a mix of Mean-Variance sophisticated optimization and a naive diversification. We show that this result is robust when short selling is allowed in the market and whatever the performance indicator chosen to gauge the relative interest of the studied investment strategies.

1 Introduction

Agent-based modelling (ABM) is widely used to study economic systems under a complexity paradigm framework. Within this research stream, financial markets have received a lot of academic and practitioners interests these last years, notably in offering an alternative to mathematical finance and financial econometrics. Among the features that can be grasped with ABM, the co-evolving aspects of stock markets (investors making decisions that affect the system, which hitherto impact their behaviour along a feedback loop) is probably one of the most critical one.

O. Brandouy
Sorbonne Graduate School of Business, Dept. of Finance & GREGOR
(EA MESR-U.Paris1 2474)
e-mail: olivier.brandouy@univ-paris1.fr

P. Mathieu
Université Lille 1, Computer Science Dept. LIFL (UMR CNRS 8022)
e-mail: philippe.mathieu@lifl.fr

I. Veryzhenko
Sorbonne Graduate School of Business, Dept. of Finance & GREGOR
(EA MESR-U.Paris1 2474)
e-mail: iryna.veryzhenko@ensam.eu

J.B. Pérez et al. (Eds.): Highlights on PAAMS, AISC 156, pp. 323–331.
springerlink.com

In this research we actually renew the analysis of a *classical* question in Finance, namely, the relative performance of investment strategies scrutinizing their behaviour in an ecological competition where populations of artificial investors co-evolve. This approach allows us to propose new results that can be compared to those of [4] or more recently [6] and [15] who did the same kind of research but within the traditional finance philosophy (no agents, no co-evolution, no complexity). In doing so, our agents populations compete each against the others with one strategy (some are based on different variations around the canonical modern portfolio theory of Markowitz [11], others on the Naive diversification principle [3] and others combine several strategies). To understand the added value of our approach compared to those of [6] and [15] one can start in summarizing their contribution before exposing the limitations of their approach.

– De Miguel and al. [6] compare several investment strategies using a *backtesting methodology*. It consists in managing a virtual portfolio of assets as if the historical prices used to run the experimentation were known. These investment strategies belong to one of the following investment rules families: i) Naive diversification or ii) Mean-Variance Diversification ("à la" Markowitz). The authors show that complicated portfolio optimization strategies not only under-perform the naive diversification, but also generate negative risk-adjusted rates of return.
– Tu and Zhou [15], extending the backtesting methodology of De Miguel and al., suggested that a combination of the 1/N strategy with the sophisticated diversification can each of its constituents taken separately. This result is proposed in an empirical framework which is extremely similar to the one of De Miguel and al.

To our opinion, the main problem with these researches is the unrealistic *"atomistic"* assumption that supports the backtesting methodology. Said simply, this assumption allows to gauge an investment strategy with historical data as if its true implementation would have not modified these prices. These assumptions in sharp contrast to analysis of [9], [7] who clearly show that prices may well be influenced by several parameters (investment strategies, the cognitive skills of investors or the market microstructure itself) that are neglected in the backtesting approach. We defend in this research that a convincing answer to the question *"among this set of investment strategies, which one outperforms the others?"*, overcoming the previously mentioned limitations, can be delivered by a multi-agent system allowing to implement ecological competitions among these strategies. We show, among others, that the best possible strategy over the long run always relies on a mix of Mean-Variance sophisticated optimization and a Naive diversification. This result reinforces the practical interests of the Markowitz framework that is strongly discussed in [6] for example.

2 Agents Behaviour

One of the advantages of ABM is that the agents are autonomous. In a mathematical model, all market participants are defined as equal-power rational entities facing homogeneous constraints. Agents actions are predetermined by strict equations describing their reaction in response to particular market conditions. ABM allows to overcome the limitations linked to that homogeneity. In this research, we design 8 agents populations, each of them following a generic strategy. These strategies are presented in subsection 2.1.

2.1 Height Populations of Agents Based on Height Different Generic Strategies

We start by introducing how a portfolio of assets is modelled and what kind of decision agents must make in a simulation. Note again that the purpose of each strategy is to allow agents to manage a diversified portfolio of financial assets over time.

- A portfolio is defined as a vector of weights over the investment universe. This vector is denoted α^{xx}, xx allowing to identify the generic strategy determining this vector.
- Depending upon the strategy definition or the empirical design, these weights can be negative or not. If this is the case, one will refer to this situation as "shorting allowed", which means that agents are allowed to sell assets they do not hold provided they will repurchase them later on.
- Each time a new portfolio is computed, the current weight vector α_t^{xx} is compared to the previous one α_{t-1}^{xx} to adjust the number of stocks to hold. This adjustment take into account the weight vectors and the corresponding assets current prices. As a result agents decide to buy or to sell certain assets they hold to reach their new (weight vector) target.
- Last but not least, these decisions must be practically implemented, that means *"translated into buy or sell orders"*, with quantities and prices in accordance to the target. One must remember that each strategy implies different parameters that may have different values within the same agents population; thus each agent has his own weight vector rolling during a simulation.

This process being the same whatever the behaviour, we can now describe at fine grain the 8 generic strategies (see Table 1).

Population 1: Naive diversification investors

The agents endowed with the naive strategy (N) ignore all information about risk and return of assets. Naive investors allocate their funds equally among the N risky assets in equal proportions $\alpha_{j,t}^{i,N} = \frac{1}{N}\ \forall j = \overrightarrow{1,N}$ the weights of wealth allocated to stock j of agent i at the moment of time t. In contrasts to sophisticated rules that are usually asymptotically unbiased but have a large (variance) estimation error in small samples, the 1/N rule is biased, but has zero estimation error.

Table 1 Strategies description

Name	Short Name	Basic definition & particularities
Naive	*N*	Equal weights, no sophisticated behaviour
Mean Variance 1	*MLong*	Markowitz optimization, long positions only
Mean Variance 2	*MShort*	Same as *MLong*, shorting allowed
Market Portfolio	*MP*	Weights according to assets capitalisation on the market,
Holders		no sophisticated behaviour
Bayesian Traders 1	*BLong*	Based on Markowitz, estimation of moments co-moments of asset returns improved, long positions only
Bayesian Traders 2	*BShort*	Same as *BLong*, shorting allowed
Strategy Combinators 1	*CLong*	Mix of *N* and *MLong*
Strategy Combinators 2	*CShort*	Mix of *N* and *MShort*

Populations 2 and 3: Mean-variance optimizers

Agents endowed with this strategy try to minimize risk for a given target return following the mean (μ) variance (σ^2)optimisation rules introduced by Markowitz [11]. An important parameter in this process is the correlation matrix V of asset returns and the investor's risk preferences (risk aversion) defined in his quadratic utility function:

$$min\frac{1}{2}\sigma_p^2 = min_\alpha \alpha' V \alpha \tag{1}$$

$$\mu_p = \alpha' \mu \tag{2}$$

$$\sum_{i=1}^{n} \alpha_i = 1, \alpha^M = (\alpha_1, \alpha_2, \ldots, \alpha_n) \tag{3}$$

where n – number of assets, μ_p – expected return of portfolio, σ_p – standard deviation of portfolio, α^M – target weights defined according to Markowitz rules. This optimisation problem provides the solutions outside the range $[0,1]$, that allows *shorting*.

From its definition, we create two agents population, one allowed to use short selling (MShort), the other not allowed to do so (long only, MLong)

Population 4: Market portfolio holders

Market portfolio (MP) holder is the type of agents with a portfolio consisting of all assets in the market with weights proportional to assets capitalisation [14]. In more realistic context, if an investor has no special insights about expectation returns and volatility of individual stocks he is supposed to hold the market portfolio (portfolio of all available stocks).

$$\alpha_{j,t}^{i,MP} = \frac{P_{j,t} \times Q_{j,t}}{C_t} \tag{4}$$

$P_{j,t}$ price of asset j at moment t, $Q_{j,t}$ number of asset j traded on the market at the moment t, C_t total market capitalisation.

Population 5 and 6: Bayesian traders

Agents within this population have a behaviour that extends the Markowitz rules described in 2.1. The Markowitz approach has been criticized due to measurement errors in the estimation of assets' moments and co-moments. To overcome these problems authors like [1], [8] or [5] propose to improve the co-moments estimation in using a factor equal to $1+\frac{1}{M}$ that reduces its estimation error and leads to more reliable investment weights. Moments and co-moments being estimated following this rule, agents use equations (1)–(3) to determine the target weights.

From this logic we define two different population, one in which short selling is allowed, Bayesian Short Selling *(BShort)* and one in which it is forbidden, Bayesian Long Only strategies *(BLong)*.

Population 7 and 8: Strategy combinators

A last population has the ability at combining the naive 1/N strategy with the sophisticated Mean-Variance optimization strategy. It has been studied by some authors who thought it could improve the overall performance of investors [4]. Mathematically the weights definition of strategies combination can be describe as follow:

$$\begin{aligned} \hat{\alpha}_{j,t}^{i,C} &= (1-\delta)\alpha_{j,t}^{i,N} + \delta\alpha_{j,t}^{i,M} \quad \delta = \frac{\varphi_1}{\varphi_1+\varphi_2} \\ \varphi_2 &= \frac{1}{A^2}\left[\frac{(T-2)(T-n-2)}{(T-n-1)(T-n-4)}\right] \end{aligned} \tag{5}$$

where $\alpha_{j,t}^{i,C}$ – weights defined by strategies combination, $\alpha_{j,t}^{i,N}$ – weights defined according to naive diversification rule, $\alpha_{j,t}^{i,M}$ –weights defined according to Markowitz rule, δ – combination parameter $0 \leq \delta \leq 1$, n – number of assets, T – memory span or the length of available historical data. "Markowitz Shorting allowed" and "Markowitz Long-only" are used for combinations, hence Combination Short *(CShort)* and Combination Long *(CLong)* populations are studied in this research.

3 Simulations and Results

As mentioned earlier in this research, we have chosen to compare the relative performance of each investment strategy using an ecological simulation. We use an empirical design that is closely related to the one developed in [2]. The basic ideas governing this approach can be summarized as follows :

- Each strategy is encoded in an initial population of NNN agents. These populations are mixed and compete in the same market, trading the same stocks. Prices are the direct result of the flow of orders sent by the agents to the central order books ruling the artificial stock exchange.

- A time step in our ecological competitions is made of several *rounds*, each of them encompassing 1000 trading days.
- For such complex experiments we use the powerful Artificial Trading Open Market *(ATOM)* [12]. ATOM allows to implement thousands of agents evolving simultaneously, with different strategies.
- Initially, we populate the ATOM environment with our 8 populations of agents. The size of each of these populations x_i for $i = \overrightarrow{1,8}$ is the same $\forall i$. The total number of agents is $X = \sum_{i=1}^{8} x_i$. Populations are updated every simulation round according to their performance $x_i = X\frac{P_i}{P_T}$, where P_i the performance of population i and P_T the overall performance of the whole soup of populations. The performance can be measured as i) the total wealth (cash + market capitalization of the stocks of all the agents in each population) or ii) the average Sharpe ratio [13] of the population, during the previous round. A population is said to be extinct if $x_i = X\frac{P_i}{P_T} < 1$

Model parametrisation

The results presented later in this paper were obtained for the following parametrization of our model :

- The investment set is made of 30 different stocks (like the 40 different families of stocks listed in the CAC40 index)
- We study the 8 populations of agents presented in table 1.
- In each population, we start with 100 agents.
- 1000 days of trading form a *simulation round* in the experiments reported below. Between rounds, both stocks and cash endowments are carried forward.
- MLong, MShort, BLong, BShort, CLong, CShort use a rolling time window T=500 to estimate the necessary moments and comoments implied in the optimization process.
- All information concerning the underlying probability distribution of security prices as well as current security prices are available continuously and at no cost for all investors.
- Contrary to [6] and [15] who consider risk aversion as 1 or 3 for Markowitz strategies and its extensions, risk aversion in our simulations is uniformly drawn between [0.5,5] $A \sim D(x|0.5,5)$ in order to test a larger variety of behaviours, from risk averse agents to risk takers.
- Agents enter the market with 50 units of each type of stocks and 1000$ cash
- All agents have the same daily trading frequency, or said differently, are equally active in the market

4 Results and Discussions

We present here the results of two different ecological competitions. In the first one, the reproduction rate of each population is linked to dollars earnings (see subsection 4.1) while in the second one, it is a function of the Sharpe ratio (see subsection 4.2).

4.1 Ecological Competition 1: Wealth

The simulations results (figure 1(a)) show that all the constrained (long-only) strategies (*MLong*, *BLong*, *CLong*), the naive (*N*) and the market portfolio strategies (*MP*) quickly disappear from the market at the end of 50 rounds. According to [10], a possible explanation of this phenomenon could be linked to the large positions (positive or negative) implied by short selling, when it is allowed : the long-only strategies have zero-positions ($\alpha_{j,t}^{i,*} = 0$) in about 50% of the traded assets. Thus, the agents with long-only strategies trade only the half of the investment set to maintain their target weights. At the same time, the agents with short-selling strategies trade the whole set of assets and increase their wealth more efficiently.
In addition, we observed that the population *CShort* are better than their individual component rules (*MShort* and *N*) which is clearly in line with the results of [15].

We also investigated a possible effect of the initial size of the population in its survival time. We therefore changed population initial distributions dynamically ($\sim D(x|20,200)$) so to get a majority of certain types of agent in the whole population soup at the beginning of each experiment. Our results indicate that even if the initial proportion of naive agents (≈ 200 individuals) is much bigger than the proportion of others (100 individuals), they cannot survive much longer in the ecological competitions where wealth rules the reproduction rate.

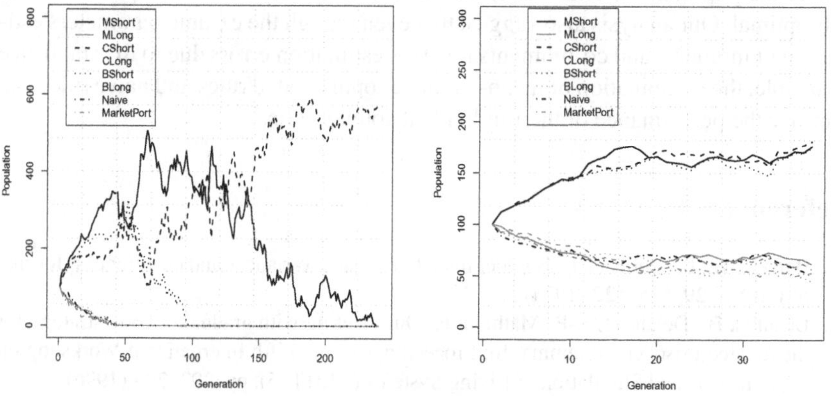

(a) Wealth (b) Sharpe ratio

Fig. 1 Ecological competition.

4.2 Ecological Competition 2: Sharpe Ratio

We measure the Sharpe ratio in order to estimate the agents ability to hedge the portfolio risk with many assets. Figure 1(b) reports the average evolution of agents proportions based on this indicator. As it can be observed in that figure, here again the unconstrained strategies outperform the constrained ones. These results confirm those of [10], who stress the importance of short selling in markets with many assets. At the same time, our results are not congruent with those of [6] who report that the Sharpe ratio of sample-based mean-variance strategy is much lower than that of naive strategy. One reason that could explain this discrepancy is that these authors use diversified portfolios with low volatility in their numerical simulations while our simulations rely on individual assets with more volatility.

5 Conclusion

Contrary to research works claiming the useless of the Markowitz theory, we show that this classical rule still outperforms the naive rules in high- and low-volatility market regimes when the traders behaviour generate such price motions. The major benefit of our method, which relies on agent-based modelling, is its flexibility comparing with approaches of [6] and [15]. Indeed, the ABM approach allows (i) any number of traders on the market (ii) combination of large variety of strategies (iii) any number of risky assets. This flexibility provides a distinct advantage over alternative approaches to the portfolio optimisation problems.

Our findings are consistent with those of [15] and [10]. The performance of unrestricted portfolio strategies outperforms the long-only and naive strategies in both ecological competitions where the Sharpe ratio or the earnings rule the reproduction rate of the populations. Thus our result show that naively diversified portfolios are sub-optimal. Our analysis also suggest that even though the ex-ante parameters estimation of moments and co-moments involves estimation errors due to the small size of sample, the combination of mean-variance sophisticated rules and naive rules can improve the performance of their individual counterparts.

References

1. Barry, C.B.: Portfolio analysis under uncertain means, variances and covariances. Journal of Finance 29, 515–522 (1974)
2. Beaufils, B., Delahaye, J.-P., Mathieu, P.: Our meeting with gradual, a good strategy for the iterated prisoner's dilemma. In: Proceedings of the Fifth International Workshop on the Synthesis and Simulation of Living Systems (ALIFE'5), pp. 202–209 (1996)
3. Benartzi, S., Thaler, R.H.: Naive diversification strategies in defined contribution saving plans. The American Economic Review 91(1), 79–98 (2001)
4. Brown, S.: Optimal portfolio choice under uncertainty: A bayesian approach. Ph.D. thesis, University of Chicago (1976)
5. Brown, S.: The effect of estimation risk on capital market equilibrium. Journal of Financial and Quantitative Analysis 14, 215–220 (1979)
6. DeMiguel, V., Garlappi, L., Uppal, R.: Optimal versus naive diversification: How inefficient is the 1/n portfolio strategy? Review of Financial Studies 22, 1915–1953 (2009)
7. Hommes, C.H.: Heterogeneous agent models in economics and finance. In: Handbook of Computational Economics. Agent-based Computational Economics, vol. 2, p. 1109 (2006)

8. Klein, R.W., Bawa, V.S.: The effect of estimation risk on optimal portfolio choice. Journal of Financial Economics 3, 215–231 (1976)
9. Levy, M., Levy, H., Solomon, S.: Microscopic simulation of the stock market: the effect of microscopic diversity. Journal de Physique I (France) 5, 1087–1107 (1995)
10. Levy, M., Ritov, Y.: Mean-variance efficient portfolios with many assets: 50% short. Quantitative Finance 11(10), 1461–1471 (2011)
11. Markowitz, H.: Portfolio selection. Journal of Finance 7, 77–91 (1952)
12. Mathieu, P., Brandouy, O.: Efficient Monitoring of Financial Orders with Agent-based Technologies. In: Demazeau, Y., Pechoucek, M., Corchado, J.M., Pérez, J.B. (eds.) 9th International Conference on Practical Applications of Agents and Multi-Agents Systems (PAAMS 2011). AISC, vol. 88, pp. 277–286. Springer, Heidelberg (2011)
13. Sharpe, W.F.: Mutual fund performance. Journal of Business 39(S1), 119–138 (1966)
14. Treynor, J.: Toward a theory of market value of risky assets. In: Korajczyk, R.A. (ed.) A Final Version was Published in 1999 in Asset Pricing and Pertfolio Performance, pp. 15–22. Risk Books, London (Fall 1962)
15. Tu, J., Zhou, G.: Markowitz meets talmud: A combination of sophisticated and naive diversification strategies. Journal of Financial Economics 99, 204–215 (2011)

Modeling Firms Financial Behavior in an Evolving Economy by Using Artificial Intelligence

Gianfranco Giulioni, Edgardo Bucciarelli, and Marcello Silvestri

Abstract. In this paper we investigate the dynamic behavior of firms capital structure. Differently from a number of existing studies, our aim is to tackle the issue by combining two approaches: experimental and computational economics. The main goal of this paper is in fact to build an artificial agent whose behavior mimics that of experimental agents. The paper shows how the heuristic approach could provide for a valid alternative micro-foundation for the financial decisions of entrepreneurs.

1 Introduction

The capital structures of the various economic agents are playing a key role in the recent negative economic and financial events. This paper focuses in particular on enterprises. Despite economic theory have dedicated them a particular attention, the existence of an optimal capital structure is still a debated question. The modern analysis of the topic initiated by Modigliani and Miller (1958) is evolving towards a dynamic framework (see for example Titman and Tsyplakov, 2007). Aiming to contribute to this recent strand of literature, in this paper we investigate how the capital structure evolves when economic conditions change. However, the approach we follow is different from that used in recent papers. We tackle the issue by combining two approaches: experimental and computational economics. The roles of these two elements can be better understood from the structure of the paper. In section 2 we briefly describe the simple microeconomic structure on which our work is based. This structure is submitted to selected people in a set of laboratory experiments whose results are reported in section 3. In section 4 we build an artificial agent whose behavioral rules are economically grounded. The parameters of the artificial

Gianfranco Giulioni · Edgardo Bucciarelli · Marcello Silvestri
Department of Quantitative Methods and Economic Theory,
Viale Pindaro 42 - 65127 - Pescara, Italy
e-mail: {g.giulioni,e.bucciarelli}@unich.it,
marcello.silvestri@yahoo.it

J.B. Pérez et al. (Eds.): Highlights on PAAMS, AISC 156, pp. 333–340.
springerlink.com

agents are endogenously established in section 5 by using the differential evolution algorithm. Section 6 gives conclusions and discusses future research opportunities.

Before going on, it is useful to have a glance at the mosaic the tile represented by this paper could belong. The final goal of our project is to build an experimentally micro-founded multi-agent artificial system made up of a large number of different types of avatars interacting with each other and with banks.[1] The model could be used for analyzing macroeconomic issues; as an example it could be used to check whether capitalism economies endogenously generate severe financial crisis as pointed out some decades ago by Hyman Minsky (see for example Minsky, 1986).

2 Background

We set up a microeconomic structure which eases the investigation of financial behavior of firms. We briefly summarize it hereafter and we point readers to Giulioni et al. (2011) for a detailed and technical presentation. The entrepreneur's task is to set the level of production as close as possible to the unknown level of demand. In fact, large deviations (both shortages and excesses) generate losses. When the loss is higher than the equity base, we require the firm to activate a bailout procedure which is costly. Bailouts can be avoided having a high level of equity base, but, on the other hand, a high level of equity base reduces the return of investment (*roi*) because the cost of equity base is assumed to be higher than the interest rate on debt. More formally, we assign to a generic entrepreneur labeled with j the goal of maximizing a score given by

$$score_{j,t} = \langle\rho\rangle_{j,t} - O_{j,t}c^O \tag{1}$$

where $\langle\rho\rangle_{j,t}$ denotes the average *roi* up to time t, $O_{j,t}$ is the number of bailouts until time t and c^O is a parameter representing the bailout cost. This goal takes care of two aspects: *i)* maximize the entrepreneurs reward and *ii)* minimize the number bailouts. A forecasting service whose reliability must be assessed is present in our model.

Our aim is to create an artificial agent in humans likeness. Toward this goal we first submit our microeconomic framework to humans in a set of laboratory experiments indexed by e. Each experiment presents a different degree of difficulty in guessing the level of demand. To graduate the guessing activity we use three variables: $\hat{y}_t$, $\tilde{y}_{e,t}$ and $y^*_{e,t}$. $\hat{y}_t$ is a very smooth series (whose value is never revealed to the subjects of the experiments) used to build $y^*_{e,t}$ which is the demand received by a subject. The latter in turn serves to build $\tilde{y}_{e,t}$ which is the information the forecasting service supplies to the entrepreneur. Experiments differ for having different combinations of σ which denotes the standard deviation of the percentage deviations between $\hat{y}_t$ and $y^*_{e,t}$, and $\tilde{\sigma}$ which denotes the standard deviation of the

[1] The caveats to this approach can be overcome following the protocols of experimental economics.

percentage deviations between $\tilde{y}_{e,t}$ and $y^*_{e,t}$. $\tilde{\sigma}$ regulates the accuracy of forecast. The latter is crucial for the performance of agents. In fact, if $\tilde{\sigma} = 0$, the experimental subjects shall set the production at the same level of the forecast. In this way s/he will always realize the maximum profit and the bailout procedure will never be activated. At high levels of $\tilde{\sigma}$, the standard deviation of demand (σ) assumes a great importance. In fact, if σ is low, good guesses for the future production could still be obtained extrapolating the trend from past values. When both $\tilde{\sigma}$ and σ are high, setting the production is a problematic task because both the forward looking and the backward looking conducts are not reliable. The left part of table 1 reports the levels of standard deviations for each experiment.

3 Human Entrepreneurs

Among the numerous subjects we have contacted, at the time of writing this paper eight of them have reached a good knowledge of the model and consequently they have been admitted to the experiments. Summary statistics from their performances are reported in table 1. In the table, j indexes experimental subjects, e experiments, $\langle\rho\rangle$ denotes the average *roi* and O the number of bailouts.

Table 1 Standard deviations and Results from experiments. j indexes experimental subjects, e experiments, $\langle\rho\rangle$ denotes the average *roi* (percentage) and O the number of bailouts.

			$j=1$		$j=2$		$j=3$		$j=4$		$j=5$		$j=6$		$j=7$		$j=8$	
e	σ	$\tilde{\sigma}$	$\langle\rho\rangle$	O	$\langle\rho\rangle$	O	$\langle\rho\rangle$	O	$\langle\rho\rangle$	O	$\langle\rho\rangle$	O	$\langle\rho\rangle$	O	$\langle\rho\rangle$	O	$\langle\rho\rangle$	O
B1a	0.02	0.03	8.25	0	8.53	0	6.26	0	8.35	0	8.11	0	8.54	0	7.02	1	7.55	0
B1b	0.02	0.05	7.25	1	8.02	0	7.29	1	7.91	0	7.44	0	7.56	0	6.9	0	5.87	0
B1c	0.02	0.1	7.65	0	7.84	0	5.78	3	8.1	0	7.7	0	7.85	1	7.23	0	2.94	5
B2a	0.03	0.04	5.48	3	7.43	1	7.33	0	7.98	0	7.53	0	6.79	1	5.63	1	6.66	1
B2b	0.04	0.05	6.61	2	7.08	1	6.43	1	5.85	3	6.13	1	4.7	2	5.6	1	4.2	4
B2c	0.05	0.06	4.23	3	5.75	3	4.16	4	5.92	2	6.58	0	4.44	2	1.5	5	5.32	1

Looking at table 1 we can say that humans' performances gradually deteriorate when the standard deviations increase. Except in a number of occasions, the average *roi* decreases, while the number of bailouts increases when standard deviations increase.

Replicating the humans' performance by an artificial agent is a task addressed in the remainder of the paper.

4 Artificial Entrepreneur

In this section we will discuss on how to built an artificial agent able to move in the same framework submitted to humans. Our aim is to make the performances of the artificial agent as close as possible to these of humans. We adopt in this paper the results of recent studies which show how the heuristic approach can be a solid foundation of the smartness and adaptability of human decision making (Gigerenzer et al., 2011).

As pointed out above, agents take two decisions: 1) the amount of production, and 2) the level of equity ratio. We analyze them in the following sections. Before starting, let us give a comment on the notation. To be rigorous, we should use $x_{j,e,t}$ to denote agent's j variable x at time t of experiment e. However, we are building a prototype artificial agent so that the j is dropped in what follows and $x_{e,t}$ intend variable x of the artificial agent at time t of experiment e.

4.1 The Production Choice

The decision on production is made by taking into account some information from the past and the forecast. We denote the information set on which the decision is based with $I_{e,t} = \{y^*_{e,t-N} \dots, y^*_{e,t-1}, \tilde{y}_{e,t}\}$. A different way to pose the problem is to use growth rates. The growth rate of a variable x at time t is $g_{x_t} := (x_t - x_{t-1})/x_{t-1}$. The agent's choice variable is now $g_{y_{e,t}}$, and s/he sets the production according to

$$y_{e,t} = (1 + g_{y_{e,t}})y^*_{e,t-1}.$$

We assume $g_{y_{e,t}}$ to be set as a linear function of the variables in $I_{e,t}$:

$$g_{y_{e,t}} = s^u_{1,e,t} g_{y^*_{e,t-N+1}} + s^u_{2,e,t} g_{y^*_{e,t-N+2}} + \dots + s^u_{N,e,t} g_{\tilde{y}_{e,t}}. \tag{2}$$

The agent's problem has been transformed into the one of choosing the $\mathbf{s}^u_{e,t} := \{s^u_{n,e,t}\}$ coefficients with $n \in \{1, 2, \dots, N\}$.

In our model, the agent achieves the final decision on $y_{e,t}$ by taking two steps:

- determine which one would have been the best achievable solution $\mathbf{s}^*_{e,t} := \{s^*_{n,e,t}\}$;
- use best solutions of the previous periods to decide the $\mathbf{s}^u_{e,t}$ in the current period.

We will go into details in the following paragraphs.

The Best Achievable Solution ($\mathbf{s}^*_{e,t}$). Once $g^*_{y_{e,t}}$ is known, $\mathbf{s}^*_{e,t}$ can be established. A systematic check on all the combinations of s is not possible. We use a *greedy* algorithm to identify a path in the coefficients space which realizes a sequence of improvements bringing to the best achievable solution given the available resources. The starting point of the algorithm is chosen in a set of easily computable arithmetic averages we will refer to as "focal points" (Schelling, 1960; Zuckerman et al., 2007).

Given a parameter called "difference" denoted hereafter with d and the number of addends (N) of the average in (2), we use the notation $\mathbf{F}^d_{R\times N}$ to identify the matrix containing the focal points. R here denotes the number of rows of this matrix. The latter is implicitly chosen when the values of N and d are set. If we set $N = 2$ and $d = 0.5$ we have for example

$$\mathbf{F}^{0.5}_{3\times 2} = \begin{pmatrix} \mathbf{f}_{1\times 2,1} \\ \mathbf{f}_{1\times 2,2} \\ \mathbf{f}_{1\times 2,3} \end{pmatrix} = \begin{pmatrix} 0 & 1 \\ 0.5 & 0.5 \\ 1 & 0 \end{pmatrix}$$

When N is higher, a software able to generate evenly spaced points in the unit simplex (Giulioni, 2011) can be used to obtain **F**.

Each row of this matrix gives the coordinates of a focal point. We denote with i the row number of F at which a given focal point ($\mathbf{f}_{1\times N,i}$) can be found.

The first step consists in choosing one of the focal points. Remember this computation is done when $y^*_{e,t}$ is known. Let us denote the column vector of the known growth rates with $\mathbf{g}_{N\times 1,e,t}$. The best performing focal point ($i^*_{e,t}$) is determined by

$$i^*_{e,t} = \min_{i\in\{1,2,\ldots,R\}} \left| g_{y^*_{e,t}} - \mathbf{f}_{1\times N,i}\mathbf{g}_{N\times 1,e,t} \right|$$

where $\mathbf{f}_{1\times N,i}\mathbf{g}_{N\times 1,e,t}$ is the dot product.

Then, the local search starts by iterations which have the following steps:

1. identify a set made up of the starting point and a number of its neighbors;
2. select the best point in the set;
3. if the selected point is equal to the starting point or if the number of iterations reaches a maximum (Z), then stop the search;
   else, set the selected point as the new starting point and go to point 1).

Step 1) can be done in several ways. We take a real number ψ and we build a neighborhood by permuting with repetition the elements of $\{-\psi, 0, \psi\}$ and taking them N at a time. We arrange all these in a matrix

$$\mathbf{D}_{3^N\times N} = (\mathbf{d}_{1\times N,l}) \qquad l \in \{1,\ldots,3^N\}.$$

If we denote with $\mathbf{s}^{(z-1)}_{1\times N}$ the point selected at iteration $z-1$, the neighbors to be considered in iteration z are gathered in the matrix $(\mathbf{s}^{(z-1)}_{1\times N} + \mathbf{d}_{1\times N,l})$. Iterations go on until one of the stopping conditions at point 3) is satisfied.

This process brings us to the best solutions $\mathbf{s}^*_t$

The Used Solution ($\mathbf{s}^u_t$). To let the agent learn and evolve the most reliable focal point, we build a learning classifier system whose rules are the focal points. Based on the previous experience, a score is associated to each rule and at each time step the rule with the highest score is taken as basis to establish the s coefficients to be used in t. We call it the “reference” rule (or reference focal point).

Given a memory length m, the entrepreneur has a set $\{(i^*_{t-m}, \mathbf{s}^*_{t-m}), (i^*_{t-m+1}, \mathbf{s}^*_{t-m+1}), \ldots, (i^*_{t-1}, \mathbf{s}^*_{t-1})\}$ informing on which one was the best focal point in the past.

The reference rule, denoted with i^u_t, can be determined by maximizing a function defined on the set of the best focal points. The function we use is

$$i_t^u = \max_{i \in \{1,2,\ldots,R\}} \sum_{z=1}^{m} \delta(i, i_{t-z}^*)$$

where $\delta(\circ,\circ)$ denotes the Kronecker delta function.

Once the reference focal point has been determined, we select the best solutions which was obtained starting from the reference focal point in the previous m periods. The rule to be used in t, $\mathbf{s}_t^u$ is obtained as an average of such points:

$$\mathbf{s}_t^u = \frac{\sum_{z=1}^{m} \mathbf{s}_{t-z}^* \delta(i, i_{t-z}^u)}{\sum_{z=1}^{m} \delta(i, i_{t-z}^u)}.$$

4.2 The Equity Ratio

Being the cost of equity greater than the interest rate on debt, the best strategy concerning the level of equity ratio (which is denoted with a and is defined as the ratio between equity base and total investments) is to keep it at the minimum level. However, as we did for humans, we assign the artificial agent the goal of maximizing the score (equation 1) which establishes the trade off between the level of *roi* and the probability of run up against the bailout procedure.

Because we want the agent to be able to adapt to changing economic conditions we let m^a be the memory length the agent has for the ρ values (it could coincide or not with the memory length for focal points selection). At time t, the agent considers the following set of values $\{\rho_{t-m^a} - h, \rho_{t-m^a+1} - h, \rho_{t-m^a+2} - h, \ldots, \rho_{t-1} - h, \rho_t - h\}$ where h is a precautionary parameter.

Now consider the sum

$$\Theta_t = \sum_i \delta(-, sign(\rho_{t-i} - h))(\rho_{t-i} - h);$$

we calculate the target level of the equity ratio ($\hat{a}$) as

$$\hat{a}_t = 1 - \exp(\lambda \Theta_t)$$

where λ is a parameter. If $a_{t-1} \neq \hat{a}_t$, the agent manages to reach the target level as fast as possible.

5 Human and Artificial Entrepreneurs

In this section we let the artificial entrepreneur perform the same experiments submitted to humans. In this way we obtain results which can be compared with those of humans. However, before the comparison could be made, the artificial agent behavioral parameters must be set.

Parameters setting. The artificial agent's behavior depends on the following parameters: Z, m, m^a, h and λ. We let them to be endogenously selected by the differential evolution algorithm (Storm and Price, 1997). The result of this procedure depends on the objective assigned to the algorithm. The assigned goal in what presented hereafter is to find the combination of parameters which achieve the maximum score. In other words the algorithm task is the same of humans (maximize equation 1). The final step to setup the artificial agent concerns the **F** matrix which depends on N and d. Several trials revealed the simplest choice ($N = 2$ and $d = 1$) as the one which is most suitable to replicate the results of humans.

Table 2 Results from the differential evolution algorithm.

e	$\langle\rho\rangle\%$	O	Z	m	m^a	h	λ
B1a	8.27	0	1	1	$\forall$	$\forall$	$\forall$
B1b	7.37	0	1	5	$\forall$	$\forall$	$\forall$
B1c	7.4	0	1	9	5	4.96	0.6
B2a	6.7	0	1	1	10	5.65	0.75
B2b	6.62	1	7	13	7	2.65	9.31
B2c	4.89	0	4	27	5	3.13	5.02

Table 3 Scores of humans and artificial entrepreneur (a. e.).

$j=1$	$j=2$	$j=3$	$j=4$	$j=5$	$j=6$	$j=7$	$j=8$	a. e.
8.25	8.53	6.26	8.35	8.11	8.54	6.02	7.55	8.27
6.25	8.02	6.29	7.91	7.44	7.56	6.9	5.87	7.37
7.65	7.84	2.78	8.1	7.7	6.85	7.23	-2.06	7.4
2.48	6.43	7.33	7.98	7.53	5.79	4.63	5.66	6.7
4.61	6.08	5.43	2.85	5.13	2.7	4.6	0.2	5.62
1.23	2.75	0.16	3.92	6.58	2.44	-3.5	4.32	4.89

Table 2 reports the results obtained from running the differential evolution algorithm on the same experiments carried out by humans. Concerning the behavioral parameters of the artificial agent we can divide them in two subsets. Z and m governing the choice of production and m^a, h and λ governing the choice of financial position. Table 2 shows how Z and m increases with the standard deviations. The behavior of the parameters which regulate the financial decision also changes with the level of standard deviations. For low levels of σ and $\tilde{\sigma}$ (experiments B1a and B1b) the values of these parameters are irrelevant ($\forall$ symbol). When the standard deviations have intermediate levels (experiments B1c and B2a) the value of h is high while that of λ is low. h decreases and λ increases at high levels of the standard deviations (experiments B2b and B2c).

Results Comparison. To summarize the work done in this paper, we report the scores (equation 1) of humans and artificial agent in Table 3. The performance of the artificial agent is in line with the average score of humans in the initial four experiments. In experiments B2b and B2c, the artificial agent performs better than humans, however a number of the latter still perform similarly (or better) than the artificial agent.

6 Conclusions

In this paper we focus on how entrepreneurs manage the capital structure of their firms. A result of our analysis is that, beside the traditional approach which assumes rational maximizing subjects, the heuristic approach could also provide for a valid alternative micro-foundation for the financial decisions of entrepreneurs. A second point worth to be mentioned is that financial behaviors could mainly depend on the volatility of demand and on the accuracy of demand forecasts instead of depending on the business cycle phases.

The model presented in this paper has many potential future developments. First of all, the agents' choice under different assumptions on divisibility and reversibility of investment could be analyzed. Second, a statistical analysis of experimental data could lead us to identify a number of heterogeneous avatars. The third extension concerns interaction. It could be investigated how the results change when selected information on the other subjects are given to each experimental subject, and how the situation evolves when credit demands are selected by a real life banker.

References

Gigerenzer, G., Hertwig, R., Pachur, T. (eds.): Heuristics: The Foundations of Adaptive Behavior. Oxford University Press, New York (2011)

Giulioni, G.: A Software to Generate Evenly Spaced Points on the Unit N-Simplex. SSRN eLibrary (2011), `http://ssrn.com/paper=1822002`

Giulioni, G., Bucciarelli, E., Silvestri, M.: A model implementation to investigate firms financial decisions (2011), `http://www.dmqte.unich.it/users/giulioni/model_description.pdf`

Minsky, H.P.: Stabilizing an Unstable Economy. Yale University Press, New Haven (1986)

Modigliani, F., Miller, M.: The cost of capital, corporate finance, and the theory of investment. American Economic Review 3(48), 261–297 (1958)

Zuckerman, I., Kraus, S., Rosenschein, J.S.: Using focal point learning to improve tactic coordination in human-machine interactions. In: IJCAI 2007, pp. 1563–1569 (2007)

Storm, R., Price, K.: Differential evolution - a simple and efficient heuristic for global optimization over continuous spaces. Journal of Global Optimization 11(4), 341–359 (1997)

Titman, S., Tsyplakov, S.: A dynamic model of optimal capital structure. Review of Finance 11(3) (2007)

Zuckerman, I., Kraus, S., Rosenschein, J.S.: Using focal point learning to improve tactic coordination in human-machine interactions. In: IJCAI 2007, pp. 1563–1569 (2007)

Simulating Artificial Stock Markets with Efficiency

Philippe Mathieu and Yann Secq

Introduction

Stock markets have been studied for years in economy and finance academic departments by relying on the idea of a general homo economicus that makes rational choices. Classical approaches use equational representation to enable a global markets characterization, but they fail to explain the link between individual behaviours and the global market dynamic and trends that emerge.

Schelling seminal work on segregation models [12] has initiated a novel approach relying on an individual behaviour characterization that leads to global observations. This approach enables a finer grain behaviour modelling that allows more detailed simulations. Within this context, it is possible to have several kind of agents, with heterogenous cooperative or concurrent behaviours, and to study their actions aggregation and feedback loops which produces emergent macroscopic behaviours. The trade-off between the expressivity available with an agent-based approach and its execution speed relies on the complexity and information volume of involved processes.

Our study is focused on order driven markets and describes why scalability issues appear when using an agent-based approach. After introducing our existing simulation platform ATOM, we study the impact of the number of orders and transactions and the cost of notification on execution time. These aspects lead us to search how we could optimize our platform to be able to handle higher volumes that are close to those generated by High Frequency Trading (HFT) algorithms.

First section studies problematics linked to order-driven market simulators in a centralized setting. Second section identifies problematics that have to be tackled to scale agent-based simulations. Third section presents the two main approaches to efficiently distribute stock market simulations: network-based or GPGPU-based. Last section concludes and identifies future works.

Philippe Mathieu · Yann Secq
Université Lille 1, Computer Science Dept. LIFL
(UMR CNRS 8022)
e-mail: firstname.surname@univ-lille1.fr

J.B. Pérez et al. (Eds.): Highlights on PAAMS, AISC 156, pp. 341–348.
springerlink.com

1 Agent-Based Stock Market Simulation with ATOM Framework

This section first introduce general notions on order-driven markets. Then, principles and dynamic of agent-based market simulators are described. These elements show why scaling can be a crucial issue with this type of approach and within this application domain. Then, section 2 study which solutions can be proposed to ease scaling issues in agent-based simulations.

In order-driven markets, price formation is based on the confrontation of a formalized offer and demand. This formalization is done through the emission of orders by traders that contain an asset name, a direction (bidding or selling), a quantity (number of shares) and a price. All orders for a given asset are gathered in an order book (generally a double auction order book) that is organised in two sets sorted on price: sell orders and buy orders. Price formation is realized by confronting the best ask order price with the best sell order price. Thus, in a stock market driven by orders, actors involved in price creation are traders that send ask or bid orders, order books that stores and generate prices (one for each negotiated share on the market) and the market that can be seen as the container for all these information exchanges (orders and quotes).

In an agent-based approach of stock markets simulation, traders are represented as agents, there are several order books (as much as the number of assets) and there is the market that establish the link between traders and order books. The order processing dynamic has two stages: in the first stage, traders send orders to the market, that forward them to the concerned order book, the second stage happens if a price is fixed. When a new price is created, it is sent to the market (to store price history) and traders whose orders where involved in the last quote have their portfolio and their balance updated accordingly to the exchanged shares and fixed prices. Traders behaviours can then react to these changes in the market leading to interesting feedback loops between agents strategies. This approach allows simulations with the finest possible grain, because all entities in the system are observable. Nevertheless, this granularity which is an interesting aspect that enables experimentations and observations close to real stock markets is also an important challenge to be able to scale these simulations.

The ATOM (ArTificial Open Market, `atom.univ-lille1.fr`) platform has been developed to easily build fine-grain stocks markets simulations. This platform is made of several layers that enable the definition of markets, simulations and to allow their execution. If needed, visualization and live interaction with a running simulation can be done. The main component of ATOM is a Java API that define main entities: Market, OrderBook, Order and Agent. This component includes detailed predefined object for orders (`Limit`, `Market`, `Cancel`, `Update`, `Stop`) and trading strategies (`ZIT`, `Chartist`, ...). Thus, a simulation can easily be built as shown in figure 1 (a more detailed description of the ATOM platformis available in [6]).

```
Simulation sim = new Simulation();
for (int i=0; i<10; i++) { // Adding 10 order books
  sim.addNewOrderBook("IBM_"+i);
}
for (int i=1;i<=1000;i++) { // Adding 1000 ZIT agents
  Agent trader = new ZIT("zit_"+i);
  sim.addNewAgent(trader);
  trader.account.credit(10000);
}
// 1000 days with 100 ticks (opening), 800 (continuous) and 100 (closing)
sim.run(1000,100,800,100);
```

Fig. 1 A simple intra/extra day simulation with ATOM

2 Scaling Issues in Agent-Based Simulations

The main principle of agent-based simulation (ABS) is its focus on individual behaviours. Thus, simulation involving a high number of agents or messages implies scalability issues. These issues can be categorized in two classes: volume handling (number of messages, information to be stored for post-analysis) and information exchange efficiency (latency between an update within the system and its notification to agents).

2.1 Handling the Volume

The first problem that arise is information volume that is generated and that has to be processed to compute one simulation time step. In this specific application domain, information volume is function of the number of order books, of agents, of prices fixed and thus, of agent wealth updates. Each time that an agent send an order, it has to be processed by the market (stored and forwarded to the specified order book), then by the order book (insertion and matching), and if a price is fixed, the market must be informed (to keep price history) and then involved traders on this transaction have to be updated.

When one thousand agents and a dozen of order books are created (as in figure 1), it involves a very large number of orders, prices and portfolios to be managed. The cost is high memory but also in computation. In the above example, the number of orders is $1000 * (100 + 800 + 100) * 1000 = 10^9$, prices generated are in the order of 10^6 and the total time taken for this simulation is roughly one hour on a 2.66Ghz i7 computer. It is difficult to know the number of orders sent in a day on Euronext, but the Global Average Daily Volume is in the order of 8 to 10 millions prices fixed in a single day (Source) with an average order matching speed below 1 millisecond (Source).

2.2 Information Exchange Efficiency

The second important aspect is linked to the information exchange cost between traders and order books. There are mainly two main information flows: the incoming orders flowing from traders through the market to be handled by order books and an outgoing flow starting with a price fixed within an order book that has to be transmitted to the market, then to traders involved in the transaction (and more globally to all traders). A last information exchange that can happen between traders and the market is concerned with information on volumes and order books (bid-ask spread and first rows of best bid-ask orders waiting).

Traders in agent-based simulations use several information to compute their next action (do nothing or send an order): market (quotes, orders waiting in orderbooks, bid-ask spread) and social (news or messages from others agents). In function of their trading strategy (chartist, arbitragist ...), agents need to access more or less information and thus, the mechanism used to transmit them to traders becomes a bottleneck for the simulator. When the number of orders increases, the delay between an order emission and its execution notification (if another order can match it) increases also.

For real markets, we have not been able to find quantified information on the duration of such round trip message. But, with a matching process that is below 1 millisecond and a publicized data message latency given under 5 milliseconds (Source). we can infer that the whole process should not take more than 10-100 milliseconds. But in an artificial markets, this information exchange of a new quote generates more computation costs because agents portfolio have to be updated.

3 Agent-Based Simulator Distribution

To handle scalability issues two main approaches can be used: distribute the load among a computer network or on a General Purpose Graphical Processor Unit (GPGPU). The first approach enables a scalability linked to the number of available computers while the second is highly dependant on the GPGPU design to be able to chain several GPGPU on a given host. A third hybrid approach using a network of computers using GPGPU is possible but raises others issues (see [13] for internet scale and [4] for local scale networks).

3.1 Distributing an Artificial Stock Market

To be able to scale agent-based stock market simulations, we have seen that two main issues are involved: data volume storage and querying and information exchange speed between traders and order books. Visualisation has also to be taken into account because it involves a process that gather an important information set. Even if visualisation can be deported, it involves a high cost in information transfer.

Several strategies can be used to distribute and parallelize market computations but one has to understand that there are three distinct problematics: computation costs involved by trading strategies, computation costs implied by order/price handling (emission, logging, insertion and matching, portfolio updating), communication latency occurring because of information exchange between traders and the market.

These problematics correspond to different kind of computations: heterogenous behaviours for trading strategies, homogenous computations for orders matching process and network infrastructure and libraries for communications. In the following sections, we explain how network-based distribution should be used to enable concurrent execution of heterogenous computations (ie. traders strategies) while GPGPU are fitted to intense data parallelism, and thus interesting for homogenous computations (ie. price fixing mechanism).

3.2 Should Quotes Be Pushed or Pulled?

The second problematic related to information exchange becomes critical in a distributed setting. Two main schemes can be used to transmit information between traders and the market: pulling or pushing. In a pulling scheme, traders initiate a request to the market to retrieve some information, while in a pushing scheme, the market sends the information to all traders. Pulling is efficient when trading strategies do not use a lot of information from the market, but when strategies needs information on multiple assets, with an important level of details (bid-ask spread, orders within order books ...), pushing the information is more efficient.

This choice has also important implications on reliability. Indeed when a pulling mechanism is used, denial of service attacks can be generated by malicious traders by overloading the market with information requests. Even with fair traders, if their trading strategy use an important information volume, it generates loads on the market to prepare and transfer all these information. It should be noted that in a distributed setting efficient group communication can be done thanks to publish/subscribe architectures, linked to networks broadcasting capabilities. For all these reason, we believe that pushing information is clearly more fitted in a network-based distribution. For GPGPU, this choice can be queried because memory latencies are really low.

3.3 Network-Based Distribution

Distributing computation on a computer network allow to harness power of commoditized computers network. The main idea is to enable concurrent computations on each node and to use some message passing scheme to coordinate tasks. This approach is particularly fitted to heterogenous computations that can be expressed through task parallelism or embarrassingly parallel computations (as in [2]). The main limiting factor is communication latency cost. To bypass this limitation, distributed systems try to cover communication costs with a coarser grain of computation chunk. These costs have been clearly evaluated in [8], where large scale traffic simulations lead to a factor of 10 between a simulation on computers on a LAN and on parallel machines. An interesting comparison of some of the well known platforms (Repast, Cougaar, Aglets and AAA) can be found in [3].

Because heterogenous computations can be distributed, it is efficient to distribute traders among the network. Even if some agents share the same trading strategy, there are several strategies used during a simulation. This approach enables a concurrent computation of traders decisions. Difficulties that arise with this approach is the guaranty of equity between agents. In a centralized setting, equity can be easily enforced with a turn of speak. In a distributed setting it becomes harder, but it can be enforced by waiting that all traders have transmitted their action before integrating orders in the market. This synchronisation barrier decreases performances because some computers are idle while the market is waiting for all traders answers. An interesting use case that can leverage embarrassingly parallel computation is the execution of several simulations. Experimentations done in [7] to compute social welfare in function of orders permutations requires *orders*! simulations that are all independent from each other. This is typically a context where network-based distribution is clearly fitted.

3.4 GPGPU-Based Distribution

To understand General Purpose Graphical Processing Unit (GPGPU) principles, concepts of SISD, SIMD and streams have to be detailed. A Single Instruction Single Data (SISD) can be seen has the default computation model, with one instruction applied to one data in a time step. Early in computer hardware history, the concept of Single Instruction Multiple Data (SIMD) has been developed in order to gain computation power through a parallel execution of the same instruction on multiple data.

Recently, these graphical processors that were tailored to graphical computations have been redesigned to allow also some more general computation scheme by relying on a stream processing approach. GPGPU processors allow scalability through the use of a large number of SIMD unit that can handle several parallel computation chunks (called *kernels*).

GPGPU computation distribution is fitted to data intensive processing, but on homogenous computations. The same algorithm can be applied in parallel to a large number of data chunks. The main difficulty is then to redefine computation in a stream oriented design to reach interesting performance gains. Works done on agent-based simulation on GPU (as [10] and [9]) demonstrate the speedup that can be achieved but also illustrate difficulties and challenges to transform agent-based computations in a stream parallelism model.

In stock markets, we can identify several mechanisms that could benefit a data parallelism model: market trends characterization (as done in [5]) and price fixing mechanism. The first computation type, for example assets mean value on a given period or time series analysis, the benefit would come from being computed once and shared by all trading strategies. The same process being applied to several assets fit nicely with the stream processing model. The second computation type is harder to parallelize. Indeed, price are fixed by heterogenous information chunks (orders which varies in quantities, prices and assets). Nevertheless, an important time slice is taken by orders sorting in ask and bid data structures. This point could benefit from parallelized sorting algorithms.

A last important issue implied by GPGPU distribution is reproducibility. As demonstrated in [11], if computations are not rigorously checked when using floating point arithmetics, errors can grow rapidly and results become totally biased. This could be crucial if algorithms linked to market data aggregation are parallelized in order to provide on the shelves results to trading strategies.

4 Conclusion

This paper has described the main challenges that appear while simulating large order-driven stock markets with an agent-based approach. These problems have been identified through the use of our ATOM platform whose core components are executed on a single computer. We have identified two fundamental problematics that have to be tackled in order to scale simulations to a large number of orders: information volume and information exchange. These problematics are particularly critical to study the impact of High Frequency Trading algorithms that generate a high number of orders in short time slice. The first approach, distributing computation on a computer network; is fitted to scaling heterogeneous computations. These computations are typically trading strategies used by artificial agents. The main limiting criteria of this kind of distribution is network communication costs that have to be compensated by a high computation cost. The second approach relies on GPGPU optimization and thus is particularly fitted for homogeneous computations. We have seen that data intensive computation that can be parallelized can leverage GPGPU computation power, but as with the network-based approach, the cost of information transfer between host and graphic card memory is the limiting factor.

We have begun experimentations to implement an hybrid approach that tries to leverage network and GPGPU approaches (in the same spirit as [1] but not focused on spatial based issues). We focus on distributing only trading strategies on a computer network and to using GPGPU cards to manage the market. The main difficulties are the network latency hiding, the parallelization of price fixing mechanism and determining an efficient price notification mechanism.

Network	GPGPU
heterogeneous computation	homogeneous computation
actor/concurrent model	stream processing model
network communication cost (high latency)	host/card communication costs (low latency)
high scalability (number of machines)	scalability limited to GPGPU chaining
language agnostic	C-99 (and OpenCL / CUDA)

Fig. 2 Network vs. GPGPU distribution

References

1. Aaby, B., Perumalla, K., Seal, S.: Efficient simulation of agent-based models on multi-gpu and multi-core clusters. In: 3rd International ICST Conference on Simulation Tools and Techniques, SIMUTools 2010, pp. 29:1–29:10 (2010)
2. Chiara, R.D., Mancuso, A., Mazzeo, D., Scarano, V., Spagnuolo, C.: A framework for distributing agent-based simulations. In: HeteroPar 2011 jointly published with Euro-Par 2009. Springer, Heidelberg (2011)
3. Gorton, I., Haack, J., McGee, D., Cowell, A., Kuchar, O., Thomson, J.: Evaluating Agent Architectures: Cougaar, Aglets and AAA. In: Lucena, C., Garcia, A., Romanovsky, A., Castro, J., Alencar, P.S.C. (eds.) SELMAS 2003. LNCS, vol. 2940, pp. 264–278. Springer, Heidelberg (2004)
4. Hamada, T., Narumi, T., Yokota, R., Yasuoka, K., Nitadori, K., Taiji, M.: 42 tflops hierarchical n-body simulations on gpus with applications in both astrophysics and turbulence. In: Conf. on High Performance Computing Networking, Storage and Analysis, pp. 62:1–62:12 (2009)
5. Lee, M., Hong Jeon, J., Kim, J., Song, J.: Scalable and parallel implementation of a financial application on a gpu: With focus on out-of-core case. In: 2010 IEEE 10th Int. Conf. on Computer and Information Technology, pp. 1323–1327 (2010)
6. Mathieu, P., Brandouy, O.: A Generic Architecture for Realistic Simulations of Complex Financial Dynamics. In: Demazeau, Y., Dignum, F., Corchado, J.M., Pérez, J.B. (eds.) Advances in PAAMS. AISC, vol. 70, pp. 185–197. Springer, Heidelberg (2010)
7. Mathieu, P., Brandouy, O.: Efficient Monitoring of Financial Orders with Agent-based Technologies. In: Demazeau, Y., Pechoucek, M., Corchado, J.M., Pérez, J.B. (eds.) PAAMS 2011. AISC, vol. 88, pp. 277–286. Springer, Heidelberg (2011)
8. Nagel, K., Rickert, M., Barrett, C.: Large Scale Traffic Simulations. In: Palma, J.M.L.M., Dongarra, J. (eds.) VECPAR 1996. LNCS, vol. 1215, pp. 380–402. Springer, Heidelberg (1997)
9. Perumalla, K., Aaby, B.: Data parallel execution challenges and runtime performance of agent simulations on gpus. In: Spring Simulation Multiconference, pp. 116–123. Society for Computer Simulation International (2008)
10. Richmond, P., Coakley, S., Romano, D.: Cellular level agent based modelling on the gpu. In: High Performance Computational Systems Biology (2009)
11. Saponaro, P., Taufer, M.: Improving numerical reproducibility and stability in large-scale simulations on gpu. Master's thesis, University of Delaware (2010)
12. Schelling, T.: Models of segregation. American Economic Review 59(2) (1969)
13. Spurzem, R., Berczik, P., Berentzen, I., Nitadori, K., Hamada, T., Marcus, G., Kugel, A., Männer, R., Fiestas, J., Banerjee, R., Klessen, R.: Astrophysical particle simulations with large custom gpu clusters on three continents. Comput. Sci. 26, 145–151 (2011)

Adaptive Learning in Multiagent Systems: A Forecasting Methodology Based on Error Analysis*

Tiago M. Sousa, Tiago Pinto, Zita Vale,
Isabel Praça, and Hugo Morais

Abstract. Electricity markets are complex environments, involving a large number of different entities, playing in a dynamic scene to obtain the best advantages and profits. MASCEM is a multi-agent electricity market simulator to model market players and simulate their operation in the market. Market players are entities with specific characteristics and objectives, making their decisions and interacting with other players. MASCEM provides several dynamic strategies for agents' behaviour. This paper presents a method that aims to provide market players strategic bidding capabilities, allowing them to obtain the higher possible gains out of the market. This method uses an auxiliary forecasting tool, e.g. an Artificial Neural Network, to predict the electricity market prices, and analyses its forecasting error patterns. Through the recognition of such patterns occurrence, the method predicts the expected error for the next forecast, and uses it to adapt the actual forecast. The goal is to approximate the forecast to the real value, reducing the forecasting error.

Keywords: Adaptive Learning, Electricity Markets, Error Analysis, Forecasting Methods, Information Theory, Multiagent Systems.

Tiago M. Sousa · Tiago Pinto · Zita Vale · Isabel Praça · Hugo Morais
GECAD – Knowledge Engineering and Decision-Support Research Center
Institute of Engineering – Polytechnic of Porto (ISEP/IPP)
Rua Dr. António Bernardino de Almeida, 431
4200-072 Porto
e-mail: {tmsbs,tmp,zav,icp,hgvm}@isep.ipp.pt

* The authors would like to acknowledge FCT, FEDER, POCTI, POSI, POCI, POSC, POTDC and COMPETE for their support to R&D Projects and GECAD Unit.

J.B. Pérez et al. (Eds.): Highlights on PAAMS, AISC 156, pp. 349–357.
springerlink.com

1 Introduction

The recent restructuring of the energy markets, characterized by an enormous increase of the competition in this sector, led to relevant changes, and consequently new challenges in the participating entities operation [1]. In order to overcome these challenges, it is essential for professionals to fully understand the principles of the markets, and how to evaluate their investments in such a competitive environment [2]. The necessity for understanding those mechanisms and how the involved players' interaction affects the outcomes of the markets, and thus, the revenues of the investments, contributed to the need of using simulation tools, with the purpose of taking the best possible results out of each market context for any participating entity. Multi-agent based software is particularly well fitted to analyze dynamic and adaptive systems with complex interactions among its constituents [3, 4, 5].

To explore and study such approaches is the main goal of this research, and for that the multi-agent system MASCEM (Multi-Agent System for Competitive Energy Markets) [5, 6] is used. This system simulates the electricity market, considering all the most important entities that take part in such operations. Players in MASCEM are implemented as independent agents, with their own capability to perceive the states and changes of the world, and acting accordingly. These agents are provided with biding strategies, to try taking the best possible advantage from each market context. MASCEM main entities include: a market operator agent, a system operator agent, a market facilitator agent, buyer agents, seller agents, VPP agents, and VPP facilitators.

MASCEM allows the simulation of several market models: day-ahead pool, bilateral contracts, complex market, and balancing market. It also allows hybrid simulations, regarding players operation in a chosen combination of the mentioned market models.

This paper presents a new methodology to support the negotiating agents' bidding definition. These agents' bids must take into account the expected market price for a certain negotiation period, therefore the use of forecasting methods to predict the market price is an essential tool.

Since forecasts and predictions are always subject to some error, it is important to analyze that error properly, in order to try to overcome it [8]. When observing the forecasting errors' distribution over time, it becomes visible that many times those errors present patterns in their occurrence. In some cases forecasts fail by predicting higher or lower values than the reality, in recurrent events that may have something in common (e.g. context, such as the considered period, day of the week, or season of the year).

This strategy's goal is to analyze the forecasting errors' evolution of a forecasting method, to try finding patterns in that error sequence and provide a prediction on the next error, which will be used to adequate the initial forecast.

The proposed methodology is integrated with ALBidS (Adaptive Learning strategic Biding System) [6], a multiagent system integrated in MASCEM with the purpose of providing decision support to negotiating players' actions. ALBidS uses reinforcement learning algorithms to choose from a set of different action proposals, provided by different algorithms that present distinct approaches in defining the player's most adequate action. ALBidS global structrure is presented in Figure 1.

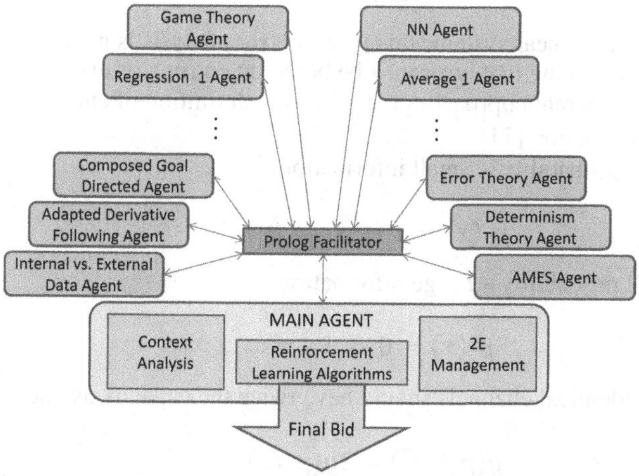

Fig. 1 ALBidS overview.

The different approaches included in ALBidS are based on alternative perspectives on the analysis of the market data, competitors' past behaviors, and market particularities [9]. One of the main strategic approaches considered by ALBidS is based on an error analysis, considering the forecasting error's tendencies over time. Details on the error analysis approach are presented and discussed in this paper.

2 Error Analysis

When a prediction is made, there will always be some error or uncertainty. For any measurement, there is a set of factors that may cause deviation from the actual (theoretical) value. Most of these factors have a negligible effect on the outcome and usually can be ignored. However, some effects can cause a significant change, or error, in the final result. In order to achieve a useful prediction, it is necessary to have an idea on the extent and direction of the errors [8].

In adaptive systems, the data input is important in order to adjust the system so that it learns the best way. The data introduced must take into account the outputs generated by the system in previous iterations, appropriate to reflect the difference between the reality and the forecast results. This is called the prediction error. To represent this error a Gaussian distribution is used most of the time [10].

The cost function, responsible for calculating the prediction error, is the most widely used Mean Square Error (MSE), because it offers the best solution, although it is not guaranteed when using a Gaussian distribution. Another measure of accuracy to qualify the error is the Mean Absolute Percentage Error (MAPE), which expresses the precision in percentage and is defined by (1):

$$M = \frac{1}{n}\sum_{t=1}^{n}\left|\frac{A_t - F_t}{A_t}\right| \tag{1}$$

However, for nonlinear systems this type of error analysis is not adequate. The criterion for determining that error may be based on entropy or divergence [10].

There are different approaches regarding the definition of entropy:

- Hartley Information [11]:
  - Large probability – Small information

$$p_x(x) = 1 \rightarrow S_H = 0 \tag{2}$$

  - Small probability – Large information

$$p_x(x) = 0 \rightarrow S_H = \infty \tag{3}$$

  - Two identical channels should have twice the capacity as one

$$g(p_x(x)^2) = 2g\big(p_x(x)\big) \tag{4}$$

  - Log2 is a natural measure for additivity

$$S_H = -\log_2 p_x(x) \tag{5}$$

- Shannon Entropy [12]:

$$H_S(X) = E[S_H] = -\sum p_x(x)\log_2 p_x(x) \tag{6}$$

- Rényi Entropy [13]:

$$H_\alpha(X) = \frac{1}{1-\alpha}\log\sum p_X^\alpha(x) \tag{7}$$

Rényi entropy becomes entropy when Shannon's entropy when:

$$\alpha \rightarrow 1 \tag{8}$$

- Fisher Entropy (local) [14]:

$$H_f(X) = \int \frac{[p'_X(x)]^2}{p_X(x)}\,dx \tag{9}$$

For divergence, there are the following approaches:

- Kullback-Leibler divergence [15]:

$$K(f,g) = \int f(x)\log(f(x)/g(x))\,dx \tag{10}$$

$$I(x,y) = \int\int f_{xy}(x,y)\log(f_{xy}(x,y)/f_x(x)f_y(y))\,dx\,dy \tag{11}$$

• Mutual information [8]

$$I(x,y) = H(x) + H(y) - H(x,y) = H(x) - H(x|y) = H(y) - H(y|x) \quad (12)$$

To address the existing error problem determined by the entropy and divergence, José C. Príncipe proposes the use of the Information Theory [8].

The Information Theory [16] is a probabilistic description of random variables that quantify the very essence of the communication process. It has been fundamental in the design and quantification of communication systems. The Information Theory provides a consistent and quantitative framework to describe processes with partial knowledge (uncertainty). Thus, it is suggested to overlap the samples in standard distributions to determine which one fits the best.

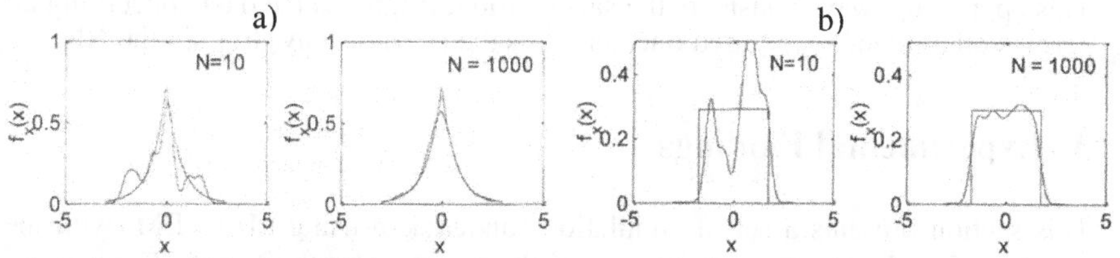

Fig. 2 Overlapping samples in: a) Laplace distributions, b) uniform distributions [8].

According to Information Theory, to choose the sample size, which is just a scale value, it is suggested to use the dynamic range:

$$3\sigma \sim 0.1 \quad (13)$$

Or use the rule of Silverman [17]:

$$\sigma = 0.9AN^{-1/5} \quad (14)$$

The proposed approach presented in this paper makes a prediction of the error value, taking into account the trends that this error has over time, depending on the circumstances and context, in order to be able to correct the value of another prediction previously made, and try to obtain a final value closer to the real one.

For this problem, a dynamic neural network (NN) [18] is used to forecast the errors, which is re-trained in each iteration, so that the forecast values are always updated according to the most recent observations. The NN receives as input the prediction error history data of the market's prices, and is trained to generate an output value, which will be the expected error. Then, this error is used to adjust the value of a prediction made by other forecasting strategy. Errors are stored in a market's history, registered in the system's database.

When defining the architecture of the NN, a matter of high importance had to be analyzed – the way the errors' sequence is "looked at" when trying to forecast it. This is a very important issue, because it does not matter how much data one has access to if that data is not used properly. For this reason three different approaches were considered. For the three is used a neural network with one value in the output layer - the value of the expected error, two intermediate nodes, and an input layer of four units, which differs in the different approaches on how the history of the error is considered:

- Error Theory A - This strategy makes a prediction along the 24 periods per day, using for the training of each period, the error of the same period for: previous day, the previous 7 days, the previous 14 days, and the previous 21 days.
- Error Theory B - This strategy makes a prediction along the days, using the error of the following periods: prior period, 2 previous periods, 3 previous periods, and 4 previous periods.
- Error Theory C - this strategy makes a prediction always in the same period (the time period in question), using the error for: the previous day, the previous 7 days, the previous 14 days, and the previous 21 days.

This approach always considers the same period. It ignores the data concerning all periods other than the required one, and uses the previous days to train the NN.

3 Experimental Findings

This section presents a set of simulations undertaken using MASCEM, with the purpose of analyzing the performance of the proposed methodology. These simulations involve 7 buyers and 5 sellers. This group of agents was created with the intention of representing the Spanish reality, reduced to a smaller summarized group, containing the essential aspects of different parts of the market, in order to allow a better individual analysis and study of the interactions and potentiality of each of those actors. The data used in this case study has been based on real data extracted from the Iberian market - OMEL [19].

The executed simulations concern 61 consecutive days with different starting dates, so that the results can be analyzed independently from the specific data of a certain time period. For that, four initial dates were chosen randomly. The selected dates were: February 13th. June 18th, September 1st, and October 29th. For each of these dates a simulation is performed for each strategy: Error Theory A, Error Theory B and Error Theory C.

As a reference value to forecast market prices, which afterwards will be adjusted by using the proposed approaches, a feed-forward Neural Network (NN) is used. The NN is trained with the historic market prices from OMEL.

Table 1 MAPE comparison of each proposed strategy with the auxiliary strategy.

		Error Theory A	*Error Theory B*	*Error Theory C*	Auxiliary forecast strategy (NN)
Starting Day	February 13th	0.0591	0.0612	0.0651	0.0687
	June 18th	0.0407	0.0425	0.0620	0.0485
	September 1st	0.0445	0.0527	0.0351	0.0496
	October 29th	0.0911	0.0963	0.1382	0.0898
Mean		0.0588	0.0632	0.0751	0.0642

Table 1 presents the MAPE results, displaying the error between the forecast and the actual market price, concerning the three approaches and the NN.

From Table 1 it is visible that the Error Theory A strategy was the approach that achieved the best overall results. This approach was able to achieve the best forecast from all the approaches in two simulations, and in the average of all. Error Theory C was able to achieve the best forecast result in one simulation. In what concerns the fourth simulation, none of the proposed approaches was able to reduce the forecasting error of the auxiliary strategy. Figure 3 presents an easier comparison between the concerned approaches performances.

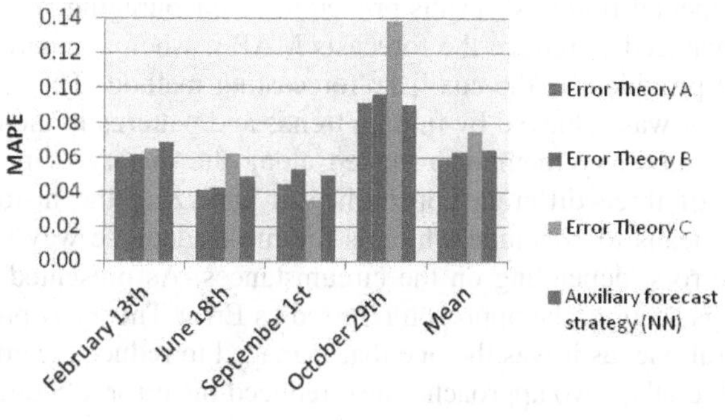

Fig. 3 Results of the test on Error Theory A strategy.

Figure 3 shows that, in what refers to the Error Theory A strategy, it was able to get smaller MAPE values when compared to the NN in all situations except from the simulation starting on October 29th. This was, in fact, an atypical situation. This simulation originated way higher MAPES from behalf of all approaches, when compared to the other simulations. Therefore, it is acceptable to globally state that the Error Theory A strategy is able to reduce the forecasting error of the NN. The other two approaches: Error Theory B and Error Theory C, present a more inconstant behavior. They are capable of reducing the NN error in some situations, while in other they are not so effective. This is especially visible in the case of Error Theory C, which in the simulation starting in the September 1st got the best result from all, while in the one starting from October 29th it got clearly the worst result from all. However, this can once more be justified by the atypicality of this simulation.

The observed results are motivating, since all three approaches were able to adjust the forecasting values, considering the prediction of the forecasting error that was expected from the NN. This is easy to see from Figure 3, which shows that the Error Theory approaches, for the presented days, are able to approximate the forecasts to the real market value, taking as basis the forecasted value from the auxiliary strategy – the NN, and the forecasting of the NN strategy's forecasting errors.

4 Conclusions

This paper presented a new approach to provide decision support to electricity markets' negotiating players. This approach is based on forecasting error analysis, to reduce the prediction error and consequently approximate the forecasts to the real observed values. This approach is integrated in the MASCEM simulator of electricity markets, through its inclusion in the ALBidS decision support system. This allowed not only to automatically take advantage on this methodology to provide support to negotiating players, but also to test the proposed method's performance using real electricity market's data.

The experimental findings' results proved to be encouraging, since all the three approaches managed to reduce the forecasts MAPE, when compared to the original prediction provided by the auxiliary forecasting method. The reduction of the forecasting error was achieved by finding trends and patterns in the log of forecast errors that the auxiliary method presented along the history of predictions. The consideration of three different approaches in analyzing the historic error data provided the means to conclude which is the most adequate way to interpret the sequence of errors, depending on the circumstances. As presented in the experimental findings section, the approach referred as Error Theory A proved to be the most successful one, as it was the one that managed to reduce the error in a higher percentage. The other two approaches also reduced the error, although in a smaller percentage.

The presented tests show the adequacy of the proposed method, and relevance in its usage. Moreover, as showed by the results of the three considered approaches, the performance of each approach is depending on the simulation days. Therefore, further analysis must be performed, in order to understand in which circumstances each of the three approaches should be used, depending on their results under different contexts. This way the ALBidS system will be able to automatically choose the most adequate from the three approaches in different situations.

References

[1] Shahidehpour, M., et al.: Market Operations in Electric Power Systems: Forecasting, Scheduling, and Risk Management, pp. 233–274. Wiley-IEEE Press (2002)

[2] Meeus, L., et al.: Development of the Internal Electricity Market in Europe. The Electricity Journal 18(6), 25–35 (2005)

[3] Li, H., Tesfatsion, L.: Development of Open Source Software for Power Market Research: AMES Test Bed. Journal of Energy Markets 2(2), 11–28 (2009)

[4] Koritarov, V.: Real-World Market Representation with Agents: Modeling the Electricity Market as a Complex Adaptive System with an Agent-Based Approach. IEEE Power & Energy Magazine, 39–46 (2004)

[5] Praça, I., et al.: MASCEM: A Multi-Agent System that Simulates Competitive Electricity Markets. IEEE Intelligent Systems 18(6), 54–60 (2003); Special Issue on Agents and Markets

[6] Vale, Z., et al.: MASCEM - Electricity markets simulation with strategically acting players. IEEE Intelligent Systems 26(2), 54–60 (2011); Special Issue on AI in Power Systems and Energy Markets
[7] Vale, Z., et al.: Electricity Markets Simulation: MASCEM contributions to the challenging reality. In: Handbook of Networks in Power Systems. Springer, Heidelberg (2011)
[8] Principe, J.: Information Theoretic Learning. Springer, Information Science and Statistics (2010)
[9] Pinto, T., Vale, Z., Rodrigues, F., Morais, H., Praça, I.: Bid Definition Method for Electricity Markets Based on an Adaptive Multiagent System. In: Demazeau, Y., Pechoucek, M., Corchado, J.M., Pérez, J.B. (eds.) Advances on Practical Applications of Agents and Multiagent Systems. AISC, vol. 88, pp. 309–316. Springer, Heidelberg (2011)
[10] Rao, S., Principe, J.: Mean Shift: An Information Theoretic Perspective. Transanctions on Pattern Analysis and Machine Intelligence 30(3) (2008)
[11] Hartley, H.: Transmission of Information. Bell System Technical Journal 7(3), 535–563 (1928)
[12] Shannon, C.: A Mathematical Theory of Communication. Bell System Technical Journal 27, 379–423, 623–656 (1948)
[13] Rényi, A.: Probability Theory. American Elsevier Publishing Company, New York (1970)
[14] Fisher, R.: On the mathematical foundations of theoretical statistics. Philosophical Transactions of the Royal Society 222, 309–368 (1922)
[15] Kullback, S., Leibler: On Information and Sufficiency. Annals of Mathematical Statistics 22(1), 79–86 (1951)
[16] Liu, W., et al.: An Information Theoretic Approach of Designing Sparse Kernel Adaptive Filters. Transactions on Neural Networks 20(12), 1950–1961 (2008)
[17] Silverman, J.: The Arithmetic of Elliptic Curves. Graduate Texts in Mathematics. Springer, Heidelberg (1986)
[18] Amjady, N., et al.: Day-ahead electricity price forecasting by modified relief algorithm and hybrid neural network. IET Generation, Transmission & Distribution 4(3), 432–444 (2010)
[19] Operador del Mercado Ibérico de Energia – homepage, http://www.omel.es (accessed on August 2011)

Author Index